PLANT BIOCHEMISTRY

PLANT BIOCHEMISTRY
AN INTRODUCTION

Reece Evans
Charlotte Roberts

Kruger Brentt
Publishers
2023

Kruger Brentt Publishers UK. LTD.
Company Number 9728962

Regd. Office: 68 St Margarets Road, Edgware, Middlesex HA8 9UU

ISBN: 978-1-78715-105-5

For information on all our publications visit our website at http://krugerbrentt.com/

PREFACE

Biochemistry, as the name implies, is the chemistry of living organisms. Living organisms, whether they are microorganisms, plants or animals are basically made up of the same chemical components. Biochemistry is the study of the way in which these components are synthesized and utilized by the organisms in their life processes. It bridges the gap between the conventional chemistry and biology. In other words, life is nothing but thousands of ordered chemical reactions or chemistry is the logic of all biological phenomena. Biochemistry deals with study of the nature of the chemical constituents of the living matter and the chemical substances produced by living things, the functions and transformations of their chemical entities in biological systems, the chemical and energetic changes associated with the transformation in the course of the activity of living matter.

Plant biochemistry examines the molecular mechanisms of plant life. One of the main topics is photosynthesis, which in higher plants takes place mainly in the leaves. Photosynthesis utilizes the energy of the sun to synthesize carbohydrates and amino acids from water, carbon dioxide, nitrate, and sulfate. Via the vascular system, a major part of these products is transported from the leaves through the stem into other regions of the plant, where they are required, for example, to build up the roots and supply them with energy. Hence, the leaves have been given the name source, and the roots the name sink. The reservoirs in seeds are also an important group of the sink tissues, and, depending on the species, act as a store for many agricultural products such as carbohydrates, proteins, and fat.

The present book contains 16 chapters covering all related disciplines. These chapters include Introduction, plant systemics: an overview, taxonomy and phylogeny systematics, the chloroplast and its role in photosynthesis, a leaf cell consists of several metabolic compartments, the calvin cycle, photosynthesis is an electron transport process, evolution and diversity of green and land plants, cell division and genetics, use of energy from sunlight by photosynthesis, nitrogen and sulfur metabolisms, respiration in plants, synthesis and metabolism of storage and structural carbohydrates, diversity of isoprenoids has multiple functions in plant metabolism, products of nitrate assimilation are deposited in plants as storage proteins, a plant cell has three different genomes. Related terminology is given at the end for ready reference. It is written in a lucid style with theoretical discussions being supplemented with illustrations, and tables for easy understanding of the subject. This book describes in details about the molecular mechanisms of plant life. The book is also supported with a comprehensive glossary and index in the end. The book is an ideal text not only for graduates but also for researchers interested in understanding the basics of plant metabolism. This book will also serve as a reference book to researchers in plant breeding, agronomy, plant physiology and plant protection.

We are grateful to all those persons as well as various books, manuals, periodicals, magazines, journals etc. that helped in the preparation of this book. In spite of the best efforts, it is possible that some errors may have occurred into the compilation and editing of the book. Further queries, constructive suggestions and criticisms for the improvement of the book are always welcome and shall be thankfully acknowledged.

Reece Evans

Charlotte Roberts

CONTENTS

CHAPTER 1
INTRODUCTION

Biochemistry is a discipline of Chemistry that deals with the chemical composition of living organisms. It deals with interactions between living organic cells and their surrounding fluids/matter and is the study of important chemical processes occurring within living organisms.

Biochemistry is the subdivision of Biology and chemistry that can further be divided into three segments, namely, metabolism, structural Biology and enzymology. By the end of the 20th century, these three variants together have been able to successfully explain the process of living.

It is the study of how living things are made, how they work, and what chemicals they have. It also looks at the human body's muscles and bones. In a Biochemistry Class, students can learn how cells are made or how blood works.

Thus Biochemistry helps in understanding the chemical basis that gives rise to the process through biological molecules that are occurring between the living cells and within the cells. This, in turn, relates to the understanding of tissues and organs as well as the structure of the organisms and their functions. Therefore, Biochemistry can alternatively be defined as the study of molecular Biology that relates to the molecular mechanisms of biological phenomena.

Biochemistry is the study of structures and the interactions of biological macromolecules. These macromolecules include protein, nucleic acids, lipids, and carbohydrates present in your body. As a result, Biochemistry is being used in research related to botany, medicine, and gene enhancement. Biochemistry is now working on finding the secret of life and how biological molecules boost the processes that occur within the living cells. This study will help in understanding the whole working organism.

Originally, it was generally believed that life was not subject to the laws of science the way non-life was. It was thought that only living beings could produce the molecules of life (from other, previously existing biomolecules). Biomonomers and biopolymers are the structural basis of the four main macromolecules classes in biochemistry: carbohydrates, lipids, proteins and nucleic acids or biopolymers. Monomers are smaller micro molecules that are put together to make macromolecules.

Biopolymers are those macromolecules that are created when monomers are synthesized together. When they are synthesized, the two molecules undergo a process called catabolization or decomposition. A biomolecule is a chemical compound that naturally occurs in living organisms. Biomolecules consist primarily of carbon and hydrogen, along with nitrogen, oxygen, phosphorus and sulfur. Other elements sometimes are incorporated but these are much less common. Biomolecules are necessary for the existence of all known forms of life. For example, humans possess skin and hair. The main component of hair is keratin, an agglomeration of proteins which are themselves polymers built from amino acids.

Amino acids are some of the most important building blocks used in nature, to construct larger molecules. Another type of building block are the nucleotides, each of which consists of three components: a purine or pyrimidine base, a pentose and a phosphate group. These nucleotides, mainly, form the nucleic acids. Besides the polymeric biomolecules, numerous small organic molecules are absorbed or synthesised by living systems.

Plant biochemistry examines the molecular mechanisms of plant life. One of the main topics is photosynthesis, which in higher plants takes place mainly in the leaves. Photosynthesis utilizes the energy of the sun to synthesize carbohydrates and amino acids from water, carbon dioxide, nitrate, and sulfate. Via the vascular system, a major part of these products is transported from the leaves through the stem into other regions of the plant, where they are required, for example, to build up the roots and supply them with energy. Hence, the leaves have been given the name source, and the roots the name sink. The reservoirs in seeds are also an important group of the sink tissues, and, depending on the species, act as a store for many agricultural products such as carbohydrates, proteins, and fat.

In contrast to animals, plants have a very large surface, often with very thin leaves in order to keep the diffusion pathway for CO_2 as short as possible and to catch as much light as possible. In the finely branched root hairs, the plant has an efficient system for extracting water and inorganic nutrients from the soil. This large surface, however, exposes plants to all the changes in their environment. They must be able to withstand extreme conditions such as drought, heat, cold, or even frost, as well as an excess of radiated light energy.

Day after day the leaves must contend with the change between photosynthetic metabolism during the day and oxidative metabolism during the night. Plants encounter these extreme changes in external conditions with an astonishingly flexible metabolism in which a variety of regulatory processes take part. Since plants cannot run away from their enemies, they have developed a whole arsenal of defense substances to protect themselves from being eaten.

Then, in 1828, Friedrich Wöhler published a paper on the synthesis of urea, proving that organic compounds can be created artificially. The dawn of biochemistry may have been in 1833 the discovery of the first enzyme, diastase (also known as amylase), by Anselme Payen. Eduard Buchner contributed to the first demonstration of a complex biochemical process at the outside of a cell in 1896: the alcoholic fermentation based on cells coming from yeast.

Although the term biochemistry seems to have been first used in 1882, it is generally accepted that the formal coinage of biochemistry was mentioned in 1903 by Carl Neuberg, a German chemist. Previously, this area would have been referred to as physiological chemistry. Since then, biochemistry has advanced, especially since the mid–20th century, with the development of new techniques such as chromatography, X-ray diffraction, dual polarisation interferometry, NMR spectroscopy, radio isotopic labelling, electron microscopy and molecular dynamics simulations.

These techniques allowed the discovery and detailed analysis of many molecules and metabolic pathways of the cell, such as glycolysis and the Krebs cycle (citric acid cycle). Another significant historic event in biochemistry is the discovery of the gene and its role in the informational transfer in the cell. This part of biochemistry is often called molecular biology. In the 1950s, James D. Watson, Francis Crick, Rosalind Franklin, and Maurice Wilkins were instrumental in solving DNA structure and suggesting its relationship with genetic transfer of information. In 1958, George Beadle and Edward Tatum received the Nobel Prize for work in fungi showing that one gene produces one enzyme. In 1988, Colin Pitchfork was the first person convicted with murder with DNA evidence, which led to growth of forensic science. More recently, Andrew Z. Fire and Craig C. Mello received the 2006 Nobel Prize for discovering the role of RNA interference (RNAi), in the silencing of gene expression.

Plant biochemistry involves the study of the biochemistry of autotrophic organisms such as photosynthesis and other biochemical processes specific to plants. Plant Biochemistry presents each topic from the cellular level to the ecological and environmental levels, placing it in the context of the whole plant. Biochemical pathways are represented as route maps, showing how one reaction follows another. These maps emphasize the dynamism and flexibility of the plant facing the environmental challenges. The unique and wide–ranging approach of this book emphasizes the importance of teaching and learning pathways within the framework of what the pathway does and why it is needed.

Plant Biochemistry illustrates the impact of plants on human activity and success, in terms of their importance as a food supply and as raw materials for industrial and pharmaceutical products, and considers that humans can benefit from exploiting plant biochemical pathways. So, Plant Biochemistry is the science which studies chemical and physico–chemical processes that take place in the living organisms having the role to establish the material substrate of the life's phenomena. Usually the development of the organisms is possible due to the biochemical processes that take place in them in different environmental conditions. The influence exerted by the environment is reflected in the morphological and chemical structure of the organisms.

The fundamental problems of the animal biochemistry cannot be broken from the vegetal biochemistry and from the unicellular organisms. The universal principle of nature's integrity is the base of the chemical composition of the organisms and of the processes that are continuously developing and whose meaning is life. The organic matter is forming, transforming and degrading in any living body (plant, animal or microorganism); it doesn't differ from an organism to other by the component substances that are always the same or related also differs through secondary and unessential features determined by the proportion and the conditions of the formation and transformation of the organic matter. This is the reason why in the organisms' life there are some formation and degradation developing processes of the live matter and they vary between certain limits determining the variability and heredity of the organisms. Animal, plant and environment are forming an indissoluble unit and the biochemistry shows the diversity of the live matter.

Plant biochemistry is related to the animal and human activity where there are similar, however in all plants each cell has achieved a much more complicated metabolic process comparing to the animal organisms which are specialized on different functions. Animal, plant and environment form an inseparable unit, and only this way, in a comparing and general way, biochemistry shows a material world and the different shapes that it can take. So, when studying biochemistry there have to be treated at the same time both the fundamental chemistry of animals and that one of chemical plants and their need.

Biochemistry has a particular role in the acquisition of agricultural raw materials for finished products with good yields, quality, and economic conditions in high yield. Biochemistry is involved in obtaining such materials in the conduct of technological processes, to conserve packaging and finished products, waste recovery in manufacturing, etc. During their industrial transformation raw materials and agricultural products suffer biochemical processes necessary or harmful, caused by enzymes, processes that must be traced, suppressed or directed.

Any living organism is an energetic system. The amount of energy received from the environment is equal to the consumed energy, plus the accumulations. If the accumulations are positive the organism will gain weight, and if the accumulations are negative it will lose weight. In a healthy organism it is desired that the accumulation should be maintained constant in time. The totality of the energetic exchanges that take place permanently between the living body and the environment, meaning that the organism is spending (losing) energy replaced by the energy contained in food is called energetic expenses.

The energy from the nutrients is the potential chemical energy and it is contained in the C–H, C–O, and the C–N links from proteins, lipids and carbohydrates, these links being able to split in the conditions of the high temperatures, the pH and the chemical environment created by cellular activities. After the usual oxidative splitting of the nutrients, the organism stocks the potential chemical energy and transforms the energy in other energy forms necessary to it. If the energetic needs of the organism are reduced and the energetic effort is increased, the glucose, the fatty acids and the amino acids are being entirely catabolised and transformed in stocking forms such as glycogen, triglucides and proteins.

The basis of all the biological processes is the exchange of matter, information and energy, this exchange process being active between the living body and the environment. The nutrients, the environmental factor indispensable for life, are crude or processed substances designated for the nutrition and satisfaction of the energy necessary for the maintaining the vital functions, for providing the growing and development of the living body and for maintaining the aptitudes for a good physical and mental activity.

The nutrients are complex products which are capable to suffer the necessary transformations in the living organism in order to maintain life and forming the tissues. Along with the oxygen, the living body also incorporates water and minerals, proteins, lipids, carbohydrates and vitamins from environment. Except oxygen, which is essential for the respiration process and the water, which is the vital environment, the other compounds are imported. The consumption of substances coming from external sources is a fundamental biological necessity of any organism. Other substances are named nutritive principles or alimentary principles and for their definition are used the terms nutrients or trophynes. All metabolic processes take place in cells where the mitochondria is the main centre of the energogenesis, and the ribosomes are the place for the proteins' synthesis.

The functions of the nutritive principles found in food are:

- Catalytic–favouring the normal development of the biological processes;
- Energetic–providing the necessary energy for the development of the vital processes;
- Plastic providing the synthesis of the substances owned by the organism; etc.

Biochemistry studies the process of the incorporation of the substances which participate in the metabolic processes, the transformation of the nutrients in soluble substances capable to be absorbed by the blood and produced in the digestive apparatus.

Food's nutritive factors have some complex combinations that cannot be used and necessitate their transformation in simple elements which can be easily assimilated by the organism.

THE NUTRITIVE PRINCIPLES, FUNDAMENTAL SUBSTANCES

Modern biochemistry agrees with the existence of nine categories of nutritive principles: water, alimentary fibres, lipids, carbohydrates, macrominerals, oligominerals, proteins, bioactive substances and vitamins.

The chemical substances which are part of these categories have precise functions useful to the organism that ingrates them. Thus, some of them provide energy and heat, others provide construction material for the organism helping with the growth or the repair of the degraded tissues, and others participate directly or indirectly in the regulation of the biochemical and physiological processes from the organism. Yet none of these nutritive principles activate independently.

For a good metabolic activity and health of the organism the alimentation should contain all those nine nutritive principles in adequate amounts. Due to their structure and properties the organic substances are the basic material of the living bodies and activate in all living cells. The organic compounds mainly consist of: carbon, hydrogen, oxygen, nitrogen, phosphorus and sulphur. Proteins, lipids and carbohydrates are part of this category and serve as plastic and energetic material with a distinct structural and functional role in the living bodies.

The living matter is formed of two main groups of substances: inorganic and organic. The inorganic substances (water, minerals) are up taken from environment and are not synthesized by plants. The organic substances are the main compounds of the living bodies. According to their role in organism they can be classified as:

- plastic substances: (carbohydrates, proteins, and lipids) with a structural and energetic role;
- substances with informational role: nucleic acids;
- substances with catalytic role: enzymes;
- substances with a role in balance: enzymes, vitamins, phytohormons, pigments, antibodies etc.;
- substances with an energetic role: macroergic compounds;

Intermediary and final substances of the metabolism: glycosides, lignin, eteric oils, alkaloids, antibiotic substances, alcohols, amides, amines, esters etc.

The living matter is formed especially of the compounds of 12 chemical elements called plastic macroelements or constitutive elements. These are: C, H, O, N, P, S,

Cl, K, Na, Ca, Mg, and Si. These plastic elements, together with other elements a.i. oligoelements or microelements: Mo, Cu, Zn, Fe, B, Mn, Co, As, F, Br, Al etc. make less than 1% of the dry matter. The main source of supply with inorganic compounds for plants is the soil. Mineral substances, component part of the food, are important in forming tissues and participate in the biological and physiological processes of the body. They are classified into macroelements (K, Ca, P, Na, Mg, Fe, Cl, S), which exist in the body in large amounts and microelements (Zn, Cu, Cr, Mn, Co, Mo, I, F, Ni etc.) found in small quantities.

The microelements have a very important functional role in the reduction reactions, in the balance of the osmotic pressure and of the acid-base balance, and in maintaining cellular integrity even if they are found in small amount. Having also an essential role in the activation of some enzymes they become catalytic elements. Studies show the presence in the vegetal organisms of more than 60 chemical elements.

From these only seven are fundamental or macroelements and together with the C, H, O and N they form about 99% of the body weight. These macroelements are: Na, K, Ca, P, Mg, Cl, and S. The other mineral elements found in very small amount are called microelements. The most important microelements are: Fe, Cu, Co, Zn, I, Mn, Cd, Se, Nil, Cr, Pb, Hg, and As. Macro and microelements are indispensable for the physiological equilibrium of the vegetal, animal and human organisms and the lack of one or more minerals results in pathological manifestations more or less severe.

Carbon is part of those organic compounds existing in the organisms and represents 48–50% of the dry substance. Hydrogen represents 6–7% of the dry matter and just as carbon, is found in all organic substances and in the water from body. It also participates in the oxidation-reduction reactions. Oxygen is found in most of the organic compounds from the living bodies in a proportion of 41–43% and along with the hydrogen, takes part in the oxidatio-reduction processes.

Nitrogen is one of the basic elements of protides, and represents 16% of plant protides. In living tissues the proportion of nitrogen reaches 1.4–1.6%. Phosphorus is part of some important biologic compounds such as: enzymes, complex lipids, ATP etc. and interferes in many biochemical processes (photosynthesis etc.). In living bodies is found in a proportion near to 0.1–1.1%. Sulphur is part of some proteins, vitamins and carbohydrates etc. and can be found as 0.02–1.8% of the dry matter. It has important role in the growth and development of the living bodies.

The basic components in the alimentation of living organisms are the essential amino acids, polyunsaturated fatty acids, all the vitamins and minerals. Some components with high physiological activity, such as phospholipids, glycoproteidele, phosphoproteinele and many other compounds, are considered essential substances. The basic principles of a balanced nutrition are the proper correlations based on substances and biologically active nutrients, like proteins, lipids, carbohydrates, vitamins, mineral substances, depending on age, gender, activity at work and lifestyles.

Nutritional factors are subjected to a series of changes before their addition to the body tissues a.i. the burning and the disposal of unused remainders, formed as a result of metabolic processes taking place in the body.

CHAPTER

PLANT SYSTEMICS: AN OVERVIEW 2

WHAT IS A PLANT?

Plants are living things that grow on land or in water. From snowy mountain slopes to dry, hot deserts, plants can survive almost anywhere on Earth. Plants can be divided into two groups: flowering plants, for example, sunflowers, orchids, and most types of tree. The other group is nonflowering plants, which includes mosses and ferns. All plants make their own food, taking energy from sunlight. Unlike animals, plants cannot move from place to place, and most are rooted in the ground.

Plants are essential for any ecosystem. They provide all the energy for the ecosystem, because they can get energy directly from sunlight. They use a process called photosynthesis to use energy from the sun to grow and reproduce. They also must get nutrients from the soil. Those nutrients get into the soil when decomposers break down waste and dead materials. Plants require space to grow and reproduce. The size of your ecodome will influence how much space your plants have.

All other organisms in the food chain get energy from plants, either by directly eating them as herbivores do, or by eating plant eaters, like carnivores do. Omnivores can get energy either by eating plants directly or by eating herbivores. Likewise, decomposers get energy either from plants or from the animals that eat them. Since all the energy in your ecosystem comes from plants, you'd better have a lot of them.

There are several different kinds of plants, and not all animals can eat all kinds of plants.

PLANTS AND THE EVOLUTION OF LIFE

As shown in Fig. 1, plants are thought to have evolved from an aquatic green alga protist. Later, they evolved important adaptations for land, including **vascular tissues**, seeds, and flowers. Each of these major adaptations made plants better suited for life on dry land and much more successful.

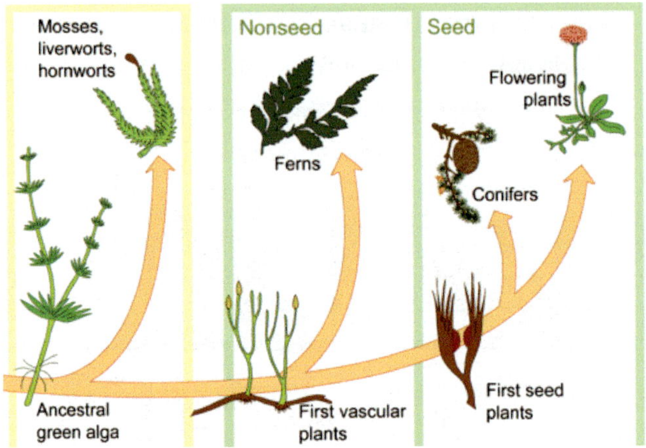

Fig. 1 From a simple, green alga ancestor that lived in the water, plants eventually evolved several major adaptations for life on land.

THE EARLIEST PLANTS

The earliest plants were probably similar to the stonewort, an aquatic algae pictured in Fig. 2. Unlike most modern plants, stoneworts have stalks rather than stiff stems, and they have hair-like structures called rhizoids instead of roots. On the other hand, stoneworts have distinct male and female reproductive structures, which is a plant characteristic. For fertilization to occur, sperm need at least a thin film of moisture to swim to eggs. In all these ways, the first plants may have resembled stoneworts.

LIFE ON LAND

By the time the earliest plants evolved, animals were already the dominant organisms in the ocean. Plants were also constrained to the upper layer of water that received enough sunlight for photosynthesis. Therefore, plants never became dominant marine organisms. But when plants moved onto land, everything was wide open. Why was the land devoid of other life? Without plants growing on land, there was nothing for other organisms to feed on. Land could not be colonized by other organisms until land plants became established.

Plants may have colonized the land as early as 700 million years ago. The oldest fossils of land plants date back about 470 million years. The first land plants probably resembled modern plants called liverworts, like the one shown in Fig. 2.

Colonization of the land was a huge step in plant evolution. Until then, virtually all life had evolved in the ocean. Dry land was a very different kind of place. The biggest problem was the dryness. Simply absorbing enough water to stay alive was a huge challenge. It kept early plants small and low to the ground. Water was also needed for sexual reproduction, so sperm could swim to eggs. In addition, temperatures on land were extreme and always changing. Sunlight was also strong and dangerous. It put land organisms at high risk of mutations.

Fig. 2. The first land plants may have been similar to liverworts like this one.

VASCULAR PLANTS EVOLVE

Plants evolved a number of adaptations that helped them cope with these problems on dry land. One of the earliest and most important was the evolution of vascular tissues. Vascular tissues form a plant's "plumbing system." They carry water and minerals from soil to leaves for photosynthesis. They also carry food (sugar dissolved in water) from photosynthetic cells to other cells in the plant for growth or storage. The evolution of vascular tissues revolutionized the plant kingdom. The tissues allowed plants to grow large and endure periods of drought in harsh land environments.

In addition to vascular tissues, these early plants evolved other adaptations to life on land, including lignin, leaves, roots, and a change in their life cycle.

- Lignin is a tough carbohydrate molecule that is hydrophobic ("water fearing"). It adds support to vascular tissues in stems. It also waterproofs the tissues so they don't leak, which makes them more efficient at transporting fluids. Because most other organisms cannot break down lignin, it helps protect plants from herbivores and parasites.
- Leaves are rich in chloroplasts that function as solar collectors and food factories. The first leaves were very small, but leaves became larger over time.
- Roots are vascular organs that can penetrate soil and even rock. They absorb water and minerals from soil and carry them to leaves. They also anchor a plant in the soil. Roots evolved from rhizoids, which nonvascular plants had used for absorption.
- Land plants evolved a dominant diploid sporophyte generation. This was adaptive because diploid individuals are less likely to suffer harmful effects of mutations. They have two copies of each gene, so if a mutation occurs in

one gene, they have a backup copy. This is extremely important on land, where there's a lot of solar radiation.

With all these advantages, it's easy to see why vascular plants spread quickly and widely on land. Many nonvascular plants went extinct as vascular plants became more numerous. Vascular plants are now the dominant land plants on Earth.

PLANT SYSTEMATICS

Plant systematics is a science that includes and encompasses traditional taxonomy; however, its primary goal is to reconstruct the evolutionary history of plant life. It divides plants into taxonomic groups, using morphological, anatomical, embryological, chromosomal and chemical data. However, the science differs from straight taxonomy in that it expects the plants to evolve, and documents that evolution. Determining phylogeny - the evolutionary history of a particular group - is the primary goal of systematics.

Classification Systems For Plant Systematics

Approaches to classifying plants include cladistics, phenetics, and phyletics.

- Cladistics: Cladistics relies on the evolutionary history behind a plant to classify it into a taxonomic group. Cladograms, or "family trees", are used to represent the evolutionary pattern of descent. The map will note a common ancestor in the past, and outline which species have developed from the common one over time. A synapomorphy is a trait that is shared by two or more taxa and was present in their most recent common ancestor but not in earlier generations. If a cladogram uses an absolute time scale, it is called a phylogram.
- Phenetics: Phenetics does not use evolutionary data but rather an overall similarity to characterize plants. Physical characteristics or traits are relied upon, although the similar physicality can reflect evolutionary background as well. Taxonomy, as brought forth by Linnaeus, is an example of phenetics.
- Phyletics: Phyletics is difficult to compare directly with the other two approaches, but it may be considered as the most natural approach, as it assumes new species arise gradually. Phyletics is closely linked to cladistics, though, as it does clarify ancestors and descendants.

TAXONOMY AND SYSTEMATICS

The word systematics comes from the word 'systema', meaning the systematic arrangement of the organisms. It takes into consideration the evolutionary relationship of the organisms. Plant systematics deals with interrelation between plants and their evolutionary descent. Systematics studies biological diversity and organises the information into a classification.

Organisms are classified on the basis of similarities, closeness or relationship between them. It shows the phylogenetic relationship between different organisms and shows their line of descent. The similarities among individuals show that they might have developed from the common ancestor. It shows the evolutionary pathway of modern living organisms. Closely related organisms are included in a group, which share a common gene pool.

Organisms are categorised into different taxonomic categories according to the similarities and specific features. The different taxonomic categories in their hierarchical order are:

- Kingdom
- Phylum
- Class
- Order
- Family
- Genus
- Species

The number of common characteristics decreases as we move from species to the kingdom, where species having fundamental similarities and organisms in the same kingdom having least common features.

List of Systems of Plant Taxonomy

The earliest system of classification considered only a few vegetative characters. Modern taxonomic studies have been more elaborate and taken into consideration various morphological, cellular and molecular characteristics, e.g. cellular and reproductive features, mode of nutrition, habitat, evolutionary relationships, etc. along with morphological features.

There are three main types of systems for plant classification. Here is the list of systems of plant taxonomy:

- Artificial system of Classification
- Natural system of Classification
- Phylogenetic system of Classification

Artificial systems

Artificial systems were the earliest systems, which attempted to classify organisms based on a few superficial characters.

These were important in the history of biological classification as this was a novel attempt to organise living organisms. The demerit was that it didn't consider morphological details and the evolutionary relationship. They gave equal importance to vegetative and sexual characters but it is not true. Vegetative characters are greatly influenced by the environment. As a result, the closely related species were kept apart.

Aristotle classified plants more than 2000 years ago on the basis of simple morphological characters into herb, shrub and trees.

Theophrastus in his book "Historia Plantarum or Enquiry into plants" attempted to arrange plants in various groups based on how plants reproduce and its uses. He is called "Father of Botany".

Carl Linnaeus is known as the "Father of Modern Taxonomy". In his book "Systema Naturae" (1735), he gave the hierarchical system of classification of the natural world into the plant kingdom, the animal kingdom and mineral kingdom.

He understood the importance of floral characters and classified plants based on the number of stamens present in them. It is also known as the sexual system of classification.

Linnaeus kept on adding new work to his publications. In "Species Plantarum" (1753), he gave a brief description of all the species known to him. He described around 7,300 plant species in it. He divided the plant kingdom into 24 classes based on the structure, union, length and the number of stamens. E.g. Monoandria (1 stamen), Diandria (2 stamens), Polyandria (more than 12 stamens), Monoadelphia (stamens united in a single bundle), Monoecia, Dioecia, Polygamia (polygamous plants), Cryptogamia (flowerless plants), etc.

He gave the Binomial nomenclature system. In "Philosophia Botanica", he had given rules for naming every species. It is called binomial because each name has two components, genus name and species name, e.g. Solanum melongena (brinjal), Solanum tuberosum (potato) having the same genus but different species name.

Natural systems

In this system of classification, more characters were considered while classifying. It was based on the natural similarities of vegetative and floral characters among the organisms. It took into consideration various external and internal features like the anatomy of a cell, types of embryo and phytochemistry.

Bentham and Hooker system of classification was the most important natural system of classification of flowering plants. They classified plants into Cryptogams (non-flowering plants) and Phanerogams (flowering plants).

In the Bentham and Hooker classification system, 97,205 species of seed plants were classified into 7569 genera and 202 families. The flowering plants were further classified into three classes namely Dicotyledon, Gymnosperm and Monocotyledon.

It helped to determine relationships between the various groups of plants but failed to identify phylogenetic relationships among different groups of plants. It wrongly placed gymnosperms between monocotyledons and dicotyledons.

The Phylogenetic system of Classification

This system is based on evolutionary sequence and genetic relationship. The phylogenetic system of classification was developed after the publication of Darwin's theory of evolution. Apart from the morphological characteristics found from fossil records, genetic constituents were also considered. It has been widely accepted by biologists all over the world. According to this system, all the organisms belonging to the same taxa originated from the common ancestor.

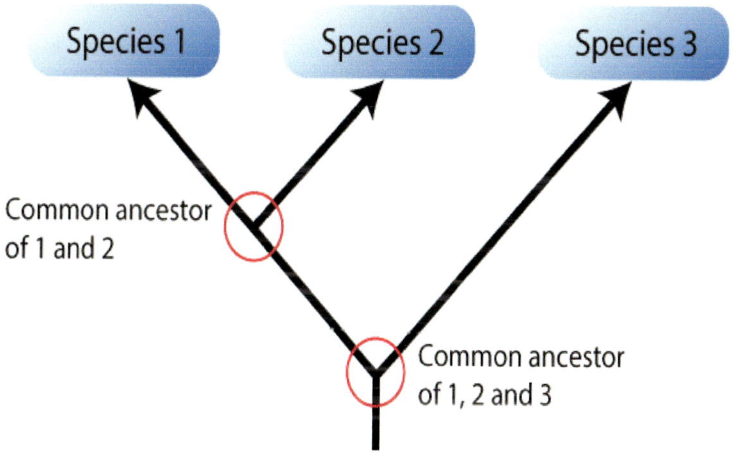

Fig. 3. The Phylogenetic system of Classification

Various scientists namely Engler and Prantl, Hutchinson, Takhtajan, Cronquist, Rolf Dahlgren and Robert F Thorne contributed to the phylogenetic system of classification.

The two main phylogenetic systems of classification are:

- Engler and Prantl system of classification: In this type of phylogenetic system of classification, floral characters like single whorl or no perianth, unisexual flowers pollinated by wind were considered primitive characters as compared to two whorls in the perianth and bisexual flowers pollinated by insects. They arranged plants based on the increasing complexity of the flower morphology.

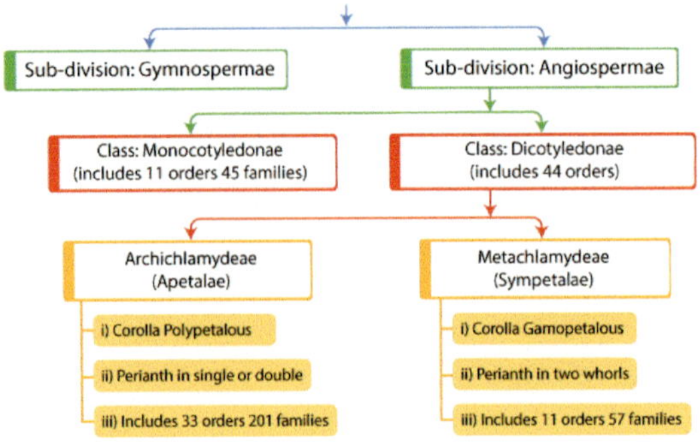

Fig. 4. Division: Embryophyta (Siphonogama)

- The plant kingdom was divided into 13 divisions:
  - 11 are **Thallophytes**
  - 12th **Embryophyta Asiphonogama**, i.e. plants having embryos but pollen tubes are absent (bryophytes and pteridophytes)
  - 13th **Embryophyta Siphonogama**, i.e. plants with embryo and pollen tubes (seed plants)
- **Hutchinson's classification: John Hutchinson,** a British botanist gave the phylogenetic system of classification in his book named **"The Families of Flowering Plants"** in two volumes in 1926 and 1934. In his classification, Hutchinson divided angiosperms into monocotyledons and dicotyledons.
- Dicotyledons were further divided into two divisions, namely, Lignosae (woody plants) and Herbaoae (herbaceous plants)

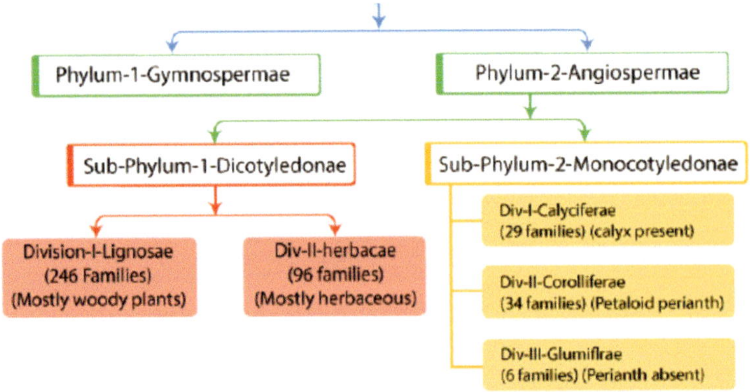

Fig. 5. Hutchinson's classification: Flowering Plants

- Monocots were divided into 3 divisions on the basis of flower morphology, namely, Calyciflerae (calyx present), Corolliferae (petaloid perianth) and Glumiflorae (perianth absent)

MODERN TAXONOMIC ADVANCEMENTS

With the advent of molecular biology, many techniques to identify genetic materials have been developed. This has equipped us to compare individuals at different taxonomic levels and resolve the difficulties of classifying them even if there is no fossil evidence.

- **Numerical taxonomy:** It is done by using computers and all the observable characters are taken into consideration. Each character is assigned with a code and a number. Hundreds of characters can be considered together and given equal importance.
- **Cytotaxonomy:** it utilises cytological information like chromosome number, shape size, etc. to understand the taxonomy.
- **Chemotaxonomy:** Use of chemical constituents of plants for taxonomic studies is known as chemotaxonomy. Proteins, amino acids, nucleic acids and peptides, starch grains, wax, fat, oil, phenols are studied in chemotaxonomy.

IMPORTANCE OF PLANT TAXONOMY

- It gives a detailed overview of various morphological and anatomical structures of a plant species
- It organises all the information of plants into an orderly fashion
- It indicates the phylogenetic relationship between species and its ancestry
- Plant taxonomy enables to identify any unknown species and its place in the classification by comparing with known species
- Analysis of genetic constituents can be done on the basis of systematics
- It is used to scientifically name any species, which helps in the uniformity of the name around the world and avoids confusion
- It helps to understand the biodiversity present at a place
- It helps in recording all the living species known until now
- Taxonomy is widely used in agriculture, medicine and forestry

PLANT SYSTEMATICS IN THE AGE OF GENOMICS

As plant biologists enter a new era in which comparative genomics promises to address fundamental questions in botany, such as unraveling metabolic and regulatory networks, the inestimable value and usefulness of robust systematic studies quickly become clear. In simplest terms, systematic studies can indicate which genomes in the plant kingdom to search, sample, and study for the answers to questions relating to the evolution of chemical and physical structures and their synthesis or ontogeny. After several model and crop species have been sequenced, the next phase of plant genomics will necessarily build on new phylogenies that are greatly assisted by molecular techniques and whose interpretation and applications will be guided by "traditional" botanical knowledge.

Plant systematics was long considered to be an "artful science," but well before the application of molecular techniques to systematics, semisubjective authority was supplanted by rapidly developing analytical methods and the computers that run them. In the age of genomics, the art of modern plant systematics lies in its applications and its links to other disciplines; conversely, the applications of genomics to an expanding array of plant species will be grounded in plant systematics, itself still based largely on field work and knowing the plants.

Much is new—and much is not—for plant systematics in the age of genomics. Molecular techniques have introduced vast and numerous independent data sets, and there are continual advances in preparing DNA, sequencing genes, aligning sequences, and designing software for interpreting the data. As a consequence of this increased accessibility, mainstream plant systematics has been able to incorporate molecular approaches, which no longer occupy a separate domain but rather constitute part of the normal repertoire of skills for systematic botanists.

The special usefulness of molecular approaches in analyzing phylogenetic relationships at higher ranks has resulted in still unresolved but clearly better and dramatically new classifications. These developments, coupled with other advances in phylogenetic analysis, now place systematics in a key position among other disciplines in biology, with increasingly diverse and powerful applications in investigations of biosynthetic and developmental pathways, natural products, origins and migrations of evolutionary lineages, and conservation. Whether it occupies the hub or spokes, more than ever plant systematics is needed to make the genomics wheel roll.

The biggest non-news is that molecular techniques have in fact not revolutionized methodologies in systematics. Instead, molecular data have rather rapidly been accommodated in existing analytical methods whose revolution—cladistics—had come and for some time had been the new order. Once computers could be harnessed to execute complex pattern analysis and resample data thousands of times for statistical rigor, phylogenetic systematics or cladistics had overruled the authority represented by a few great intellects in favor of greater objectivity and more reproducible results. Systematists were already wrestling with issues about adequate sampling and about merging data sets before DNA sequences began flooding the market.

News perhaps for non-systematists, but not for systematists, is a greater need than ever before for traditional botanical knowledge and activities. Field work, collections, diversity surveys, floras, monographs, and conservation efforts still provide the primary means for working with the physical materials needed for investigations in systematics and genomics (e.g. fresh tissue for RNA extraction and synthesis of cDNA expressed sequence tag libraries), as well as for formulating hypotheses, interpreting the results, and making useful applications of those results, thereby linking the genes to the whole plants and the world outside them.

Phylogenetic Analysis

The field of plant systematics has undergone a renaissance during the last 20 years. This is due primarily to the incorporation of cladistic methodology and subsequently DNA sequence data into phylogenetic reconstruction, which is predicated on the recognition of monophyletic ("natural") groups. Cladistics represents an approach whereby the data pertaining to terminals are analyzed using the criterion of parsimony to minimize the number of ad hoc hypotheses needed to explain the pattern inherent in the data. The results are represented as "trees" that are termed "unrooted networks." If the same data set is again analyzed by parsimony, the same answer will be obtained. When used for the purposes of phylogenetic systematics, the tree is rooted between the ingroup and the outgroup, i.e. the group with which the ingroup is compared. This tree becomes a phylogenetic tree that constitutes a hypothesis of the genealogical relationships of the taxa under study and can be subjected to further tests with additional data.

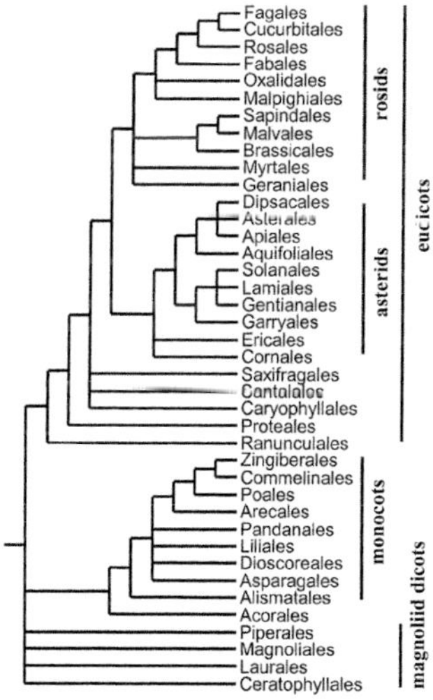

Fig. 6 A conservative estimate of phylogenetic relationships among the orders of flowering plants (modified fromAngiosperm Phylogeny Group, 1998). Those plants with a single cotyledon (monocots) are a monophyletic clade nested within at least two major lineages of dicots; the eudicots are further divided into rosids and asterids. Arabidopsis is a rosid eudicot classified within the order Brassicales; rice is a member of the monocot order Poales.

Systematists use cladistics to produce phylogenetic trees that can be used for purposes of phylogenetic classifications. Such trees must remain as hypotheses because there is only one "true" evolutionary tree, which is based upon a long history of life on Earth, and even if we obtained that true tree we would not know it definitively because we cannot directly observe that history. We can assume, however, that such trees from phylogenetic analyses are nearer to the true tree than would be a randomly generated tree, and these hypotheses of relationships provide our best estimate of genealogy and may be used for starting points for further research in systematics and for understanding the biology of the organisms under study.

New Alignments

These phylogenetic studies and the resulting trees have had a substantial impact on systematists' hypotheses about plant relationships and ultimately on our systems of classification. The new cladograms of higher ranks, based largely on combined gene sequence data from an international team of botanists, differ in a number of fundamental ways from past classification systems.

Perhaps more than any in the previous half century, the comprehensive (down to the rank of family) classification system of flowering plants by Cronquist (1981) has been the most widely followed. That system divided the flowering plants into two major classes that will be familiar to most readers: Magnoliopsida (including all the dicotyledons or "dicots") and Liliopsida (including all the monocotyledons or "monocots").

The long-held notion that the flowering plants are first and foremost separated into monocots and dicots was not supported by cladistic analyses of morphological data, and the need for change was later confirmed by molecular phylogenetic studies. The monocots continue to be recognized as a monophyletic lineage, but that lineage is embedded among separate clades that are characterized by two cotyledons. This led to the realization that two cotyledons are in fact the generalized (i.e. "primitive") condition for seed plants as they occur also in cycads, Ginkgo biloba, many conifers, and the Gnetales, while having one cotyledon is uniquely derived (autapomorphic).

It appears that a more natural (monophyletic) grouping of the angiosperms would consist of the "eudicots" (see Fig. 6), with typically three pollen apertures; the monocots; and the magnoliid dicots, comprising several diverging lineages. This is harmonious with the formal higher rank classification proposed recently by an international team of plant systematists calling themselves the Angiosperm Phylogeny Group (1998). Theirs was the first attempt to realign the orders and families of flowering plants into a phylogenetic system based on an analysis of molecular data. A summary of that system is presented in the form of a phylogenetic tree or cladogram (Fig. 6).

A spectrum of applications

Phylogenetic studies already have direct applications in agriculture, natural products chemistry, biomedical sciences, and other disciplines. In the age of genomics, the breadth and power of these applications will increase as interpretation of phylogenies helps to identify those taxa for which genomic treatments will answer fundamental and in some cases long-standing questions about metabolic and regulatory networks in all the evolutionary corners of the plant kingdom.

The evolutionary origin of symbiosis between plants and nitrogen-fixing bacteria living in root nodules represents an example of applications in agriculture. Known to occur in only 10 different flowering plant families, it had traditionally been thought that these families were only distantly related to one another. In the new system of angiosperm classification, however, all nitrogen-fixing angiosperms are members of only four ordinal clades (Fabales, Cucurbitales, Rosales, and Fagales) that are linked as a monophyletic group. This information suggests that the predisposition for root-nodule symbiosis and nitrogen fixation in angiosperms evolved only once. From a phylogenetic perspective, therefore, the investigation of genes that contribute to nodule formation would be best studied within other genomes from this clade, rather than randomly throughout the angiosperms.

Phylogenetic trees are useful also in directing research in natural products chemistry. Taxol is a compound known only from the yew family (Taxaceae). If we wish to look for additional sources of taxol and other taxanes, rather than randomly sample the plant kingdom, we can focus on the nearest relatives as understood from a phylogenetic tree of the seed plants. This in turn leads us to investigate the Podocarpaceae because current analyses indicate that this is the sister group to the yew family, i.e. these two conifer families share a more recent common ancestor with each other than either does to any other plant family. That is indeed the case, and taxanes have recently been reported in the Podocarpaceae.

Chemical information is often used in data sets for phylogenetic analysis. The compound β-N-methylamino-L-Ala (BMAA, a non-protein alpha amino acid) was first discovered in the cycad genus Cycas; it binds to Glu receptors and is implicated in the ailment Guam Dementia. Subsequently, it has been demonstrated to occur in all other cycads examined and in no other plants, so the shared derived compound BMAA is termed a "synapomorphy" for the cycads. It is also highly toxic to most herbivores. After Glu receptors were discovered in plants, treatment of Arabidopsis with BMAA led to the discovery that a plant-derived neurotoxin may act as a signaling molecule in plants, involved in the control of morphogenesis.

New collaborative studies on cycad genomics initiated by The New York Botanical Garden, Cold Spring Harbor Laboratories, and New York University are being used to identify cycad genes involved in the synthesis of this neurotoxin and to identify its potential target genes in plants, among other applications. In addition, because of the group's basal phylogenetic position, sequences from a cycad genome will be valuable for evolutionary studies.

Mustard oil glucosinolates are known to be produced in at least 15 different families of plants. Depending on the system of classification, the biochemical pathway for these secondary metabolites was thought to have arisen and then to have been lost on multiple occasions. Molecular phylogenetic information, however, indicates that all mustard oil-producing families are part of the same ordinal clade, Brassicales, with the exception of the genus Drypetes (a member of Malpighiales), indicating only two evolutionary origins for mustard glucosinolates. Moreover, these mustard oils are derived from two different biosynthetic pathways, thus showing that even though the final product may be the same, the ways they were synthesized are not; hence they are not the same compounds in an evolutionary, i.e. historical, sense. This new information clearly has profound implications for the plant biologist interested in the genetic basis for mustard oil glucoside synthesis in the plant kingdom.

RECIPROCAL ILLUMINATION AND THE APPLICATIONS OF PHYLOGENETIC STUDIES

When morphological characters are used in a data matrix, they are coded as states of a character based upon presumed evolutionary homology, and phylogenetic analysis can be used to examine their developmental equivalence. The example of the two origins of the mustard oils demonstrates the problems of character coding based upon descriptors and of divorcing descriptors from process. A descriptor may be accurate at a structural level but not at the developmental level, and this can be detected through phylogenetic analysis and subsequent "reciprocal illumination", the process of using the phylogenetic tree to test the notion that a character reappearing independently is not really the same (i.e. evolutionarily homologous) despite its apparent similarity.

An example of reciprocal illumination concerns the origin of storage products in the seeds of some monocotyledonous plants. There is a storage tissue derived from cells of the nucellus occurring in some monocots that is termed "perisperm" and is formed as storage tissue instead of the more typical endosperm. Phylogenetic analyses of the monocots using both morphological and molecular sequence data have shown that perisperm has evolved three times within the monocots, thus indicating that not all perisperm is the same or minimally had the same history.

Perisperm has been described for Acorus, the putative sister taxon to the rest of the monocots, for Hydatella, nested within the monocots, and for Cyanastrum, also nested within the monocots in a different clade from Hydatella. Recent work has demonstrated that the perisperm of Acorus is derived from the epidermis of the nucellus and that of Hydatella is derived from subepidermal layers of the nucellus, while Cyanastrum in fact does not even have perisperm but rather its seed storage tissue is derived from the chalazal tissue of the ovule. Thus, as the tree indicated, these are not homologous but different in origin, despite having a similar function and appearance. The implication is that perhaps different genes are controlling storage tissue development in these three taxa.

Clearly, the construction of phylogenetic trees, tempered by testing with reciprocal analysis, is a powerful tool for understanding the biology of plants. This is taking on increasing importance for our understanding of development and processes; in turn, this knowledge can be applied to various purposes that include improving agriculture, exploring natural products chemistry, and contributing to biomedical science. In the current age of genomics, the applications of phylogenetic studies constitute one of several major criteria for deciding which plant genomes to tackle next.

CHOOSING CLADES

For future comparative genomics studies, the importance of classification systems based on solid phylogenetic information cannot be overemphasized. Zoologists have already recognized this fact and have recently begun to debate which mammalian species should be targeted for sequencing now that the human genome has been nearly completed. Candidates for new model taxa include the mouse (Mus musculus), rat (Rattus norvegicus), chimpanzee (Pan troglodytes), and rhesus monkey (Macaca mulatta).

As researchers pointed out, however, these candidate species may not necessarily be the best from a phylogenetic perspective, as they are all members of the same evolutionary lineage. Instead, the authors proposed a phylogenetically more even sampling that includes organisms rather distant from as well as quite close to human beings, while factoring in other criteria such as economic value, biomedical relevance, and genome size.

The same approach should be applied to plants. With the Arabidopsis genome sequenced, and that of rice (*Oryza sativa*) nearing completion, we will soon be able to make direct comparisons between these two species, which are phylogenetically distant. Subsequent genome sequencing of other model plant species, such as tobacco (*Nicotiana tabacum*), snapdragon (*Antirrhinum majus*), corn (*Zea mays*), tomato (*Lycopersicon esculentum*), or Gerbera jamesonii is warranted, but it should be noted that these are all derived from the same common ancestor as Arabidopsis and rice. Selection of species for genomic studies must consider phylogenetic position, such that the sampling allows comparisons among closely related genomes as well as among distantly related ones throughout the plant kingdom.

We propose that the next wave of model plants should include species from these three categories: (a) species not too distantly related to currently used model taxa (e.g. other members of the order Brassicales close to Arabidopsis); (b) representatives of species several lineages derived from the same common ancestor as Arabidopsis and rice but close to neither (e.g. *Lamiales* and *Caryophyllales*); and (c) angiosperms that are not descendants of the last common ancestor of rice and Arabidopsis, i.e. the point of divergence between the monocots (which include rice) and the eudicots (which include Arabidopsis). Examples of the latter are the Piperales and Laurales.

Back to basics

In the genomics age, molecular techniques represent an immensely powerful resource that, surprising to some, greatly increases rather than obviates the need for non-molecular data sets, the botanists who can interpret them, and the means for generating them, i.e. field work, floristic studies, herbarium collections, and the building blocks of basic taxonomy, morphology and anatomy. Sequences and cladograms are merely tools, and one has to know the plants in order to pose the questions, structure the sampling, select the characters (including which genes), and interpret the results. As the examples of applications discussed above show, cladograms often pose as many questions as they answer; there is no reciprocal illumination if there is no light on the other end.

Plant systematics provides molecular and phylogenetic investigations with the physical and informational materia prima for the data sets that make them tick, and this can be achieved only through greatly increased activity in basic taxonomic and floristic research, and vastly accelerated and better-coordinated programs of field work and plant collections. A useful relationship between systematics and those molecular and phylogenetic investigations depends heavily on accurately identified specimens, well-prepared ancillary collections (silica-dried material and others), and carefully planned sampling of taxa and/or populations representing the fullest possible range of geography, variation, and phylogeny in the group. Monographs and floras are the most efficient way of generating the needed information and specimens, as well as putting into practice the extensive phylogenetic realignments now under way. Taxonomists are needed to execute the numerous nitty-gritty nomenclatural changes involved.

Returning to the example of the cycads, participation in long-term floristic projects and years of field work ultimately powered what has followed. Field work has documented the diversity of the group, including the discovery of a new genus in this conspicuous group in recent years. It produces the physical material that has formed the basis not only for monographic work in the family but also for chemical and subsequently for molecular investigations.

Work on the labyrinthine nomenclature of this group, which contains a number of ornamentals, established which legitimate genus and species names are linked to which historic reference specimens. In turn, morphology- and anatomy-based taxonomic work, culminating in monographs and revisions, has provided the means for identifying the specimens being used for anything from floras to genomics.

The data sets produced by the taxonomic work are the raw material for generating the cladograms that are making it possible to thoroughly sample the group (i.e. one taxon from each clade) for any subsequent investigation, including natural products and genomic and other molecular investigations. Thorough sampling of the cycads and related groups for the presence of BMAA subsequent to its discovery in Cycas circinalis (Vega and Bell, 1967) tested and confirmed the hypothesis that this compound is unique to the cycads, resulting in the focus on that group by the current genomic investigations of cycads mentioned above.

Conservation

Floristic work, built on taxonomy and phylogenetic analyses that often incorporate geographic data, drives plant conservation. Data on the distributions, habitat preferences, and population structures of organisms identify centers of endemism and of diversity, as well as rare and/or threatened species; this constitutes the first layer of information for setting conservation priorities.

Sound phylogenetic trees can establish another layer of information and analysis for maximizing the conservation of genetic diversity, the basis of biodiversity. Habitat destruction and threats of extinction continue while the major realignments of taxa highlight how much of the plant kingdom is still a biological black hole in many ways: as pointed out above, to date most plant taxa used in some way as model taxa fall in one clade, so entire sectors of plant diversity (e.g. families or genera endemic to a small region) could go extinct before we have any notion of significant differences in chemistry, developmental processes, or even gene regulation. Because we cannot realistically save all areas from habitat destruction, phylogenetic studies can help us make intelligent choices that will allow us to conserve areas with the highest genetic and hence taxic diversity. It should also allow us to identify evolutionarily significant species whose DNA should be stored in banks; this will complement existing efforts to create DNA banks for rare and threatened species (e.g. Randell and Morden, 1999). It is also necessary to conserve the engines of evolution: phylogenetic analyses incorporating molecular data sets can help guide the protection of evolutionary processes by illuminating historical migrations and geographic and ecological origins.

Conclusions

During the 1980s, plant systematics underwent a decade-long revolution in methodology that has enabled the discipline to adapt itself to absorb and analyze the subsequent avalanche of molecular data sets, and now molecular techniques are part of the mainstream of plant systematics. In the age of genomics, the horizons of plant systematics have vastly expanded. New phylogenetic realignments and diversifying links to other disciplines give systematic studies an increasingly important role in targeting species for genomic treatments and other research that will help elucidate—and conserve—the evolution of metabolic pathways and developmental processes. Execution and interpretation of the entire effort will continue to be grounded in botanical field work and traditional botanical knowledge.

CHAPTER

TAXONOMY AND PHYLOGENY SYSTEMATICS 3

Modern plant taxonomy is an extensive feld that is increasingly benefting from achievements in many other felds of the biological sciences, such as genetics, cytology, molecular biology, ecology, phytogeography, and paleobotany. Te infuence of these branches of science on plant taxonomy is expressed not only through a signifcant multiplication of information and detailed data but also through new phylogenetic concepts. Reconstructing phylogenetic compounds that unite all lines of evidence is a great challenge. Te process of describing a new species is considered equivalent to generating a new hypothesis in phylogenetics. Hence, taxonomy and phylogenetics are inseparably connected with each other.

Systematics today is a vigorous and exciting field that has been given great impetus by the discoveries of molecular biologists, who now are describing organisms at their most fundamental level—the DNA sequences of the cells—and providing the systematists new data on which to base their phylogenetic trees. Phylogenetic trees are the graphic representation of the evolutionary divergences of organisms that put together on the same branches the organisms most closely related, with oldest ancestors near the base, youngest descendants near the top. The trees obtained from the DNA sequences basically trace the history of how the genes have changed through time.

Even if organisms did not 'naturally' occur in groups it would be important for biologist to define groups; there is simply so much biological diversity that some sort of 'filing system' is required to organize it all. Thus, there are two aspects to organizing organisms into groups: (1) a practical aspect, producing a way to manage all the variety of organismal life, to arrange its vast diversity, (2) a mechanistic (process) aspect, producing a system that will allow biologists to understand mechanisms that result in the patterns of diversity, e.g., why there are groups. To a large extent these two aspects can be satisfied simultaneously, i.e., there are systems that categorize life easily and also allow users to study the processes (evolutionary mechanisms) that lead to the categories. However, the biological landscape is vast; clusters range from groups (species, or perhaps subspecies) that might have less than 1000 organisms that all 'look' very similar to groups (families, orders) that are comprised of millions of individuals, with members that do not 'look' very similar yet do possess some 'fundamental similarities.'

With such a span in scale one should appreciate that there may be situations where a system that is useful to consider the evolutionary relationships (i.e., the phylogeny) between groups of organisms may not very handy as a filing system. Or, looking at it from a different perspective, classification schemes that easily organize life's diversity may not do so in a way that reflects the phylogeny of different groups.

An example familiar to most biology students would be 'reptiles', a handy group in terms of classification (reptiles are vertebrates that aren't amphibians, birds or mammals), but a group that does not represent phylogeny.

We will be dealing with scientific names, and the groups that they describe, throughout the book and it is important to appreciate the basis and the limitations of the naming. Shakespeare asks "what's in a name?" and answers with the implication that names are basically trivial. But names are profoundly significant in ways that often are not appreciated. Like water to a fish and air to humans, we are so immersed in names that we rarely stop to consider them.

But names say much about humans, about how we think and what we think. Indeed, it is probably the case that names not only reflect how we think, they may actually dictate how we think. Names reflect the organization by which we view things and the way we process the information that we receive.

While classification is useful whenever one is faced with a large number of variable entities, we need to consider to what extent our classification is a reflection of our thought (i.e., we are imposing order on something that isn't really ordered), or a reflection of reality (i.e., there is an order that we are describing), or something in between. Stated another way, names reflect an organization and it is important to consider whether the organization is inherent to what is being named or inherent to our minds.

Keep in mind that naming is a grouping process, i.e., it is a mechanism to put 'things' (in our case biological things) together. Faced with diversity, humans lump things together into categories, putting similar things together into groups; this makes the diversity more manageable and this is what classification (naming) is all about.

At the same time, one should appreciate that any classification, in fact, the very process of 'naming', results in a loss of information to the extent that any organism varies from the norm that characterizes the group. The name 'tree' sets up a classification of living things, one that is both useful and arbitrary; not all trees are the same and describing something as a tree strips an organism its individuality. The same could be said for the terms 'sequoia', 'Sequoiadendron', and 'Sequoiadendron giganteum'.

For any named biological group, it is important to consider how 'real' the group is: do the entities naturally fall together or are we just putting them together as a means to simplify the system. If the groups are 'real' (valid) a biologist might consider what processes might relate to their validity; what process forms the group?

Thus the fundamental question to address when naming groups of things is what criteria will be used to group them. For instance, if you are classifying motor vehicles one might group them based on color, on manufacturer, or on type of vehicle. When considering organisms, deciding what criteria to group them on is a tough question: organisms are exceedingly diverse and they differ in myriad ways.

Because living things have many, many characteristics, there are many different ways that they can be grouped. Moreover, until some goal has been attached to the classification, there is not 'right' way for it to be done; it is simply an arbitrary way to simplify a diverse system. Classifying cars based on color is certainly easy and, in some cases, might be useful, but it is not very useful if one's goal is to explain the overall patterns in car variation.

Two features make a classification easier to develop and make the entities thereby defined more 'real', i.e., an accurate representation of the reality. One feature relates to the pattern of variation. Consider a group of organisms that has only one characteristic, or perhaps only one characteristic that might distinguish one organism from another, for example, a group of organisms that are all the same except for length.

In general, organisms show variation that is discontinuous and they exhibit correlation in variation of different characters, and both these features make classification easier.

These pattens of nature not only make classification of organisms simpler, by reducing the number of characters that one needs to consider, it also hints that there are better, or perhaps even a 'best', way to classify living things. This might be described as a 'natural' classification', one that is based on 'fundamental similarities'. This idea was apparent to early naturalists.

Carl Linnaeus, who developed one of the first classification schemes, recognized that his technique of classification, although useful, was flawed because it was not 'natural', i.e., it put things together based on features that did not correspond to many other features. His categories were useful because they put organisms into bins and made their diversity much more manageable, but Linnaeus appreciated that there was an organization to the diversity of living things and that this organization was not always reflected in his categories.

If there was an organization to the diversity of living things, there should be a reason for this. Three hundred years after Linnaeus, Charles Darwin, who was an excellent student of classification (of organisms as varied as beetles, orchids and barnacles), came up with an explanation for the correlated variation and for the observation that living things appeared to fall into 'natural' groupings: it was consequence of the process of evolution, the changes in the characteristics of groups of organisms through time. The fundamental similarity of groups of organisms was due to the fact that together they shared a common evolutionary ancestry. Indeed, Darwin's taxonomic acumen was highly significant to his elucidation of the process of evolution. He made two key observations that were connected to his understanding of taxonomy:

- In widely separated parts of the world where conditions are similar, for example, deserts in South America and in Africa, organisms often look similar even though they are not 'fundamentally similar';
- In one region of the world where conditions varied considerably within a relatively small geographic area, e.g., going from plains to mountains in southern South America, organisms that on first examination seem very different, upon closer study are 'fundamentally similar'.

The process of organic evolution explained both of these patterns; in the first instance, convergent evolution could cause organisms that are fundamentally different (i.e., not closely related) to look superficially similar; in the second case divergent evolution (adaptive radiation) could cause fundamentally similar (i.e., closely related) organisms to diversify and look different.

It can be seen that although evolution explains different reasons why things might look alike, it does NOT, at least initially, explain how one might group things: should it be on the basis of 'basic, i.e. fundamental, similarity' and thus combine things that may not look that similar until 'closely examined' (see Fig. 1 which shows two different members of the cactus family, a leafy cactus and a more 'normal' looking cactus); or should things be grouped on the basis of 'superficial similarity', i.e. group things if, on the surface, they look similar (see Fig. 2 which shows two plants that look superficially similar yet are 'fundamentally' different).

The distinction may seem petty (i.e. what is the difference between superficial similarity and fundamental similarity), but it has real consequences because perceptions vary. For example, people commonly group flowers on the basis of flower color, which is one of the most easily perceived characteristics of plants, but it turns out to also be one of the most superficial.

Taxonomy and Phylogeny Systematics

For one particular group of people, biologists, Darwin's theory of evolution did provide a rationale for grouping and naming living things, organisms should be classified based on their phylogeny, their evolutionary ancestry, i.e., their fundamental similarities. For those who are studying organic diversity it would desirable to group things based on 'fundamental similarities' (= evolutionary ancestry) because, among other things, it allows us to view the consequences of evolution. But we are still left with the question of how to recognize phylogenetic groups. From pre-Linnaean times through Darwin and up until the middle of the last century, scientists searched for features that they thought reflected 'fundamental similarities'. After Darwin's ideas were accepted, scientists realized that what they were looking for were features that reflected the evolutionary past of organisms. But these characteristics are elusive entities and no one really knew if they had found one because most evolutionary history is impossible to trace—the fossil record is grossly inadequate except for large-scale overviews.

Over the last fifty years molecular biology brought new approaches to taxonomy. It allows organisms to be compared on the basis of similarities in the sequences of amino acids in proteins or sequences of nucleotide bases in nucleic acids (DNA, RNA). These are more than just new features to be compared; they are features that one can argue do reflect evolutionary lineage. Through time, changes accumulate so that the longer the time since two lines diverge the more differences that accumulate in the sequence of amino acids in proteins or nucleotide bases in nucleic acids.

One might argue that this is how classification has always been done, that the assumption has always been that groups 'accumulate' more and more differences through evolutionary time, forming separate groups with divergent characteristics. Although this is true in a very general sense, classification has always involved 'character weighting', i.e., observers have always felt that some characteristics, ones that are less easily modified by natural selection should have more 'weight' in a classification than others, characteristics, e.g., flower color, that are readily shaped by evolution and thus might occur independently in two lines that are not phylogenetically related.

The formation of groups is based not just on the accumulation of differences but rather the acquisition of key differences, the differences that reflect phylogeny. Molecular biology not only provides a tremendous increase in characteristics that can be compared. It additionally provides characteristics that are unlikely to be selected for by natural selection. This is significant because if two organisms share a feature this can be explained two ways: (1) it may reflect a common ancestry ('fundamental' similarity), or (2) it may reflect a common selective force ('superficial' similarity).

Fig. 1 These two plants are closely related and both in the cactus family, yet they 'look' very different, with the Pereksia (on the left) having a more typical plant form with typical leaves and branches, while the saguaro cactus on the right has evolved a very different form with no obvious leaves, an unusual branching pattern and an abundance of spines.

Fig. 2 Convergent evolution in plant form. The plant on the left is a euphorb that is not at all closely related to the cactus on the right. Although similar in form (unbranched, lacking typical leaves, having spines), this is not the result of a close evolutionary ancestry, but is the result of convergent evolution, two groups of plants 'converging' on a form that presumably is useful under arid conditions

However, if two organisms share a feature that is unlikely to be selected for then the only explanation involves a common ancestry. Assuming that one wants a classification based on ancestry, then using molecular data provides multiple ways to trace ancestry.

One feature of biological classification from the time of Linnaeus to now, and a feature of most (but not all) classifications, is that they are hierarchical, with species grouped into genera, genera into families, families into orders, orders into classes, classes into phyla and phyla into kingdoms (and, some would add, kingdoms into domains). However, with the exception of species (whose definition we will consider later in the course), none of these levels is defined—a genus is a group of related species, but how closely related is never specified.

Thus, although it is often the case that there is agreement that a group of living things represents a taxonomic entity, exactly what level that group should be placed at, and how this group relates to other groups, may be quite controversial. Moreover, the seemingly logical idea that groups should be organized on the basis of evolutionary ancestry doesn't mesh very well with a classification that has levels–evolution doesn't necessarily operate in a way to produce levels; and there is no reason to assume that the levels produced on one branch might coincide with levels on another branch.

One can certainly devise classification schemes that more closely match the way that we believe evolution operates; but these schemes will not be as useful in pigeon-holing (categorizing) living things. As is the case with many concepts (and in fact with words themselves!) one must balance between utility and accuracy; useful concepts often distort reality but making them more real often renders them less useful.

Most biologists approach classification from a 'cladistic' viewpoint that is centered on the idea that evolution produces 'clades' (groups) as a result of the splitting of a previously existing clade.

Seen through time one would see a branching diagram. In general, this probably reflects the pattern of evolution and the development of diversity. But we have good evidence that groups not only split but sometimes merge (e.g., the endosymbiont theory for the appearance of eukaryotes; secondary endosymbiosis and the origin of multiple algal groups) and neither cladistic approaches or a hierarchical system deals well with this possibility.

For most of this book we will speak of groups that are generally considered 'real', that is, a group of organisms that are set off, in terms of phylogeny and in terms of characteristics, from other living things. Because our focus is on organisms and their diversity, we will be less concerned with the exact placement of the group in a taxonomic scheme or the exact phylogenetic relationship between one group and others. Some of this information is covered in the phylogeny sections of the 'fact sheets' for specific groups found in the *Organisms* section of the book. But note that the text as a whole is NOT organized along phylogeny/taxonomy lines. In fact, I will refer to a number of 'artificial' (i.e., non-phylogenetic) groups. These are listed below and serve as examples of groupings that are known to be artificial yet are useful for reasons of history, ecology or convenience.

- 'inanimate life'—living things that are not animals (i.e., in the Animal Kingdom). Using a five-kingdom classification this would include Monera (i.e., prokaryotes), Fungi, Plants and Protists. A more ecological classification is shown here, along with a list that shows where the 'organism fact sheets fit in this classification.

- algae—aquatic photosynthetic organisms. This category spans most of the phylogenetic universe! Yet for ecological reasons, it is useful!
- macroalgae—multicellular or large colonial algae
- green algae—aquatic, photosynthetic organisms with multiple similarities to plants (pigments, cell wall chemistry, storage carbohydrates). This group is useful to know about, but for a host of reasons is difficult to define rigorously in a phylogenetic sense.
- gymnosperms—plants that have seeds but don't have flowers. This is an historical category that is still in common use and worthwhile to be aware of. It is an example of a grouping based on the lack of a particular characteristic, something that is not generally phylogenetically sound. Other examples of artificial groups based on what they lack are:
    - prokaryotes—cells without nuclei
    - protists—eukaryotic organisms that are not animals, fungi, plants or prokaryotes
    - bryophytes—plants (mosses, liverworts and hornworts) without vascular tissue
    - fern 'allies' —vascular plants without seeds.

CHAPTER

THE CHLOROPLAST AND ITS ROLE IN PHOTOSYNTHESIS 4

INTRODUCTION

Photosynthesis is the process in which light energy is converted to chemical energy in the form of sugars. In a process driven by light energy, glucose molecules (or other sugars) are constructed from water and carbon dioxide, and oxygen is released as a byproduct. The glucose molecules provide organisms with two crucial resources: energy and fixed—organic—carbon.

Energy. The glucose molecules serve as fuel for cells: their chemical energy can be harvested through processes like cellular respiration and fermentation, which generate adenosine triphosphate—A, T, P, end text, a small, energy-carrying molecule—for the cell's immediate energy needs.

Fixed carbon. Carbon from carbon dioxide—inorganic carbon—can be incorporated into organic molecules; this process is called carbon fixation, and the carbon in organic molecules is also known as fixed carbon. The carbon that's fixed and incorporated into sugars during photosynthesis can be used to build other types of organic molecules needed by cells.

Besides introducing fixed carbon and energy into ecosystems, photosynthesis also affects the makeup of Earth's atmosphere. Most photosynthetic organisms generate oxygen gas as a byproduct, and the advent of photosynthesis—over 333 billion years ago, in bacteria resembling modern cyanobacteria—forever changed life on Earth, end superscript. These bacteria gradually released oxygen into Earth's oxygen-poor atmosphere, and the increase in oxygen concentration is thought to have influenced the evolution of aerobic life forms—organisms that use oxygen for cellular respiration.

Photosynthetic organisms also remove large quantities of carbon dioxide from the atmosphere and use the carbon atoms to build organic molecules. Without Earth's abundance of plants and algae to continually suck up carbon dioxide, the gas would build up in the atmosphere. Although photosynthetic organisms remove some of the carbon dioxide produced by human activities, rising atmospheric levels are trapping heat and causing the climate to change. Many scientists believe that preserving forests and other expanses of vegetation is increasingly important to combat this rise in carbon dioxide levels.

Leaves are sites of photosynthesis

Plants are the most common autotrophs in terrestrial—land—ecosystems. All green plant tissues can photosynthesize, but in most plants, but the majority of photosynthesis usually takes place in the leaves. The cells in a middle layer of leaf tissue called the mesophyll are the primary site of photosynthesis.

Small pores called stomata—singular, stoma—are found on the surface of leaves in most plants, and they let carbon dioxide diffuse into the mesophyll layer and oxygen diffuse out.

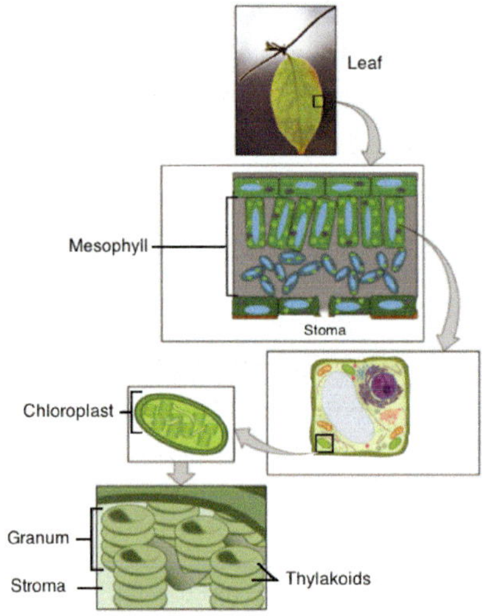

Fig. 1. Not all cells of a leaf carry out photosynthesis. Cells within the middle layer of a leaf have chloroplasts, which contain the photosynthetic apparatus.

Each mesophyll cell contains organelles called chloroplasts, which are specialized to carry out the reactions of photosynthesis. Within each chloroplast, disc-like structures called thylakoids are arranged in piles like stacks of pancakes that are known as grana—singular, granum. The membrane of each thylakoid contains green-colored pigments called chlorophylls that absorb light. The fluid-filled space around the grana is called the stroma, and the space inside the thylakoid discs is known as the thylakoid space. Different chemical reactions occur in the different parts of the chloroplast.

Plants form the basis of all life on earth and are known as producers. Plant cells contain structures known as plastids which are absent in animal cells. These plastids are double-membraned cell organelles which play a primary role in the manufacturing and storing of food. There are three types of plastids –

- Chromoplasts- They are the colour plastids, found in all flowers, fruits and are mainly responsible for their distinctive colours.
- Chloroplasts- They are green coloured plastids, which comprise green-coloured pigments within the plant cell and are called chlorophyll.
- Leucoplasts- They are colourless plastids and are mainly used for the storage of starch, lipids and proteins within the plant cell.

Chloroplast Definition

"Chloroplast is an organelle that contains the photosynthetic pigment chlorophyll that captures sunlight and converts it into useful energy, thereby, releasing oxygen from water."

What is a Chloroplast?

Chloroplasts are found in all green plants and algae. They are the food producers of plants. These are found in mesophyll cells located in the leaves of the plants. They contain a high concentration of chlorophyll that traps sunlight. This **cell organelle** is not present in animal cells. Chloroplast has its own extra-nuclear DNA and therefore are semiautonomous, like mitochondria. They also produce proteins and lipids required for the production of chloroplast membrane.

Diagram of Chloroplast

The chloroplast diagram (Fig. 2) represents the chloroplast structure mentioning the different parts of the chloroplast. The parts of a chloroplast such as the inner membrane, outer membrane, intermembrane space, thylakoid membrane, stroma and lamella can be clearly marked out.

PHOTOSYNTHETIC CELLS

Cells get nutrients from their environment, but where do those nutrients come from? Virtually all organic material on Earth has been produced by cells that convert energy from the Sun into energy-containing macromolecules. This process, called photosynthesis, is essential to the global carbon cycle and organisms that conduct photosynthesis represent the lowest level in most food chains (Fig. 3).

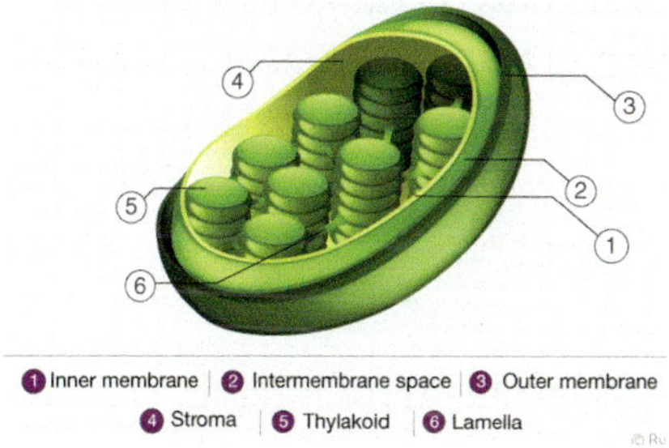

Fig. 2. The chloroplast diagram

Fig. 3: Photosynthetic plants synthesize carbon-based energy molecules from the energy in sunlight. Consequently, they provide an abundance of energy for other organisms

IMPORTANCE OF PHOTOSYNTHESIS

Most living things depend on photosynthetic cells to manufacture the complex organic molecules they require as a source of energy. Photosynthetic cells are quite diverse and include cells found in green plants, phytoplankton, and cyanobacteria. During the process of photosynthesis, cells use carbon dioxide and energy from the Sun to make sugar molecules and oxygen. These sugar molecules are the basis for more complex molecules made by the photosynthetic cell, such as glucose. Then, via respiration processes, cells use oxygen and glucose to synthesize energy-rich carrier molecules, such as ATP, and carbon dioxide is produced as a waste product. Therefore, the synthesis of glucose and its breakdown by cells are opposing processes.

The building and breaking of carbon-based material — from carbon dioxide to complex organic molecules (photosynthesis) then back to carbon dioxide (respiration) — is part of what is commonly called the global carbon cycle. Indeed, the fossil fuels we use to power our world today are the ancient remains of once-living organisms, and they provide a dramatic example of this cycle at work. The carbon cycle would not be possible without photosynthesis, because this process accounts for the "building" portion of the cycle.

However, photosynthesis doesn't just drive the carbon cycle — it also creates the oxygen necessary for respiring organisms. Interestingly, although green plants contribute much of the oxygen in the air we breathe, phytoplankton and cyanobacteria in the world's oceans are thought to produce between one-third and one-half of atmospheric oxygen on Earth.

WHAT CELLS AND ORGANELLES ARE INVOLVED IN PHOTOSYNTHESIS?

Photosynthetic cells contain special pigments that absorb light energy. Different pigments respond to different wavelengths of visible light. Chlorophyll, the primary pigment used in photosynthesis, reflects green light and absorbs red and blue light most strongly. In plants, photosynthesis takes place in chloroplasts, which contain the chlorophyll. Chloroplasts are surrounded by a double membrane and contain a third inner membrane, called the thylakoid membrane, that forms long folds within the organelle. In electron micrographs, thylakoid membranes look like stacks of coins, although the compartments they form are connected like a maze of chambers. The green pigment chlorophyll is located within the thylakoid membrane, and the space between the thylakoid and the chloroplast membranes is called the stroma (Fig. 4, Fig. 5).

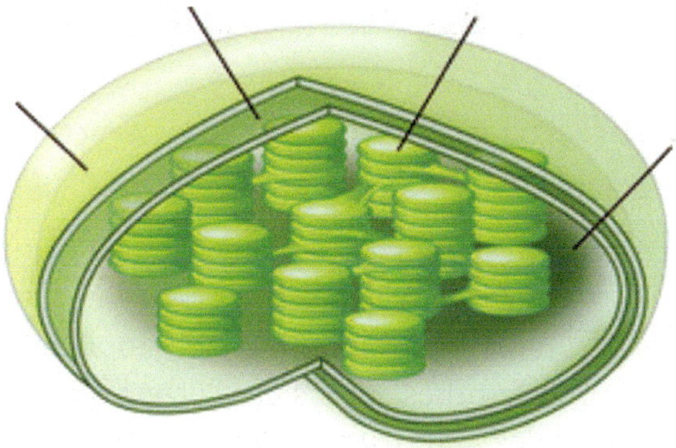

Fig. 4: Structure of a chloroplast

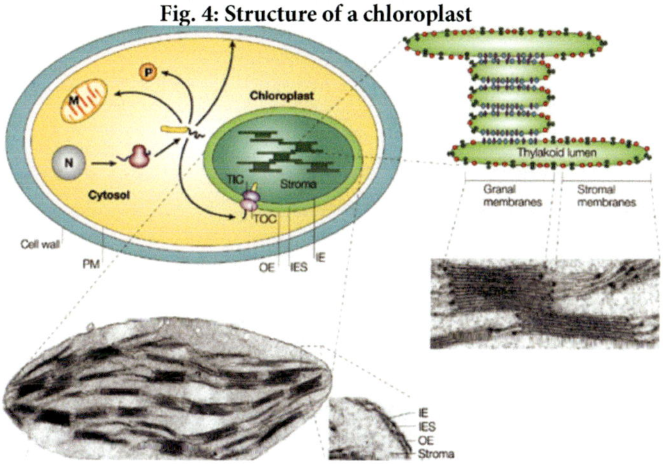

Fig. 5: Diagram of a chloroplast inside a cell, showing thylakoid stacks

STRUCTURE OF CHLOROPLAST

Chloroplasts are found in all higher plants. It is oval or biconvex, found within the mesophyll of the plant cell. The size of the chloroplast usually varies between 4-6 µm in diameter and 1-3 µm in thickness. They are double-membrane organelle with the presence of outer, inner and intermembrane space. There are two distinct regions present inside a chloroplast known as the grana and stroma.

- Grana are made up of stacks of disc-shaped structures known as thylakoids or lamellae. The grana of the chloroplast consists of chlorophyll pigments and are the functional units of chloroplasts.
- Stroma is the homogenous matrix which contains grana and is similar to the cytoplasm in cells in which all the organelles are embedded. Stroma also

contains various enzymes, DNA, ribosomes, and other substances. Stroma lamellae function by connecting the stacks of thylakoid sacs or grana.

The chloroplast structure consists of the following parts:

Membrane Envelope

It comprises inner and outer lipid bilayer membranes. The inner membrane separates the stroma from the intermembrane space.

Intermembrane Space

The space between inner and outer membranes.

Thylakoid System (Lamellae)

The system is suspended in the stroma. It is a collection of membranous sacs called thylakoids or lamellae. The green coloured pigments called chlorophyll are found in the thylakoid membranes. It is the sight for the process of light-dependent reactions of the photosynthesis process. The thylakoids are arranged in stacks known as grana and each granum contains around 10-20 thylakoids.

Stroma

It is a colourless, alkaline, aqueous, protein-rich fluid present within the inner membrane of the chloroplast present surrounding the grana.

Grana

Stack of lamellae in plastids is known as grana. These are the sites of conversion of light energy into chemical energy.

CHLOROPHYLL

It is a green photosynthetic pigment that helps in the process of photosynthesis.

Characteristics of chloroplasts

Chloroplasts are a type of plastid—a round, oval, or disk-shaped body that is involved in the synthesis and storage of foodstuffs. Chloroplasts are distinguished from other types of plastids by their green colour, which results from the presence of two pigments, chlorophyll *a* and chlorophyll *b*. A function of those pigments is to absorb light energy for the process of photosynthesis. Other pigments, such as carotenoids, are also present in chloroplasts and serve as accessory pigments, trapping solar energy and passing it to chlorophyll. In plants, chloroplasts occur in all green tissues, though they are concentrated particularly in the parenchyma cells of the leaf mesophyll.

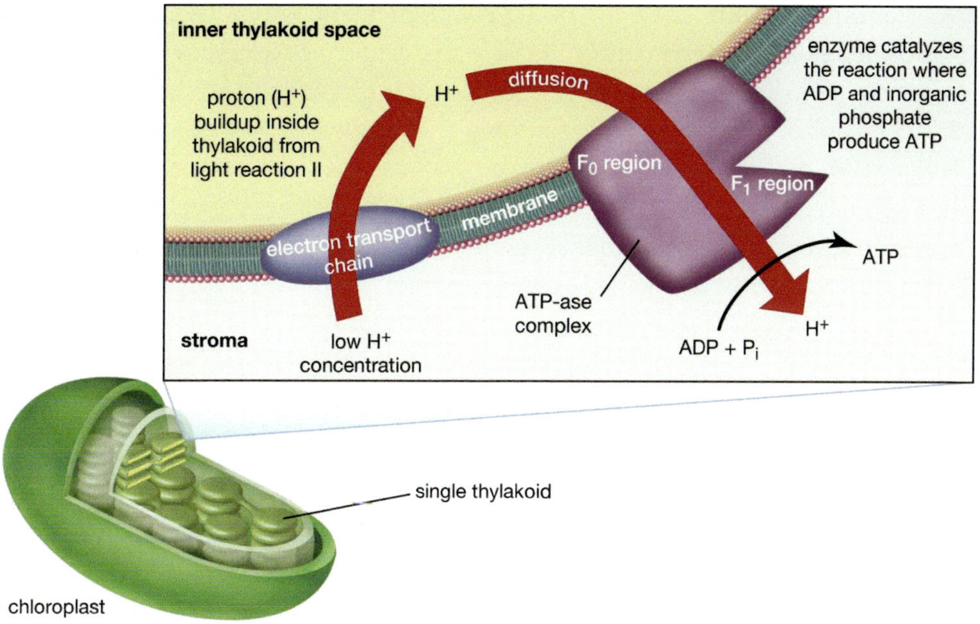

Fig. 6. Chemiosmosis in chloroplasts

Chloroplasts are roughly 1–2 μm (1 μm = 0.001 mm) thick and 5–7 μm in diameter. They are enclosed in a chloroplast envelope, which consists of a double membrane with outer and inner layers, between which is a gap called the intermembrane space. A third, internal membrane, extensively folded and characterized by the presence of closed disks (or thylakoids), is known as the thylakoid membrane. In most higher plants, the thylakoids are arranged in tight stacks called grana (singular granum). Grana are connected by stromal lamellae, extensions that run from one granum, through the stroma, into a neighbouring *granum*. The thylakoid membrane envelops a central aqueous region known as the thylakoid lumen. The space between the inner membrane and the thylakoid membrane is filled with stroma, a matrix containing dissolved enzymes, starch granules, and copies of the chloroplast genome.

The photosynthetic machinery

The thylakoid membrane houses chlorophylls and different protein complexes, including photosystem I, photosystem II, and ATP (adenosine triphosphate) synthase, which are specialized for light-dependent photosynthesis. When sunlight strikes the thylakoids, the light energy excites chlorophyll pigments, causing them to give up electrons. The electrons then enter the electron transport chain, a series of reactions that ultimately drives the phosphorylation of adenosine diphosphate (ADP) to the energy-rich storage compound ATP. Electron transport also results in the production of the reducing agent nicotinamide adenine dinucleotide phosphate (NADPH).

ATP and NADPH are used in the light-independent reactions (dark reactions) of photosynthesis, in which carbon dioxide and water are assimilated into organic compounds. The light-independent reactions of photosynthesis are carried out in the chloroplast stroma, which contains the enzyme ribulose-1,5-bisphosphate carboxylase/oxygenase (rubisco). Rubisco catalyzes the first step of carbon fixation in the Calvin cycle (also called Calvin-Benson cycle), the primary pathway of carbon transport in plants. Among so-called C_4 plants, the initial carbon fixation step and the Calvin cycle are separated spatially—carbon fixation occurs via phosphoenolpyruvate (PEP) carboxylation in chloroplasts located in the mesophyll, while malate, the four-carbon product of that process, is transported to chloroplasts in bundle-sheath cells, where the Calvin cycle is carried out. C_4 photosynthesis attempts to minimize the loss of carbon dioxide to photorespiration. In plants that use crassulacean acid metabolism (CAM), PEP carboxylation and the Calvin cycle are separated temporally in chloroplasts, the former taking place at night and the latter during the day. The CAM pathway allows plants to carry out photosynthesis with minimal water loss.

CHLOROPLAST GENOME AND MEMBRANE TRANSPORT

The chloroplast genome typically is circular (though linear forms have also been observed) and is roughly 120–200 kilobases in length. The modern chloroplast genome, however, is much reduced in size: over the course of evolution, increasing numbers of chloroplast genes have been transferred to the genome in the cell nucleus. As a result, proteins encoded by nuclear DNA have become essential to chloroplast function. Hence, the outer membrane of the chloroplast, which is freely permeable to small molecules, also contains transmembrane channels for the import of larger molecules, including nuclear-encoded proteins. The inner membrane is more restrictive, with transport limited to certain proteins (e.g., nuclear-encoded proteins) that are targeted for passage through transmembrane channels.

Fig. 7. Chloroplast Function in Photosynthesis

FUNCTIONS OF CHLOROPLAST

Following are the important chloroplast functions:

- The most important function of the chloroplast is to synthesise food by the process of photosynthesis.

- Absorbs light energy and converts it into chemical energy.
- Chloroplast has a structure called chlorophyll which functions by trapping the solar energy and is used for the synthesis of food in all green plants.
- Produces NADPH and molecular oxygen (O_2) by photolysis of water.
- Produces ATP – Adenosine triphosphate by the process of photosynthesis.
- The carbon dioxide (CO2) obtained from the air is used to generate carbon and sugar during the Calvin Cycle or dark reaction of photosynthesis.

CHLOROPLAST FUNCTION IN PHOTOSYNTHESIS

In photosynthesis, the sun's solar energy is converted to chemical energy. The chemical energy is stored in the form of glucose (sugar). Carbon dioxide, water, and sunlight are used to produce glucose, oxygen, and water. Photosynthesis occurs in two stages. These stages are known as the light reaction stage and the dark reaction stage.

The **light reaction stage** takes place in the presence of light and occurs within the chloroplast grana. The primary pigment used to convert light energy into chemical energy is **chlorophyll a**. Other pigments involved in light absorption include chlorophyll b, xanthophyll, and carotene. In the light reaction stage, sunlight is converted to chemical energy in the form of ATP (free energy containing molecule) and NADPH (high energy electron carrying molecule). Protein complexes within the thylakoid membrane, known as photosystem I and photosystem II, mediate the conversion of light energy to chemical energy. Both ATP and NADPH are used in the dark reaction stage to produce sugar.

The **dark reaction stage** is also known as the carbon fixation stage or the Calvin cycle. Dark reactions occur in the stroma. The stroma contains enzymes which facilitate a series of reactions that use ATP, NADPH, and carbon dioxide to produce sugar. The sugar can be stored in the form of starch, used during respiration, or used in the production of cellulose.

Chloroplast Function Key Points

- Chloroplasts are chlorophyll-containing organelles found in plants, algae, and cyanobacteria. Photosynthesis occurs in chloroplasts.
- Chlorophyll is a green photosynthetic pigment within the chloroplast grana that absorbs light energy for photosynthesis.
- Chloroplasts are found in plant leaves surrounded by guard cells. These cells open and close tiny pores allowing for the gas exchange needed for photosynthesis.

- Photosynthesis occurs in two stages: the light reaction stage and the dark reaction stage.
- ATP and NADPH are produced in the light reaction stage which occurs within chloroplast grana.
- In the dark reaction stage or Calvin cycle, ATP and NADPH produced during the light reaction stage are used to generated sugar. This stage occurs in plant stroma.

CHAPTER

A LEAF CELL CONSISTS OF SEVERAL METABOLIC COMPARTMENTS 5

In higher plants photosynthesis occurs mainly in the **mesophyll,** the chloroplast-rich tissue of leaves. Fig. 1 shows an electron micrograph of a mesophyll cell and Fig. 2 shows a schematic presentation of the cell structure. The cellular contents are surrounded by a **plasma membrane** called the plasmalemma and are enclosed by a **cell wall**. The cell contains organelles, each with its own characteristic shape, which divide the cell into various compartments (subcellular compartments). Each compartment has specialized metabolic functions. The largest organelle, the vacuole, usually fills about 80% of the total cell volume. Chloroplasts represent the next largest compartment, and the rest of the cell volume is filled with mitochondria, peroxisomes, the nucleus, the endoplasmic reticulum, the Golgi bodies, and, outside these organelles, the cell plasma, called **cytosol**. In addition, there are oil bodies derived from the endoplasmic reticulum. These oil bod-ies, which occur in seeds and some other tissues (e.g., root nodules), are storage organelles for triglycerides.

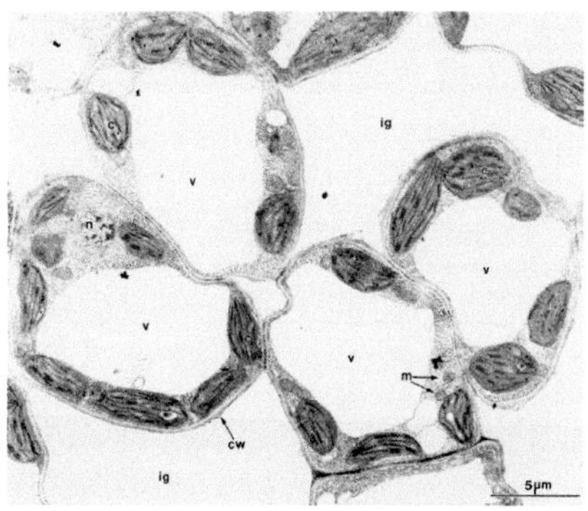

Fig. 1. Electron micrograph of mesophyll tissue from tobacco. In most cells the large central vacuole is to be seen (v). Between the cells are the intercellular gas spaces (ig), which are somewhat enlarged by the fixation process. c: chloroplast, cw: cell wall, n: nucleus, m: mitochondrion

The nucleus is surrounded by the nuclear envelope, which consists of the two membranes of the endoplasmic reticulum. The space between the two membranes is known as the perinuclear space. The nuclear envelope is interrupted by nuclear pores with a diameter of about 50 nm. The nucleus contains chromatin, consisting of DNA double strands that are stabilized by being bound to basic proteins (histones). The genes of the nucleus are collectively referred to as the nuclear genome. Within the nucleus, usually off-center, lies the nucleolus, where ribosomal subunits are formed. These ribosomal subunits and the messenger RNA formed by transcription of the DNA in the nucleus migrate through the nuclear pores to the ribosomes in the cytosol, the site of protein biosynthesis. The synthesized proteins are distributed between the different cell compartments according to their final destination

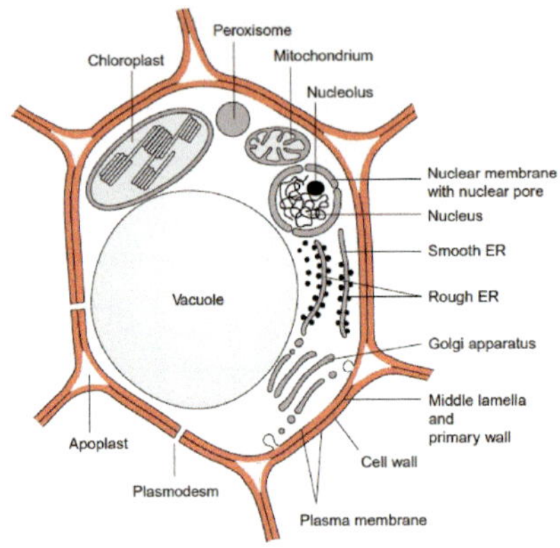

Fig. 2 Diagram of a mesophyll cell.

The nucleus is surrounded by the nuclear envelope, which consists of the two membranes of the endoplasmic reticulum. The space between the two membranes is known as the perinuclear space. The nuclear envelope is interrupted by nuclear pores with a diameter of about 50 nm. The nucleus contains chromatin, consisting of DNA double strands that are stabilized by being bound to basic proteins (histones). The genes of the nucleus are collectively referred to as the nuclear genome. Within the nucleus, usually off-center, lies the nucleolus, where ribosomal subunits are formed. These ribosomal subunits and the messenger RNA formed by transcription of the DNA in the nucleus migrate through the nuclear pores to the ribosomes in the cytosol, the site of protein biosynthesis. The synthesized proteins are distributed between the different cell compartments according to their final destination.

The cell contains in its interior the cytoskeleton, which is a three-dimen-sional network of fiber proteins. Important elements of the cytoskeleton are the microtubuli and the microfilaments, both macromolecules formed by the aggregation of soluble (globular) proteins. Microtubuli are tubular structures composed of á and b tubuline monomers. The microtubuli are con-nected to a large number of different motor proteins that transport bound organelles along the microtubuli at the expense of ATP. Microfilaments are chains of polymerized actin that interact with myosin to achieve movement.

Actin and myosin are the main constituents of the animal muscle. The cytoskeleton has many important cellular functions. It is involved in the spatial organization of the organelles within the cell, enables thermal stabil-ity, plays an important role in cell division, and has a function in cell-to-cell communication.

The cell wall gives the plant cell mechanical stability

The difference between plant cells and animal cells is that plant cells have a cell wall. This wall limits the volume of the plant cell. The water taken up into the cell by osmosis presses the plasma membrane against the inside of the cell wall, thus giving the cell mechanical stability. The cell walls are very complex structures; inArabidopsis about 1,000 genes were found to be involved in its synthesis. Cell walls also protect against infections.

The cell wall consists mainly of carbohydrates and proteins

The cell wall of a higher plant is made up of about 90% carbohydrates and 10% proteins. The main carbohydrate constituent is cellulose. Cellulose is an unbranched polymer consisting of D-glucose molecules, which are con-nected to each other by b-1,4 glycosidic linkages (Fig. 3A). Each glucose unit is rotated by 180° from its neighbor, so that very long straight chains can be formed with a chain length of 2,000 to 25,000 glucose residues. About 36 cellulose chains are associated by interchain hydrogen bonds to a crystalline lattice structure known as a microfibril. These crystalline regions are impermeable to water. The microfibrils have an unusually high tensile strength, are very resistant to chemical and biological degradations, and are in fact so stable that they are very difficult to hydrolyze. However, many bacteria and fungi have cellulose-hydrolyzing enzymes (cellulases). These bacteria can be found in the digestive tract of some animals (e.g., ruminants), thus enabling them to digest grass and straw. It is interesting to note that cellulose is the most abundant organic substance on earth, repre-senting about half of the total organically bound carbon.

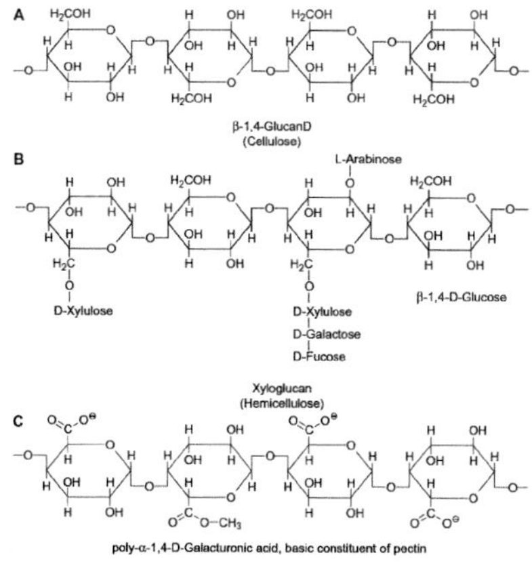

Fig. 3 Main constituents of the cell wall. 1.3A. Cellulose

Hemicelluloses are also important constituents of the cell wall. They are defined as those polysaccharides that can be extracted by alkaline solutions. The name is derived from an initial belief, which later turned out to be incor-rect, that hemicelluloses are precursors of cellulose.

Hemicelluloses con-sist of a variety of polysaccharides that contain, in addition to D-glucose, other carbohydrates such as the hexoses D-mannose, D-galactose, D-fucose, and the pentoses D-xylose and L-arabinose. Figure 1.3B shows xylogly-can as an example of a hemicellulose. The basic structure is a β-1,4-glucan chain to which xylose residues are bound via á-1,6 glycosidic linkages, which in part are linked to D-galactose and D-fucose. In addition to this, L-arabinose residues are linked to the 2' OH group of the glucose.

VACUOLES HAVE MULTIPLE FUNCTIONS

The term "vacuole" means "empty space". They help in the storage and disposal of various substances. They can store food or other nutrients required by a cell to survive. They also store waste products and prevent the entire cell from contamination.

The vacuoles in plant cells are larger than those in the animal cells. The plant vacuoles occupy more than 80% of the volume of the cell. The vacuoles may be one or more in number.

Let us have a detailed look at the structure and function of vacuoles.

Structure of Vacuole

A vacuole is a membrane bound structure found in the cytoplasmic matrix of a cell. The membrane surrounding the vacuole is known as tonoplast. The components of the vacuole, known as the cell sap, differ from that of the surrounding cytoplasm. The membranes are composed of phospholipids. The membranes are embedded with proteins that help in transporting molecules across the membrane. Different combinations of these proteins help the vacuoles to hold different materials.

Functions of Vacuole

Unlike animal cells, plant cells typically contain only one vacuole per cell (often referred to as a "central vacuole"), and the vacuole they contain is much larger than those in animal cells. Plant cell central vacuoles take up an enormous percentage of the cell, sometimes over 90% of cell space, although 30-50% is more common.

Surrounding the vacuoles in mature plant cells is an additional thin membrane called a tonoplast. **The tonoplast helps the vacuole hold its structure so that the vacuole can retain its shape.** Vacuoles in plant and fungi cells perform very similar functions, however; fungi cell vacuoles are typically much smaller than plant cell vacuoles, and each fungi cell can contain more than one vacuole (similar to animal cells).

Vacuoles in plant and fungi cells **perform more functions than vacuoles in other types of cells**; they're a critical part of keeping the plant/fungi alive and healthy. Because fungi and plant cells don't contain lysosomes, vacuoles in these cells also break down more materials than they do in animal cells. In addition to the functions listed in the previous section, vacuoles in plant and fungi cells also:

- **Maintain proper pH:** The vacuole keeps the cytoplasm in the cell acidic so that enzymes can break down different molecules. The vacuole lowers pH by moving protons from the cell cytosol into the vacuole.
- **Store water:** The vacuole can use proton motive force, a chemical gradient used to move materials in an out of the cell, to store water which allows the plant to survive longer in periods of drought.

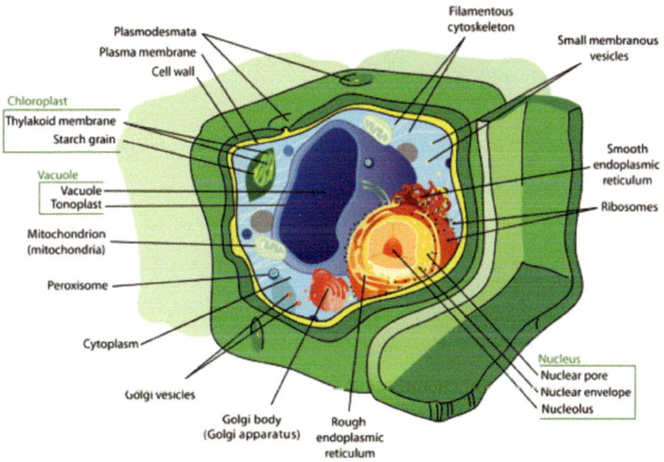

Fig. 4. Functions of Vacuole: Large blue structure

- **Maintain turgor pressure:** Turgor pressure is the pressure of the main area of the cell against the cell wall. It's one of the ways plants and trees avoid being limp and grow tall and strong. Think of fresh, crisp salad greens vs. limp ones. The former have high turgor pressure. Tonoplasts in vacuoles control turgor pressure by maintaining a particular balance of ions, which causes the vacuole to swell against the cell wall.
- **Adjust size of the cell:** Because vacuoles in plant cells can be so large, they are a key part in determining how large or small a certain plant cell is, which can in turn affect the size of different parts of the plant.

Storage

A vacuole stores salts, minerals, pigments and proteins within the cell. The solution that fills a vacuole is known as the cell sap. The vacuole is also filled with protons from the cytosol that helps in maintaining an acidic environment within the cell. A large number of lipids are also stored within the vacuoles.

Turgor Pressure

The vacuoles are completely filled with water and exert force on the cell wall. This is known as turgor pressure. It provides shape to the cell and helps it to withstand extreme conditions.

Endocytosis and Exocytosis

The substances are taken in by a vacuole through endocytosis and excreted through exocytosis. These substances are stored in the cells, separated from the cytosol. Lysosomes are vesicles that intake food and digest it. This is endocytosis and it varies in different cells.

INVESTIGATING COMPARTMENTATION IN PLANT METABOLISM

Cell fractionation

The traditional way to investigate metabolic compartmentation is by cell fractionation, which is usually a targeted approach that asks the question: 'In which compartment is enzyme X located?' To answer this question, cells or protoplasts are gently disrupted in an osmotically balanced medium, and the cell lysate is then fractionated by differential centrifugation and/or centrifugation on sucrose or Percoll density gradients. Each fraction is then assayed for activities of enzyme X and marker enzymes for each compartment, with co-localization across different fractions indicating co-localization within the cell. For enzymes that co-localize with organelle marker enzymes, further experiments on highly purified organelle preparations are carried out to see if the enzyme is really located within the organelle, or just stuck to the outside.

These can include latency measurements, in which activity is compared in intact and lysed preparations of the organelle, and protease protection assays, where resistance to proteolytic degradation until the organelle is broken indicates that the enzyme is located inside. Such assays are also used to determine sub-organellar compartmentation in multi-compartment organelles such as mitochondria and chloroplasts. Cell fractionation studies rely on accurate measurements of the target and marker enzymes, and careful accounting of all the fractions is needed to check that all of the activity in the initial cell lysate is recovered after fractionation.

This is particularly important where an enzyme is found in more than one compartment and the isoforms from different compartments show differential stability. A limitation of classical cell fractionation techniques is that a cytosolic location for an enzyme can only be inferred in a negative way, i.e. if it does not co-localize with any known organelle or membrane compartment, and cytosol-enriched fractions are almost invariably contaminated by proteins from broken organelles.

Cell fractionation studies have been carried out mostly on a few experimentally tractable systems, such as spinach leaves, castor bean seedlings, and cell suspension cultures, and information obtained from these model systems is generally assumed to be valid for comparable tissues from other species. However, such extrapolation can sometimes lead us astray, with the enzyme ADP-glucose pyrophosphorylase providing a good cautionary example. This enzyme synthesizes the substrate (ADP-glucose) for starch synthesis, which occurs only in plastids.

In leaves and most non-photosynthetic organs, ADP-glucose pyrophosphorylase is similarly restricted to plastids—it has even been used as a plastidial marker enzyme in some studies. However, it has recently been shown that in the developing endosperm of cereals (e.g. barley, wheat, maize, and rice) and wild grass species the majority of the ADP-glucose pyrophosphorylase activity is in fact cytosolic, and most of the ADP-glucose needed for starch synthesis is produced in the cytosol and imported by the amyloplasts via a specific transporter in the amyloplast envelope.

The generally accepted plastidial location for ADP-glucose synthesis in other species and organs has recently been challenged by the proposal that ADP-glucose is synthesized in the cytosol, not by ADP-glucose pyrophosphorylase, but by sucrose synthase. However, it is difficult to reconcile this proposal with the large body of biochemical and genetic evidence that plastidial ADP-glucose pyrophosphorylase does indeed supply the vast majority of ADP-glucose for starch synthesis in most species and organs. Therefore, although sucrose synthase can undoubtedly synthesize ADP-glucose *in vitro*, the weight of evidence at present indicates that it makes little, if any, direct contribution to ADP-glucose synthesis *in vivo*.

In addition to locating enzymes, cell fractionation can also be used to investigate the intracellular compartmentation of metabolites. However, this is technically more challenging than enzyme localization because the turnover times of many metabolites are much shorter than the time needed for the fractionation. One way to overcome this problem is to remove water from the tissue to arrest metabolic activity, and carry out the fractionation under non-aqueous conditions.

In a typical experiment, the tissue is first rapidly frozen in liquid nitrogen, lyophilized to remove all water, homogenized, and then centrifuged on gradients of organic solvents, such as *n*-heptane and carbon tetrachloride. The gradient is divided into fractions, which are then assayed for marker enzymes and the metabolites of interest. The separation of organelles by non-aqueous fractionation is much less effective than with aqueous systems, but it is still possible to partially resolve the cytosolic, plastidial, and vacuolar compartments of leaf tissue).

One important result from non-aqueous fractionation experiments was the finding of a significant pool of inorganic pyrophosphate (PPi) in the cytosol. This pool of PPi allows UDP-glucose pyrophosphorylase to catabolize the UDP-glucose produced by cleavage of sucrose via sucrose synthase, and thus conserve much of the energy in the glycosidic bond of sucrose.

Cytosolic PPi can also contribute to energization of the tonoplast via the H^+-pumping pyrophosphatase, and PPi can be transported into chloroplasts to replenish Pi pools in the stroma. Non-aqueous fractionation of potato tubers was also instrumental in revealing the redox regulation of ADP-glucose pyrophosphorylase and starch synthesis.

Immunohistochemistry

Another technique that has been used widely to investigate compartmentation is immunohistochemistry. Specific antibodies are used to immuno-decorate tissue sections, and then visualized by light or electron microscopy after labelling of the antibodies with fluorescent, heavy-atom, or enzymatic markers. Immunohistochemical techniques can be used for proteins that do not have enzymatic activity (e.g. transporters) or proteins of unknown function, and have also been used to locate non-protein targets such as cell wall polysaccharides and amino acids.

Immunohistochemical methods also have the advantage that cytosolic proteins can be distinguished from those that are associated with intracellular structures that are too fragile or too low in abundance to isolate by simple cell fractionation techniques, such as the endoplasmic reticulum and Golgi bodies. However, a major limitation of this technique is the need for highly specific antibodies for each target of interest, because even weak cross-reaction with a particularly abundant proteins (e.g. Rubisco in leaves) can lead to mistaken conclusions. Furthermore, antibodies do not usually distinguish between active and inactive forms of their target protein, such as unprocessed pre-proteins or partially degraded proteins. Therefore, detection of an enzyme protein in a particular compartment does not necessarily indicate the presence of activity in that location.

Proteome analysis by mass spectroscopy

Traditional cell fractionation studies and immunohistochemistry are low-throughput, targeted approaches for investigating metabolic compartmentation. In the last few years, new technologies for identifying proteins by mass spectrometry have been combined with classical cell fractionation techniques to catalogue the proteomes of various plant cell compartments, particularly in the model species *Arabidopsis thaliana.*

Such proteomic surveys have not only confirmed the locations of proteins identified in previous studies, but more importantly revealed the locations of numerous other proteins, including many whose functions are still unknown. Indeed, this has been one of the driving forces for such studies, as knowing the intracellular location of an 'unknown' protein can be a very useful clue to identifying its function.

To date, most proteomic studies in plants have focused on *Arabidopsis thaliana*, because the availability of the genome sequence makes protein identification relatively straightforward. A list of proteomic studies that have been undertaken in this species is given in Table 1. Proteomic analyses have also been carried out on the inner and outer envelope membranes of spinach chloroplasts, bundle sheath and mesophyll cell chloroplasts from maize leaves, and rice etioplasts.

Analysis of the chloroplast proteome confirmed that many of the enzymes involved in biosynthesis of fatty acids, lipids, amino acids, nucleotides, hormones, alkaloids, and isoprenoids are found in these organelles, in addition to the Calvin cycle enzymes and proteins belonging to the light-harvesting apparatus and photosynthetic electron transport chain.

Table 1. Proteomic studies of organelles and sub-organellar compartments in *Arabidopsis thaliana*

Organelle	Tissue	Proteins identified
Chloroplasts	Rosette leaves	690
Chloroplasts	Rosette leaves	426
Chloroplasts (thylakoid lumen)	Rosette leaves	81
Chloroplasts (thylakoid lumen)	Rosette leaves	36
Chloroplasts (thylakoid membrane)	Rosette leaves	154
Chloroplasts (envelope)	Rosette leaves	106
Chloroplasts (envelope)	Rosette leaves	392
Mitochondria	Stem, leaves, and cell culture	52
Mitochondria	Cell culture	91
Mitochondria	Cell culture	114
Mitochondria (protein complexes)	Cell culture	14
Mitochondria	Cell culture	416
Peroxisomes	Greening cotyledons	29
Peroxisomes	Etiolated cotyledons	19
Vacuoles	Rosette leaves	402
Vacuoles (tonoplast)	Cell culture	163
Vacuoles (tonoplast)	Cell culture	70
Nucleus	Rosette leaves	158
Nucleus	Cell culture	36
Nucleus (nucleolus)	Cell culture	217
Plasmalemma	Cell culture	97

In mitochondria, about 20% of the proteins identified have completely unknown functions. Although many of the others have been assigned to general protein classes such as protein kinases and phosphatases, their specific functions are also unknown. This large number of poorly characterized proteins suggests that there are many metabolic pathways and regulatory processes still waiting to be discovered in these organelles.

Proteomic studies have greatly increased our knowledge about compartmentation of enzymes and other proteins in plant cells, but current technologies do have some limitations. Whatever the approach used to identify proteins, the reliability of the results is absolutely dependent on the purity of the original organelle preparation, and the level of contamination by proteins from other compartments often sets the limit of sensitivity. Mass spectrometric techniques can also be limited by their dynamic range, with high abundance proteins swamping the analysis with redundant information, making it more difficult to find peptide signals from rare proteins. As with immunohistochemical techniques, mass spectroscopic methods do not distinguish between active and inactive proteins, so metabolic activities can only be inferred from such data.

Analysis of very hydrophobic proteins is still technically challenging, and the difficulties of preparing highly purified membrane preparations have held back proteomic analyses of the plasmalemma, endoplasmic reticulum, and Golgi body. Another limitation is that proteins can still only be assigned to the cytosol in a negative way, by not being present in other compartments. New technologies that might offer solutions to some of these problems are isotope tagging and laser micro-dissection and capture.

FLUORESCENT PROTEIN TAGGING AND OTHER IN VIVO IMAGING TECHNIQUES

Plant tissues generally contain more than one cell type. For example, it has been estimated that mesophyll, epidermal, and vascular cells occupy 58%, 3%, and 1%, respectively, of the volume of spinach leaves, and 51%, 27%, and 4% of the volume of wheat leaves, with most of the remainder being air spaces. As a result of this heterogeneity, any organelle preparation is likely to come from a mixture of cell types; for example, a mitochondrial preparation from leaves could include organelles from both photosynthetic and non-photosynthetic cells.

These might have very different metabolic functions and protein compositions; for example, the photorespiratory glycine decarboxylase complex would be a major component only in the mitochondria from photosynthetic cells. In the worst case scenario, preferential extraction of organelles from a minor cell type could give a very misleading impression of the metabolism in the tissue as a whole, so results obtained from cell fractionation studies on heterogeneous plant tissues must be interpreted with caution. Attempts to avoid this issue by using homogeneous cell suspension cultures still face the question of how to relate the data to whole plants.

An alternative approach to tackle this problem is to express fluorescently tagged versions of the protein of interest in the plant. These can be visualized by laser confocal microscopy in vivo to show the intracellular location of the protein, and also its distribution between different cell types and organs if the native promoter and other regulatory elements are used to drive expression of the tagged protein. In the beginning, this technique was used to locate just one or two proteins of special interest, but several larger-scale investigations have now been carried out.

These were essentially pilot studies for high-throughput approaches, using transient or stable expression systems to test different strategies for making the fusion protein constructs, and several different fluorescent proteins (e.g. GFP, YFP, and CFP). Some studies have used constitutive and/or strong promoters to drive expression of the fluorescently tagged proteins, but some authors have argued that it is preferable to use native promoters and regulatory elements, because ectopically expressed proteins may not be properly targeted.

For example, a protein that is normally expressed only in photosynthetic cells and located in the chloroplasts could be effectively trapped if expressed in the cytosol of non-photosynthetic cells with few or no plastids. Similarly, it is argued that expression of a protein at an abnormally high level could also lead to artefactual compartmentation by overloading intracellular trafficking mechanisms, or disrupting protein complexes whose integrity is needed for proper targeting. Nevertheless, at least one study has found that addition of CaMV 35S enhancer sequences led to higher expression of the tagged proteins and stronger fluorescence signals, but did not change the intracellular location of the proteins.

A very important consideration in such studies is where to attach the fluorescent tag to the target protein. Clearly the N-terminus of the protein should be avoided, as plastidial and mitochondrial transit peptides, and endomembrane signal peptides, are all located in this region. However, the C-terminus can also contain important targeting information, such as endoplasmic reticulum retention signals. As a default strategy, researchers inserted the YFP tag about 10 amino acids upstream of the C-terminus of the protein, flanked by six- or nine-amino acid Gly- or Ala-rich peptides, with the aim of maximizing the integrity of the target protein and minimizing the chance of disrupting targeting regions.

Using this strategy, every single one of the peroxisomal, tonoplast, plasmalemma, cell wall, plasmodesmatal, cytoskeletal, nuclear, proplastidial, and cytosolic proteins tested was located in the expected compartment, demonstrating that it would be feasible to use this approach to locate proteins in most compartments. For proteins of particular interest, a useful test for the reliability of the results is to check whether the fluorescently tagged fusion protein can complement a loss-of-function mutant. A good example is the YFP-tagged MEX1 maltose transporter, which was found to be targeted to the chloroplast envelope and was able to restore the wild-type phenotype to the mex1 mutant.

There are several other imaging techniques that can be used to study metabolism in vivo, although these are used more to detect metabolites than proteins. Nuclear magnetic resonance (NMR) spectroscopy has been used to determine the intracellular distribution of inorganic orthophosphate, and can detect changes in the intracellular pH and energy charge of living cells. NMR has also been used to measure rates of sucrose or water transport in phloem and xylem vessels, and for network flux analysis.

A major disadvantage of in vivo NMR is its rather poor sensitivity, as it is only able to detect metabolites that are present in millimolar concentrations, and this has so far limited its usefulness for studies of plant metabolism. A more sensitive technique for in vivo imaging is positron emission tomography (PET), which detects short-lived isotopes (11C or 18F), that have been used to label molecules of interest, for example, sugars and polyphenols. Unfortunately, the necessary equipment for PET is very expensive, and so far this technique has found little application outside the medical field.

COMPUTATIONAL APPROACHES

Proteomic and fluorescent protein-labelling studies have revealed the locations of over 4000 proteins in Arabidopsis thaliana, but the whole proteome of this species is estimated to contain another 12,000–24,000 proteins whose intracellular locations are unknown. To tackle this huge problem, various computational approaches have been developed to predict intracellular location from the primary protein sequence.

Most of these programs aim to identify N-terminal transit or signal peptide sequences that target proteins to plastids, mitochondria, or the endomembrane system (endoplasmic reticulum, plasmalemma, vacuoles) and apoplast. To assess the specificity and sensitivity of various prediction programs, several comparisons of predicted and experimentally determined sets of organelle proteins have been made. In general, the programs seem to be more successful at predicting plastidial proteins than mitochondrial or endomembrane/apoplastic proteins, although even the most favourable assessments found that at least 15% of plastidial proteins were mis-identified.

Combining results from several prediction programs was found to improve specificity, i.e. there were fewer false-positive predictions, but there was a trade-off in sensitivity, with a greater number of chloroplast proteins being wrongly rejected as non-plastidial. A similar finding was made for the mitochondrial prediction programs, which individually fail to recognize more than half of the proteins known to be located in the mitochondria.

There are several reasons why prediction programs can give both false-positive and false-negative results. One trivial explanation is that the N-terminal sequence of the protein has been incorrectly identified. The primary sequences of transit and signal peptides are often poorly conserved between species, and the open reading frame (ORF) prediction programs used for genome annotation can misidentify a conserved downstream region as the start of the ORF if no full-length cDNA sequence is available. False negatives can also arise if the protein in question is imported into an organelle by an unconventional route, although this is probably quite rare.

Perhaps one of the main reasons for the rather mixed success of the first generation of prediction programs is that they were trained on relatively small datasets, usually fewer than 100 proteins of known location. Presumably, the availability of much larger training sets will eventually allow the reliability of transit and signal peptide prediction programs to be improved, but at present they can only give a rough guide to a protein's intracellular location. Post-translational modifications, such as prenylation or addition of glycosyl phosphatidylinositol, can also play a major role in protein targeting, especially to membranes (e.g. plasmalemma and Golgi bodies). Prediction programs have been developed to identify which proteins have potential sites for such modifications, and are therefore likely to be targeted to these compartments, but in most cases the results have yet to be confirmed experimentally.

NEW PATHWAYS IN PLANT METABOLISM, AND SOME OLD ONES REVISITED

The compartmentation of plant metabolism has been painstakingly investigated over the last 40 years using the techniques described above. Although the locations of most of the pathways of primary metabolism appear to be well established, recent findings have revealed new details, or even shown that maps of some pathways need to be revised, and in this section an overview of a few examples is given. Also the compartmentation of some newly discovered metabolic pathways in plants will be discussed.

Glycolysis

During the 1970s and 1980s considerable effort was invested in understanding the metabolism of chloroplasts and non-photosynthetic plastids (e.g. amyloplasts). These studies showed that plastids play a central role not only in carbohydrate metabolism, but also in the biosyntheses of fatty acids, amino acids, nucleotides, and many secondary metabolites (e.g. alkaloids). Some of the metabolic pathways found in plastids may be wholly or partially duplicated in other compartments; for example, separate isoforms of key enzymes in the pathway of sulphate assimilation and cysteine biosynthesis are found in the chloroplasts, cytosol, and mitochondria in spinach leaves.

One of the more unexpected discoveries was the extensive duplication of glycolysis and the oxidative pentose-phosphate pathway in the cytosol and plastid stroma, providing both energy and carbon skeletons for other metabolic pathways in these two compartments. A more recent finding was that many of the glycolytic enzymes in the cytosol are closely associated with the outer mitochondrial membrane. It has been suggested that this microcompartmentation allows the product of glycolysis, pyruvate, to be directly channelled into the mitochondria to fuel the Krebs cycle. In the chloroplasts, several Calvin cycle enzymes appear to form multi-enzyme complexes that are also associated with membranes, in this instance the thylakoid membranes where NADPH and ATP are synthesized, and this association might also permit channelling of metabolites. In yeast and animals, several glycolytic enzymes act as metabolic sensors and transcription factors, and are found in the nucleus as well as the cytosol. It seems probable that some plant enzymes also have such dual metabolic and regulatory functions.

Gluconeogenesis

One of the classical examples of a highly compartmented pathway in plants is the conversion of lipids to sugars in germinating oilseeds via â-oxidation and the glyoxylate cycle. This pathway involves reactions in the peroxisomes (glyoxysomes), cytosol, and mitochondria. Although the locations of the enzymes in the pathway have been known for many years as a result of a series of studies on germinating castor bean seedlings, questions remained about the ways in which carbon is transported between the three compartments and the exact role of the peroxisomal malate dehydrogenase.

Various schemes for the pathway have been proposed by different authors, some of which show the peroxisomal malate dehydrogenase (MDH) oxidizing malate to oxaloacetate, even though researchers pointed out that the high NADH:NAD^+ ratio in the peroxisome, arising from â-oxidation, would be unfavourable for the MDH reaction to proceed in the oxidative direction. Recent analysis of knockout mutants that lack the peroxisomal MDH has shown that the reaction proceeds in the other direction, converting oxaloacetate to malate, and re-oxidizing the NADH produced by â-oxidation of fatty acids. Other genetic studies on peroxisomal citrate synthase mutants have shown that this enzyme is essential for β-oxidation of fatty acids, and that citrate is exported from the peroxisomes for conversion to isocitrate in the cytosol or oxidation via the citric acid cycle in the mitochondria, unlike in yeast where the peroxisomes export acetylcarnitine.

Also related to fatty acid–sugar interconversions was the surprising discovery of a role for Rubisco in the synthesis of lipids in developing oilseeds. Imported sucrose is a major source of carbon for storage-product synthesis in most developing seeds. In oilseeds, it was generally thought that sucrose was metabolized via glycolysis to acetyl CoA, the substrate for fatty acid synthesis. However, enzyme activity measurements and stable-isotope labelling studies in oilseed rape (*Brassica napus*) led to the proposal of an alternative pathway, in which hexose-phosphates derived from sucrose are metabolized via the non-oxidative steps of the pentose-phosphate pathway in conjunction with Rubisco. The proposed pathway would allow synthesis of 20% more acetyl CoA than the glycolytic route, but with a 40% reduction in the amount of carbon lost as CO_2.

C4 photosynthesis

Another classical example of compartmentation in plant metabolism is C_4 photosynthesis. The C_4 cycle maintains a high concentration of CO_2 at the site of fixation by Rubisco in the bundle sheath cells, and so helps to prevent photorespiration in C_4 plants. One of the key features of this pathway is that it is compartmented between two different cell types—mesophyll and bundle sheath cells—as well as between chloroplasts, cytosol, and mitochondria. CO_2 is initially fixed in the mesophyll cell cytosol by PEP carboxylase, forming the C_4-acid oxaloacetate, which is converted to malate and/or aspartate and transported via the plasmodesmata into the bundle sheath cells.

There, the malate/aspartate is decarboxylated in the chloroplasts (NADP-malic enzyme species), mitochondria (NAD-malic enzyme species), or cytosol and mitochondria (PEP carboxykinase species), releasing CO_2 to be refixed by Rubisco. The residual C_3 molecule (pyruvate or PEP) is transported back to the mesophyll cells, where pyruvate is converted back to PEP, the initial CO_2 acceptor, by pyruvate, Pi dikinase (PPDK) in the chloroplasts. Other pathways are also compartmented between mesophyll and bundle sheath cells in C_4 plants; for example, sucrose is synthesized preferentially in the mesophyll cells, whereas starch synthesis is usually restricted to the bundle sheath cells. A recent proteomic study of mesophyll and bundle sheath cell chloroplasts from maize leaves suggests that synthesis of lipids, tetrapyrroles, and isoprenoids is also preferentially located in the mesophyll cells, as is nitrogen assimilation, whereas sulphur assimilation takes place preferentially in the bundle sheath cells.

For many years it was believed that C_4 photosynthesis in terrestrial plants was inseparably linked to the possession of Kranz anatomy, where the bundle sheath cells are surrounded by a characteristic wreath of mesophyll cells. However, researchers discovered an exception to this rule in the plant *Borszczowia aralocaspica*, from the family Chenopodiaceae. This plant, and another species from the Chenopodiaceae, *Biernertia cycloptera*, both lack Kranz anatomy, but otherwise show all the characteristics of C_4 photosynthesis.

The photosynthetic chlorenchyma cells of *Borszczowia aralocaspica* have spatially separated photosynthetic enzymes, and possess two types of chloroplasts and other organelles arranged in separate parts of the cell. The chlorenchyma cells of *Biernertia cycloptera* have an unusual central cytoplasmic compartment containing mitochondria and granal chloroplasts. This is connected by cytoplasmic channels through the vacuole to the peripheral cytoplasm, which appears to lack mitochondria and has only grana-deficient chloroplasts.

Although quite different, the intracellular compartmentation in both of these species allows spatial separation of the carboxylation and decarboxylation phases of the C_4 cycle within the same cell, showing that compartmentation of C_4 cycle enzymes between two cell types is not absolutely essential for C_4 photosynthesis. Nevertheless, these two interesting species represent rare exceptions to the general rule that terrestrial C_4 plants have Kranz anatomy.

Phylogenetic studies of the occurrence of C_4 photosynthesis have shown that this pathway evolved independently at least 31 times, within 18 separate plant families. In evolutionary terms, the appearance of C_4 photosynthesis in all of these families seems to have occurred over a short period of time, when atmospheric CO_2 levels were much lower than today and would favour C_4 over C_3 photosynthesis. At first sight, it seems quite surprising that such a complex pathway should have apparently arisen so readily in so many different plant lineages, but some new observations suggest that several features of C_4 plants were already present in their C_3 ancestors. Researchers found that photosynthetic stems and petioles of some C_3 plants contain high activities of the C_4-acid decarboxylating enzymes, and that these could form part of a C_4 cycle. They proposed that C_4 acids (e.g. malate and aspartate) are synthesized in other organs, such as roots, and then transported via the phloem or xylem to the stems where they are decarboxylated to supply CO_2 for photosynthesis. In this way, cells in the stem that do not have direct access to atmospheric CO_2 via stomata can still contribute to net carbon fixation. The presence of such a C_4 cycle in C_3 plants suggests that it is not such a giant leap from C_3 to C_4 photosynthesis, and therefore the polyphyletic evolution of C_4 photosynthesis is less surprising.

THE PLASTIDIAL PATHWAY OF ISOPRENOID SYNTHESIS

It is not often that a major new metabolic pathway in plants is discovered, but a recent example is the 1-deoxy-D-xylulose-5-phosphate (DOXP) pathway of isoprenoid biosynthesis. This pathway is located in the plastids, and is responsible for the synthesis of carotenoids, phytol, plastoquinone-9, and isoprene, as well as mono- and diterpenes. Before the discovery of this pathway, it was thought that all isoprenoids, including photosynthetic pigments and electron transport chain components, were synthesized via the acetate/mevalonate pathway, which is located in the cytosol.

An obvious advantage of having the DOXP pathway in the chloroplasts is that the photosynthesis-related isoprenoids can be synthesized *in situ*, and do not need to be imported from the cytosol. Perhaps the most surprising thing about the DOXP pathway is that its existence in plants was overlooked for so many years. The discovery of the DOXP pathway in bacteria provided the stimulus to re-examine data from some earlier radiolabelling and inhibitor studies in plants. These were found to be inconsistent with the accepted view that all isoprenoids were synthesized via the acetate/mevalonate pathway, and led the way to finding the DOXP pathway in plants. As so often in science, those little inconsistencies in the data turned out to be telling us something important, if only we cared to look.

VITAMIN AND COENZYME SYNTHESIS

The biosynthetic pathways of several vitamins and coenzymes represent one of the final frontiers of plant primary metabolism. Fluxes through many of these pathways can be vanishingly small, making it difficult to measure either enzymatic activities or pathway intermediates. However, the discovery and analysis of auxotrophic mutants of *Arabidopsis thaliana* with lesions in the pathways of vitamin biosynthesis, together with new information from studies in bacteria, is slowly uncovering these pathways. A good example is the synthesis of biotin (vitamin H), the coenzyme of acetyl CoA carboxylase, and several other carboxylases, decarboxylases, and transcarboxylases.

The last enzyme in the pathway of biotin biosynthesis, biotin synthase, catalyses the conversion of dethiobiotin to biotin, and this enzyme was found to be located in the mitochondrial matrix, whereas the first committed step in the pathway, the conversion of pimeloyl CoA to 7-keto-8-aminopelargonic acid (KAPA) by KAPA synthase, takes place in the cytosol. The enzymes catalysing the two intermediate steps in the pathway, DAPA aminotransferase and dethiobiotin synthase, have not yet been characterized in plants, but *Arabidopsis thaliana* proteins showing similarity to the bacterial enzymes are predicted to be located in the cytosol and mitochondria, respectively.

The L-galactose pathway of L-ascorbic acid (vitamin C) synthesis is similarly divided between the cytosol and mitochondria, with the terminal enzyme, L-galactono-1,4-lactone dehydrogenase, being located in the inner mitochondrial membrane, in association with complex I of the mitochondrial electron transport chain. The biosynthesis of pantothenate (vitamin B_5) is also split between cytosol and mitochondria, but in the reverse orientation, starting in the mitochondria (ketopantoate hydroxymethyltransferase) and ending in the cytosol (pantothenate synthase).

In the thiamine (vitamin B_1) biosynthetic pathway, the mRNA from the *THI1* gene encoding the thiazole biosynthetic enzyme contains two alternative AUG translation initiation codons. The first of these appears to be the preferred start codon and gives rise to a chloroplast-targeted protein, while the second AUG produces a slightly smaller protein that is targeted to the mitochondria. This suggests that the thiazole moiety of thiamine is synthesized predominantly in the chloroplasts, but can also be made in the mitochondria. It is not yet known to what extent the rest of the thiamine biosynthetic pathway is duplicated between these two organelles or other compartments. The biosynthesis of folic acid (vitamin B_9) is split between three compartments: cytosol, chloroplasts, and mitochondria.

Folic acid is composed of three moieties: a pterin ring, *p*-aminobenzoic acid (*p*-ABA), and a polyglutamate chain (Glu_{1-8}). Dihydropterin and *p*-ABA are synthesized in the cytosol and chloroplasts, respectively, and then transported into the mitochondria where they are condensed with up to eight glutamate molecules to produce folic acid.

By contrast to these multi-compartmented pathways, the biosynthesis of the vitamin B_6 complex (pyridoxal, pyridoxine, pyridoxamine, and their derivatives) appears to be located entirely within the cytosol, and riboflavin (vitamin B_2) synthesis is probably restricted to the plastids. Except for the last two steps of haem biosynthesis, which are duplicated in the mitochondria, the synthesis of tetrapyrrole coenzymes (chlorophyll, haem, sirohaem, and phytochromobilin) is also restricted to the plastids. Lipoic acid, the coenzyme of several oxidative decarboxylases in the mitochondria, is made in the mitochondria.

It is interesting to note that not all coenzymes are synthesized where their partner enzymes are located; for example, most of the biotin synthesized in the mitochondria probably ends up bound to the plastidial and cytosolic isoforms of acetyl CoA carboxylase. Presumably some of these coenzymes must be transported between different intracellular compartments, and perhaps between cells. One of the sucrose-H^+ symporters in *Arabidopsis thaliana* (AtSuc5) can also transport biotin, and a folate transporter has been identified in the chloroplast envelope. Apart from these examples, little is known about vitamin and coenzyme transporters in plants at present.

The apparent importance of the mitochondria in so many vitamin biosynthetic pathways is somewhat unexpected, given the dominance of the plastids in many other biosynthetic pathways. Some of these vitamin biosynthetic pathways could have been directly inherited from the bacterial endosymbiont that gave rise to the mitochondria, whereas others might have been transferred from the cyanobacterial ancestor of the chloroplasts. The dual targeting of the thiazole biosynthetic enzyme to both chloroplasts and mitochondria suggests one possible mechanism by which enzymes, and indeed whole pathways, could be shuffled between these organelles.

COMMUNICATION BETWEEN COMPARTMENTS

One of the consequences of metabolic compartmentation is the need for a whole host of transporters to move metabolites around the cell. These are usually integral membrane proteins that are less amenable to experimental investigation than soluble proteins, so only a few transporters have been thoroughly characterized in plants. Multi-compartmented pathways also need regulatory mechanisms to balance the fluxes through the separate branches of the pathway. A good example is photosynthetic carbon metabolism, where the rate of sucrose synthesis in the cytosol is co-ordinated with the rate of CO_2 fixation and starch synthesis in the chloroplasts.

These pathways are linked by the triose-phosphate translocator in the chloroplast envelope, which imports Pi from the cytosol in exchange for triose-phosphates. The transported metabolites also act as signals that convey information about metabolic fluxes in the originating compartment; triose-phosphates exported from the chloroplast influence the level of the signal metabolite Fru2,6P_2 in the cytosol, which regulates the cytosolic fructose-1,6-*bis*phosphatase, and Pi released by sucrose synthesis in the cytosol is needed in the chloroplast for photophosphorylation. Pi is an allosteric inhibitor of the plastidial ADP-glucose pyrophosphorylase, and so the stromal concentration of Pi also influences the rate of starch synthesis. These and other mechanisms, including post-translational regulation of several enzymes, co-ordinate sucrose synthesis with the rate of CO_2 fixation, preventing excessive withdrawal of triose-phosphates from the Calvin cycle, but maintaining an adequate supply of Pi in the chloroplast for photophosphorylation.

There is emerging evidence that trehalose 6-phosphate (Tre6P), the intermediate of trehalose synthesis, could also be involved in communication between cytosol and chloroplast. Tre6P appears to be synthesized in the cytosol in plant cells, and the amount of this metabolite is strongly correlated with the level of sucrose. It has been found that supplying Tre6P to isolated chloroplasts promotes the redox activation of ADP-glucose pyrophosphorylase, although the mechanism of this effect is not yet understood. From these observations it has been proposed that accumulation of sucrose in the leaf leads to a rise in the level of Tre6P in the cytosol, which in some way leads to activation of ADP-glucose pyrophosphorylase in the chloroplasts, and so diverts photosynthate away from sucrose and into starch while maintaining a steady rate of CO_2 fixation. Thus, one of the functions of Tre6P in plants could be to act as a messenger service between the cytosol and the chloroplasts.

Phylogenetic analysis of some of the enzymes of sucrose and starch synthesis has revealed some interesting clues to the evolutionary history of their compartmentation. Most of the enzymes in these pathways appear to have a cyanobacterial origin, presumably stemming from the endosymbiosis of the chloroplast ancestor. Therefore, it seems likely that both sucrose and starch would have been made in the chloroplasts of the earliest photosynthetic eukaryotes. During the evolution of plants, most (90–95%) of the ancestral cyanobacterial genes were lost from the chloroplast genome. Many were transferred to the nucleus, and in *Arabidopsis thaliana* these account for about %18 of the protein-coding genes in the nuclear genome.

Despite their origin, the majority of these genes encode proteins that end up being located outside the plastids. These include the key enzymes of sucrose synthesis—sucrose-phosphate synthase and sucrose-phosphatase—as well as neutral/alkaline invertase, which hydrolyses sucrose in the cytosol. Interestingly, the acid invertases in the vacuole and apoplast appear to have a different evolutionary origin.

Although the key enzymes of starch synthesis (ADP-glucose pyrophos-phorylase, starch synthases, isoamylases, starch branching enzymes) are all encoded by nuclear genes, the proteins themselves are located in the plastids. Surprisingly, phylogenetic analysis showed that not all of these plastidial enzymes have a cyanobacterial origin. The starch branching enzymes are more closely related to enzymes from non-photosynthetic eukaryotes, and have entirely replaced any cyanobacterial form of the enzyme that might once have been present.

Thus, the pathway of starch synthesis appears to have been reconstituted in plastids by combining eukaryotic and cyanobacterial elements. Curiously, in red algae, whose plastids are derived from the same endosymbiotic event that gave rise to chloroplasts, the pathway for synthesis of floridean starch appears to have been reconstituted in a similar way, but in the cytosol, not the plastids. These pathways, and the adaptation of pre-existing elements to form the C_4 photosynthetic pathway, both indicate that there has been extensive shuffling of genes and proteins between compartments during the evolution of plants.

Concluding comments

Identifying pathways and mapping their intracellular locations are the essential first steps for understanding plant metabolism. Genome sequencing has led to the discovery of new metabolic pathways in plants, and new technologies such as protein mass spectrometry and fluorescent protein tagging are providing fresh insights into the compartmentation of plant metabolism. However, despite the rapid advances of recent years, there are still some major gaps in our knowledge of primary metabolism in plants. For example, the biosynthetic pathways for several coenzymes have not yet been fully resolved, and there is only limited understanding of the synthesis, transport, and assembly of many cell wall components. Knowledge of plant secondary metabolism is even more fragmentary. Collectively, plants synthesize a huge range of secondary metabolites (e.g. isoprenoids, alkaloids, and phenylpropanoids), but for many of these only the early steps in the biosynthetic pathways are known.

It is not only in biosynthetic plant metabolism where there are gaps in our knowledge. Plants often grow under nutrient-limited conditions, and recycling of nutrients from turnover of cell constituents (e.g. nucleotides, amino acids, membrane lipids, and photosynthetic pigments) allows them to reallocate limited resources to maximize their chances of survival or their ability to reproduce, for example, by degrading photosynthesis-related proteins in the leaves to supply nitrogen to developing seeds. Such reallocation can also have an important influence on the yield of crop plants, yet few of the catabolic and salvage pathways involved in nutrient recycling in plants have been fully characterized, and some are still a matter of guesswork based on information from other organisms.

Metabolite transport is yet another aspect of metabolic compartmentation where our knowledge is incomplete. Although the transport of metabolites can be inferred from the locations of the enzymes in the pathway, many of the proteins involved in membrane transport and intracellular trafficking have not yet been identified. In conclusion, it appears that there is still some way to go before there is a complete map of plant metabolism and, without this, efforts to identify regulatory networks or to understand plant metabolism as an integrated whole are hampered.

CHAPTER

THE CALVIN CYCLE 6

After the energy from the sun is converted and packaged into ATP and NADPH, the cell has the fuel needed to build food in the form of carbohydrate molecules. The carbohydrate molecules made will have a backbone of carbon atoms. Where does the carbon come from? The carbon atoms used to build carbohydrate molecules comes from carbon dioxide, the gas that animals exhale with each breath. The Calvin cycle is the term used for the reactions of photosynthesis that use the energy stored by the light-dependent reactions to form glucose and other carbohydrate molecules.

The Calvin cycle is a set of light independent redox reactions that occur during photosynthesis and carbon fixation to convert carbon dioxide into the sugar glucose. These reactions occur in the stroma of the chloroplast, which is the fluid-filled region between the thylakoid membrane and inner membrane of the organelle. Here is a look at the redox reactions that occur during the Calvin cycle.

OTHER NAMES FOR THE CALVIN CYCLE

You may know the Calvin cycle by another name. The set of reactions also is known as the dark reactions, C3 cycle, Calvin-Benson-Bassham (CBB) cycle, or reductive pentose phosphate cycle. The cycle was discovered in 1950 by Melvin Calvin, James Bassham, and Andrew Benson at the University of California, Berkeley. They used radioactive carbon-14 to trace the path of carbon atoms in carbon fixation.

CARBON FIXATION

In carbon fixation, a CO_2 molecule from the atmosphere combines with a five-carbon acceptor molecule called ribulose-1,5-bisphosphate (RuBP).

The resulting six-carbon compound is then split into two molecules of the three-carbon compound, 3-phosphoglyceric acid (3-PGA).

This reaction is catalyzed by the enzyme RuBP carboxylase/oxygenase, also known as RuBisCO. Due to the key role it plays in photosynthesis, RuBisCo is probably the most abundant enzyme on Earth.

Reduction

In the second stage of the Calvin cycle, the 3-PGA molecules created through carbon fixation are converted into molecules of a simple sugar – glyceraldehyde-3 phosphate (G3P).

This stage uses energy from ATP and NADPH created in the light-dependent reactions of photosynthesis. In this way, the Calvin cycle becomes the way in which plants convert energy from sunlight into long-term storage molecules, such as sugars. The energy from the ATP and NADPH is transferred to the sugars.

This step is called "reduction" because NADPH donates electrons to the 3-phosphoglyceric acid molecules to create glyceraldehyde-3 phosphate. In chemistry, the process of donating electrons is called "reduction," while the process of taking electrons is called "oxidation."

Regeneration

Some glyceraldehyde-3 phosphate molecules go to make glucose, while others must be recycled to regenerate the five-carbon RuBP compound that is used to accept new carbon molecules.

The regeneration process requires ATP. It is a complex process involving many steps. Because it takes six carbon molecules to make a glucose, this cycle must be repeated six times to make a single molecule of glucose.

To accomplish this equation, five out of six glyceraldehyde-3 phosphate molecules that are created through the Calvin cycle are regenerated to form RuBP molecules. The sixth exits the cycle to become one half of a glucose molecule.

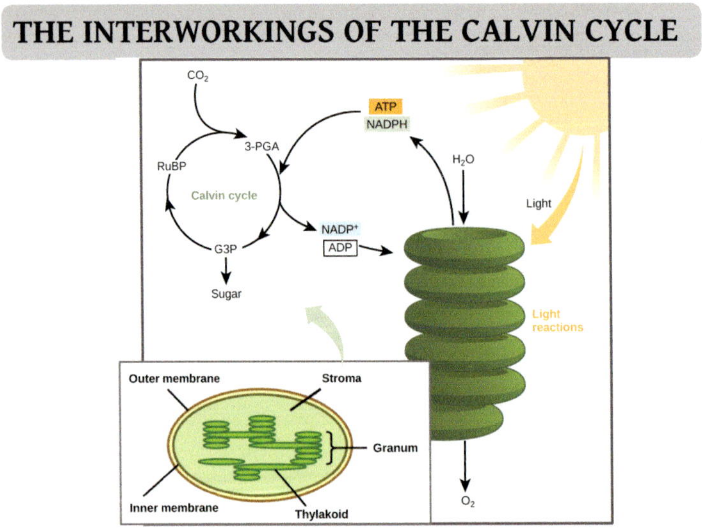

Fig. 1. Light-dependent reactions harness energy from the sun to produce ATP and NADPH. These energy-carrying molecules travel into the stroma where the Calvin cycle reactions take place.

In plants, carbon dioxide (CO_2) enters the chloroplast through the stomata and diffuses into the stroma of the chloroplast—the site of the Calvin cycle reactions where sugar is synthesized. The reactions are named after the scientist who discovered them, and reference the fact that the reactions function as a cycle. Others call it the Calvin-Benson cycle to include the name of another scientist involved in its discovery (Fig. 1).

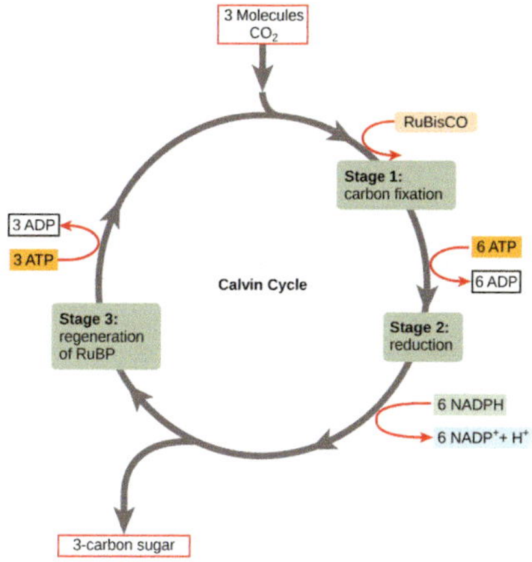

Fig. 2. The Calvin cycle has three stages. In stage 1, the enzyme RuBisCO incorporates carbon dioxide into an organic molecule. In stage 2, the organic molecule is reduced. In stage 3, RuBP, the molecule that starts the cycle, is regenerated so that the cycle can continue.

The Calvin cycle reactions (Fig. 2) can be organized into three basic stages: fixation, reduction, and regeneration. In the stroma, in addition to CO_2, two other chemicals are present to initiate the Calvin cycle: an enzyme abbreviated RuBisCO, and the molecule ribulose bisphosphate (RuBP). RuBP has five atoms of carbon and a phosphate group on each end.

RuBisCO catalyzes a reaction between CO_2 and RuBP, which forms a six-carbon compound that is immediately converted into two three-carbon compounds. This process is called carbon fixation, because CO_2 is "fixed" from its inorganic form into organic molecules.

ATP and NADPH use their stored energy to convert the three-carbon compound, 3-PGA, into another three-carbon compound called G3P. This type of reaction is called a reduction reaction, because it involves the gain of electrons. A reduction is the gain of an electron by an atom or molecule. The molecules of ADP and NAD^+, resulting from the reduction reaction, return to the light-dependent reactions to be re-energized.

One of the G3P molecules leaves the Calvin cycle to contribute to the formation of the carbohydrate molecule, which is commonly glucose ($C_6H_{12}O_6$). Because the carbohydrate molecule has six carbon atoms, it takes six turns of the Calvin cycle to make one carbohydrate molecule (one for each carbon dioxide molecule fixed). The remaining G3P molecules regenerate RuBP, which enables the system to prepare for the carbon-fixation step. ATP is also used in the regeneration of RuBP.

In summary, it takes six turns of the Calvin cycle to fix six carbon atoms from CO_2. These six turns require energy input from 12 ATP molecules and 12 NADPH molecules in the reduction step and 6 ATP molecules in the regeneration step.

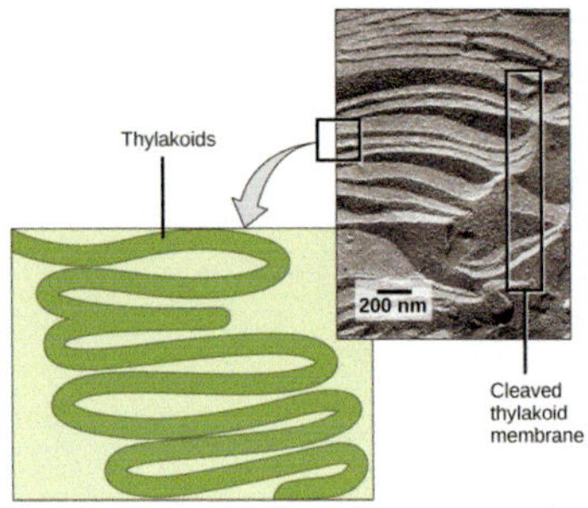

Fig. 3 A photosynthetic prokaryote has infolded regions of the plasma membrane that function like thylakoids. Although these are not contained in an organelle, such as a chloroplast, all of the necessary components are present to carry out photosynthesis

PHOTOSYNTHESIS

The shared evolutionary history of all photosynthetic organisms is conspicuous, as the basic process has changed little over eras of time. Even between the giant tropical leaves in the rainforest and tiny cyanobacteria, the process and components of photosynthesis that use water as an electron donor remain largely the same. Photosystems function to absorb light and use electron transport chains to convert energy. The Calvin cycle reactions assemble carbohydrate molecules with this energy.

However, as with all biochemical pathways, a variety of conditions leads to varied adaptations that affect the basic pattern. Photosynthesis in dry-climate plants has evolved with adaptations that conserve water. In the harsh dry heat, every drop of water and precious energy must be used to survive. Two adaptations have evolved in such plants. In one form, a more efficient use of CO_2 allows plants to photosynthesize even when CO_2 is in short supply, as when the stomata are closed on hot days.

The other adaptation performs preliminary reactions of the Calvin cycle at night, because opening the stomata at this time conserves water due to cooler temperatures. In addition, this adaptation has allowed plants to carry out low levels of photosynthesis without opening stomata at all, an extreme mechanism to face extremely dry periods.

PHOTOSYNTHESIS IN PROKARYOTES

The two parts of photosynthesis—the light-dependent reactions and the Calvin cycle—have been described, as they take place in chloroplasts. However, prokaryotes, such as cyanobacteria, lack membrane-bound organelles. Prokaryotic photosynthetic autotrophic organisms have infoldings of the plasma membrane for chlorophyll attachment and photosynthesis (Fig. 3). It is here that organisms like cyanobacteria can carry out photosynthesis.

THE ENERGY CYCLE

Living things access energy by breaking down carbohydrate molecules. However, if plants make carbohydrate molecules, why would they need to break them down? Carbohydrates are storage molecules for energy in all living things. Although energy can be stored in molecules like ATP, carbohydrates are much more stable and efficient reservoirs for chemical energy. Photosynthetic organisms also carry out the reactions of respiration to harvest the energy that they have stored in carbohydrates, for example, plants have mitochondria in addition to chloroplasts.

You may have noticed that the overall reaction for photosynthesis:

$6CO_2+6H_2O'!C_{612}O_6+6O_2 6CO_2+6H_2O'!C_6H_{12}O_6+6O$

is the reverse of the overall reaction for cellular respiration:

$6O_2+C_6H_{12}O_6'!6CO_2+6H_2O_6O_2+C_6H_{12}O_6'!6CO_2+6H_2O$

Photosynthesis produces oxygen as a byproduct, and respiration produces carbon dioxide as a byproduct.

In nature, there is no such thing as waste. Every single atom of matter is conserved, recycling indefinitely. Substances change form or move from one type of molecule to another, but never disappear.

CO_2 is no more a form of waste produced by respiration than oxygen is a waste product of photosynthesis. Both are byproducts of reactions that move on to other reactions. Photosynthesis absorbs energy to build carbohydrates in chloroplasts, and aerobic cellular respiration releases energy by using oxygen to break down carbohydrates. Both organelles use electron transport chains to generate the energy necessary to drive other reactions. Photosynthesis and cellular respiration function in a biological cycle, allowing organisms to access life-sustaining energy that originates millions of miles away in a star.

Summary

Using the energy carriers formed in the first stage of photosynthesis, the Calvin cycle reactions fix CO_2 from the environment to build carbohydrate molecules. An enzyme, RuBisCO, catalyzes the fixation reaction, by combining CO_2 with RuBP. The resulting six-carbon compound is broken down into two three-carbon compounds, and the energy in ATP and NADPH is used to convert these molecules into G3P. One of the three-carbon molecules of G3P leaves the cycle to become a part of a carbohydrate molecule. The remaining G3P molecules stay in the cycle to be formed back into RuBP, which is ready to react with more CO_2. Photosynthesis forms a balanced energy cycle with the process of cellular respiration. Plants are capable of both photosynthesis and cellular respiration, since they contain both chloroplasts and mitochondria.

CHAPTER

PHOTOSYNTHESIS IS AN ELECTRON TRANSPORT PROCESS 7

INTRODUCTION

Photosynthetic control of electron transport is a fundamental concept in the regulation of photosynthesis. It stems from the observation that the proton motive force (pmf) exerts a feedback control over electron flux, forming a regulatory circuit between ATP synthesis and electron transport in energy-transducing membranes. In this way, electron transport rates can be restrained when ATP is abundant and the pmf is high. The pmf is comprised of the transthylakoid voltage and the pH gradient (ΔpH).

As electrons pass through the photosynthetic electron transport (PET) chain the thylakoid lumen becomes positively charged and acidic relative to the stroma. Lumen acidification controls plastoquinol (PQH_2) oxidation by the cytochrome b_6/f (cytb_6f) complex. The plastoquinone (PQ)/PQH_2 pool is an integrating regulatory redox hub that controls both the efficiency of electron transport and the activation of gene expression. PQH_2 oxidation also controls post-translational regulation of photosynthetic proteins by phosphorylation and also the rate of transcription of genes encoding the reaction centre apoproteins of photosystem I (PSI) and PSII. In this way, the stoichiometries of the photosystems are balanced to optimize the relative efficiency of operation of each photosystem. Gene expression within the chloroplasts and nucleus is controlled through the activation of a network of kinases that are involved in chloroplast to nucleus retrograde signalling pathways.

The fundamental concepts and hypotheses concerning the photosynthetic control hub are considered here in detail, with particular emphasis on the central role of reduction–oxidation (redox) metabolism, homeostasis, and signalling in the regulation of gene expression in response to changes in atmospheric CO2 levels.

THE LIGHT-DEPENDENT REACTIONS OF PHOTOSYNTHESIS

How can light be used to make food? When a person turns on a lamp, electrical energy becomes light energy. Like all other forms of kinetic energy, light can travel, change form, and be harnessed to do work. In the case of photosynthesis, light energy is converted into chemical energy, which photoautotrophs use to build carbohydrate molecules. However, autotrophs only use a few specific components of sunlight.

WHAT IS LIGHT ENERGY?

The sun emits an enormous amount of electromagnetic radiation (solar energy). Humans can see only a fraction of this energy, which portion is therefore referred to as "visible light." The manner in which solar energy travels is described as waves. Scientists can determine the amount of energy of a wave by measuring its wavelength, the distance between consecutive points of a wave. A single wave is measured from two consecutive points, such as from crest to crest or from trough to trough (Fig. 1).

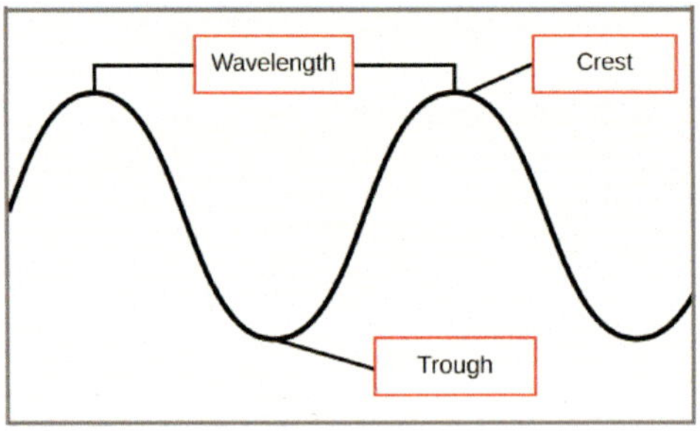

Fig. 1: The wavelength of a single wave is the distance between two consecutive points of similar position (two crests or two troughs) along the wave.

Visible light constitutes only one of many types of electromagnetic radiation emitted from the sun and other stars. Scientists differentiate the various types of radiant energy from the sun within the electromagnetic spectrum. The electromagnetic spectrum is the range of all possible frequencies of radiation. The difference between wavelengths relates to the amount of energy carried by them.

CONCEPTS OF ENERGY PRODUCTION AND UTILIZATION

The basic function of the light-harvesting and electron transport systems is to convert the free energy of absorbed light as efficiently as possible into forms [ATP, reduced ferredoxin (Fd), and NADPH] that can be used to drive metabolism in the chloroplast and cytosol. When the supply of free energy from light is in excess of the needs of metabolism, a secondary function of light harvesting and electron transport emerges, namely the safe dissipation of the excess absorbed free energy. A third function arises from the fluctuating ratio of demand for reductant and ATP. The photosynthetic apparatus must be able to adjust the relative rates of synthesis of ATP and reductant in order to prevent limitations on the operation of photosynthetic energy transduction arising from an imbalance in the availability of the two forms of assimilatory power. The integrated operation of light-limited photosynthesis, energy dissipation, and the balance between the synthesis of ATP and reductant lies at the heart of photosynthetic regulation and its responses to the environment.

LIGHT ABSORPTION AND CHARGE SEPARATION BY PSI AND PSII

An essential property of any light-driven system is its ability to absorb light. In the case of photosynthesis in higher plants, light absorption results from the presence in leaves of significant quantities of chlorophylls (chls) a and b and carotenoids, of which â-carotene, lutein, violaxanthin, and neoxanthin are the most common. The photosynthetically active wavelengths within the spectrum are normally in the range 400–700 nm, though wavelengths shorter or longer than this can still have some photosynthetic activity. For a typical leaf, the absorption of photosynthetically active radiation (PAR) is ~85%, implying that leaves are good light traps.

At the chl and carotenoid concentrations normally found in leaves, the absorptivity for PAR is relatively insensitive to changes in pigment level, though at wavelengths of <560 nm and >700 nm, where absorption is low, the effect of increased pigment levels is more noticeable. It is important to note that not only photosynthetic pigments absorb PAR; flavonoids and other phenolic substances are present in most leaves to varying degrees and absorb wavelengths <560 nm. This partly accounts for the lower quantum efficiencies for CO_2 fixation measured under light-limiting conditions at wavelengths <560 nm. The relatively inefficient energy transfer from carotenoids to chls (only 70–90%) also contributes to the lowered light use efficiency of photosynthesis at wavelengths <560 nm.

When a photosynthetic pigment (chl or carotenoid) absorbs a quantum of PAR, an excited state is formed. These excited states are ultimately transferred to chl a to form the first singlet (S1) excited state (1chl*) and it is in this form that the excitation energy migrates, by resonance transfer from pigment to pigment, to the reaction centres. 1Chl* is inherently unstable and, in PSII, for example, has a maximum lifetime of only 3 ns so excited states cannot be accumulated by the photosystems. An important property of the reaction centres of both photosystems is their ability to use the fugitive, high-energy 1chl* to drive an otherwise endothermic charge separation reaction, a stable chemical transformation, with a high efficiency, implying very rapid chemistry.

The efficiency by which this charge separation reaction is driven by PAR can be assessed in different ways. Firstly, it can be determined as a quantum efficiency, which is a measure of what proportion of quanta, on either an incident or an absorbed light basis, produce a specific change. The quantum efficiencies of photosynthesis are frequently assessed on a relative basis. They can also be measured in terms of the proportion of the incident or absorbed light energy that is conserved in free energy of the products of photosynthesis, but this parameter is less easy to measure.

The maximum (i.e. dark-adapted) quantum yield for charge separation in PSII (Φ_{PSII}) obtained from chl fluorescence measurements is ~0.83. However, Φ_{PSII} measurements derived from fluorescence analysis can be subject to error. Researchers estimate the maximum Φ_{PSII} to be 0.88, while Vasili'ev and Bruce estimated the maximum quantum yield for charge separation in PSI (Φ_{PSI}) and Φ_{PSII} to be ~0.95.

Other estimates of maximum Φ_{PSI} place values close to 1.0. However, both photosystems can undergo rapidly reversible losses of efficiency as part of the regulation of light use by photosynthesis. Imperfect excitation transfer from the carotenoids to chl will decrease efficiency, as will light-induced damage. Slowly reversible down-regulation of efficiencies has also been reported.

ELECTRON AND PROTON TRANSPORT

The PET system incorporates linear and cyclic pathways. Linear (or non-cyclic) electron transport follows the classic Z-scheme model from water to the terminal electron acceptor of PSI, Fd. Cyclic electron transport operates around PSI and is much less well defined than the linear pathway, and various components may be involved. Proton transport coupled to electron transport results in the formation of a transthylakoid proton potential (DμH+), or pm. This pmf is important for two reasons.

First it drives the formation of ATP. The gating of protons, driven by the DμH+, through the ATPase results in the otherwise energetically unfavourable synthesis of ATP from ADP and Pi in the chloroplast stroma, thus linking electron transport to ATP formation.

Secondly, the DμH+ is comprised of two components: a pH difference between the chloroplast stroma and the thylakoid lumen and a voltage difference across the thylakoid membrane, with the thylakoid lumen becoming more acid and electrically positively charged when electron transport occurs. Acidification of the thylakoid lumen is important in regulating electron transport and modulating the non-photochemical quenching (NPQ) mechanism of PSII. An important consequence of these regulatory effects is that one of the results of electron transport, lumen acidification, is a down-regulation of electron transport and an up-regulation of photoprotection in PSII. Linear electron transport results in the formation of reductant and ATP, while cyclic electron transport results only in the formation of ATP with no net formation of reductant. The flux of electrons through these pathways is regulated at the level of the cytb6f complex. While some cyclic fluxes operating around PSII have been reported, these are not considered to be significant in vivo so will not be discussed further here.

Cyclic electron flow around PSI in higher plants consists of at least two partially redundant pathways known as the ferredoxin quinone oxidoreductase (FQR)- and NAD(P)H dehydrogenase (NDH)-dependent pathways. Despite intensive efforts, FQR has not been found associated with any known chloroplast protein or protein complex. In contrast, it is now established that the NDH complex associates with PSI, a process that is considered to be important in facilitating NDH functions such as cyclic electron flow and chlororespiration.

The NDH complex can interact with multiple PSI units to form NDH–PSI supercomplexes in the thylakoid membranes, an association that requires two minor LHCI proteins, Lhca5 and Lhca6. The thylakoid PROTON GRADIENT REGULATION 5 (PGR5) and PGR 5-like 1 (PGRL1) proteins also associate with PSI in a manner that is considered to facilitate the operation of PSI cyclic electron transport. The NDH–PSI association may re-direct electron flow and so prevent over-reduction of the stroma in mutants such as pgr5, which are defective in PSI cyclic electron transport. In the NDH-dependent cyclic electron flow pathway, electrons are transported from PSI to the PQ pool via NADPH or reduced ferredoxin and the NDH complex.

Cyclic electron transport around PSI only predominates in some special cases, such as in the bundle sheath cells of some C4 leaves, though some mutants show clear up-regulation of cyclic electron transport. The majority of PET follows the linear (or non-cyclic) pathway passing from water through PSII, the cytb6f complex, and PSI to Fd, and thereafter, for most electrons, to NADP. However, it is generally accepted that cyclic electron flow makes an important contribution to the overall regulation of electron transport in response to environmental variation and metabolic cues, even though the nature of these pathways and their precise functions remain controversial.

Cyclic electron transport serves only to support the production of the pmf and thus allows ATP generation without the net formation of reductant, such as reduced Fd or NADPH. It is therefore believed that switching between cyclic and non-cyclic pathways provides a degree of flexibility in the ratio of ATP and NAPDH production and plants can hence adjust the ratio of ATP to NAPDH to meet the needs of metabolism.

Flexibility in the formation of pmf and reductant is not only important in meeting the needs of the Benson–Calvin cycle and other metabolic pathways, but it is also vital to the activation and regulation of photosynthesis in the transition from darkness to light, in response to rapid (i.e. on the time scale of seconds) fluctuations in the light environment and in atmospheric CO2 availability, and the acclimation to environmental stress. The operation of cyclic electron flow around PSI is considered to protect PSII by enhancing protonation of the lumen, thus limiting electron transport and triggering NPQ. By regulating electron transport into PSI via the cytb6f complex, cyclic electron flow also minimizes the probability of the formation of reactive oxygen species (ROS) such as superoxide (O2·–) on the acceptor side of PSI. The mid-point potential for the O2/O2·– couple is –160 mV while the redox potential of the stroma is close to –330 mV. Many of the stromal electron acceptors such as Fd, thioredoxin (TRX), and ferredoxin-NADP+ oxidoreductase (FNR) are thus able reduce O2 to O2–.

This process, which is called the Mehler reaction or pseudocyclic electron flow, is important in photosynthetic control because it contributes to the prevention of over-reduction of the PET carriers and it produces oxidative signals that regulate gene expression. The addition of H_2O_2 to plant cells leads to substantial transcriptome re-programming. Similarly, the expression of numerous genes was changed in mutants lacking a chloroplastic superoxide dismutase (SOD.

THE ASSIMILATION (A)/INTERCELLULAR CO2 (CI) CURVES OF PHOTOSYNTHESIS

The variable relationship between the PET chain and carbon metabolism can be illustrated by the A/C_i curves of photosynthesis, which are considered to show two main zones or phases of limitation. The first phase is observed at low C_i values and corresponds to a limitation by ribulose-1, 5-bisphosphate carboxylase/oxygenase (Rubisco)-catalysed carboxylation (Fig. 2). Under these conditions, which would be experienced by a leaf with closed or partially closed stomata, the capacity of electron transport could exceed that of metabolism. The second phase is observed at higher C_i values and corresponds to a limitation of photosynthesis by regeneration of ribulose 1, 5-bisphosphate (RuBP), which is believed to be substantially due to a limitation of electron transport capacity. Under these conditions, down-regulation of electron transport would be minimal.

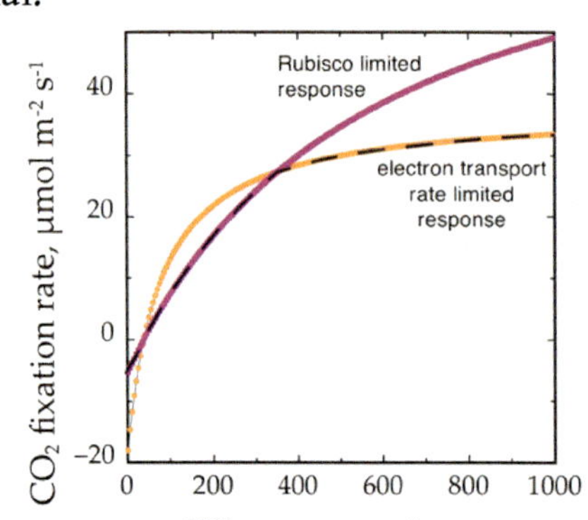

Fig. 2. The relationship between CO2 concentration at the site of carboxylation (i.e. in the hloroplast stroma) and the rate of CO2 fixation, commonly referred to as an *A*/Ci curve, for a leaf in an atmosphere containing 21% (21 kPa) oxygen. The rate of CO2 fixation is the minimum of the either the Rubisco-limited rate (the Ac line) or the ribulose-1,5-phosphatelimited rate (the Aj line). The supply of ribulose-

1,5-bisphosphate is strongly influenced by the rate of electron transport and thus irradiance, so for any leaf there are an infinite number of possible Aj lines, corresponding to different irradiances (only one is illustrated). The Ac line is determined by the amount of Rubisco, its activation state and temperature, so if the Rubisco is fully activated and temperature is constant there is only one Ac line.

The capacities for electron transport and metabolism are closely balanced by regulation of chloroplast structure and composition and by physiological regulation that controls the rate constants for electron and proton transport. For example, the Ci value of the transition point between the Rubisco- and electron transport-limited zones is offset relative to the growth Ci values. Similarly, the light-saturated, maximum rate of CO2 fixation is proportional to the leaf content of the Rubisco and cytb6f proteins.

Although the rate constants for electron or transport processes such as $P700^+$ reduction by electrons from the PQH_2 pool can vary in response to increasing irradiance, the rate constants do not change significantly (i.e. there is little need for regulation) under steady-state conditions in the absence of stress, implying that the capacity of electron transport is well balanced with the capacity (or demand) of metabolism.

REGULATION OF ELECTRON AND PROTON TRANSPORT

The supply of ATP and reductant NADPH by the thylakoid membranes must be precisely matched to their use in metabolism. This coordination minimizes singlet oxygen production in PSII and superoxide production in PSI. Considered in isolation from metabolism, electron transport is generally limited by the availability of light (i.e. the rate of charge separation in PSI and PSII) or by the intrinsic capacity of the electron transport chain (particularly the intersystem electron transport chain) to transport electrons, while metabolism, which generates demand for ATP and reductant, is limited by the extent of enzyme activation.

The key to the regulation of electron transport is the pH sensitivity of the reaction between PQH2 and the Rieske FeS of the cytb6f complex. PQH2 is able to form an H-bond with a specific histidine residue adjacent to the binding pocket of the Rieske FeS protein. The formation of an H-bond between a hydroxyl group proton on the quinol head group of PQH2 and the lone pair of a nitrogen in the imidazole ring of the histidine residue is the first step in oxidation of the bound PQH2.

Protonation of the lone electron pair on the histidine residue blocks H-bond formation. The pH optima for cytf reduction lie in the range 6.5–7.25, with the highest value being found in the alga Chlamydomonas. The pH sensitivity of the PQH2–cytb6f reaction combined with decreases in lumenal pH offers a model for the regulation of linear and cyclic electron transport that explains the observed changes in the oxidation and reduction of components of the electron transport chain during regulation. This model, however, does not explain the mechanism that brings about the changes in lumen pH required for adjustment of the kinetics of the reaction between PQH2 and cytb6f. This mechanism will be discussed once the problem of regulating the ratio of ATP and reductant formation has been introduced because the mechanisms that regulate electron transport and the ratio of ATP to NADPH have much in common.

The ATP produced/reductant produced ratio is largely determined by two parameters: the number of protons translocated to the thylakoid lumen per unit electron transport (H+/e–), and the number of protons required to synthesize ATP (H+/ATP). These parameters, acting together, will determine the yield of ATP per unit linear electron transport. Studies on the structure and operation of the thylakoid ATPase, in particular the ratio of the F0 and F1 subunits, gave an estimate of ~4.7 H+ per ATP. Although the structure-based estimate for the H+/ATP ratio is currently widely accepted, some recent measurements of the H+/ATP stoichiometry made on the ATPases from Escherichia coli and spinach have cast some doubt on the plausible 4.7 H+ per ATP derived from structural studies.

The ATPases from E. coli and spinach have different stoichiometries for the F0 subunits (10 for E. coli and 14 for spinach), but in both cases the H+/ATP ratio is 4, a value which is consistent with earlier experimental determinations of the H+/ATP ratio. Whatever the H+/ATP of the ATPase, the effective H+/ATP ratio of the thylakoid will be increased by processes that allow the uncoupled dissipation of the DμH+. Several such processes can operate. They include slippage (the passage of protons through the ATPase without ATP formation), uncoupling of the ATPase, and H+ leakage through the thylakoid membrane.

All have been demonstrated in vitro and would decrease the H+/ATP stoichiometry. Slippage and leakage are expected to increase as the pmf increases, and uncoupling of the ATPase occurs when the stromal pH falls below ~7.5 (the nominal value of the stromal pH is ~7.5). These DμH+ dissipation mechanisms, which do not lead to ATP synthesis, are expected to occur when the electron transport chain is subject to down-regulation and the intrathylakoid pH decreases, though the impact of such processes on overall regulation remains to be quantified.

In particular, the occurrence of slippage reactions under physiological conditions has been disputed. The minimum DμH+ and, by implication, the transthylakoid pH difference required to drive ATP synthesis will depend on the H+/ATP. Based on an H+/ATP of 4.7, researchers estimated that if the stromal pH was 7.5 and the DμH+ was divided equally between voltage and pH differences, then a lumen pH of no less than 6.5–7.0 would be enough to support ATP synthesis.

Uncertainty in the stoichiometry of proton transport into the thylakoid lumen coupled to electron transport hinges on the presence or absence of the Q-cycle in the cytb6f complex. Current concepts incorporate a Q-cycle around the cytb6f complex that is always in operation. If this is the case, the number of protons translocated into the thylakoid lumen per electron (H+/e–) is three for linear electron transport. The scale of any uncoupled dissipation accompanying this DμH+ formation is unknown and, although some leakage seems inevitable, there is no evidence that in vivo uncoupled dissipation of the DμH+ is controlled so as to regulate the DμH+. Given an H+/e– of 3 and an H+/ATP of 4.7, the ATP/e– is 0.64 at most (equal to an ATP/NADPH ratio of <"1.3), and this value will decrease as leakage increases.

Balancing the varying energy requirements of CO_2 fixation and photorespiration

The basic ATP/NADPH ratio required by CO2 fixation is 1.5 in the steady state (i.e. no photorespiration with stable Calvin cycle metabolite concentrations so that the rate of CO2 assimilated equals the rate of product formed). However, CO2 fixation is only one of many reductant- and ATP-requiring metabolic reactions occurring in the stroma. Many of these reactions such as nitrite and ammonia assimilation occur at much lower rates than CO2 fixation. They will therefore have only modest effects on the ATP/reductant demand of the stroma. Moreover, considerable energetic as well as metabolic cross-talk occurs between the energy-producing reactions of the chloroplasts and mitochondria, which are highly interdependent. These organelles cooperate in the coordinated regulation of photosynthesis, sucrose synthesis, and primary nitrogen assimilation.

Mitochondria also dissipate excess reducing equivalents produced by the chloroplasts and thus mutations in mitochondrial proteins can adversely affect chloroplast processes. The metabolic and energy dialogue that occurs between the energy-producing organelles is also important in the regulation of gene expression. As discussed later, organelle energy metabolism impacts on the elaborate network of anterograde and retrograde signalling pathways between the chloroplasts, mitochondria, and nucleus that regulates organelle biogenesis and composition.

The oxygenation of RuBP, which occurs alongside CO2 fixation, drives photorespiration. Use of RuBP for oxygenation in parallel with CO2 fixation changes the ATP/reductant requirements associated with the Calvin cycle. First, unlike carboxylation, metabolism driven by oxygenation does not require reductant (in the chloroplast), but it does require ATP for the final reaction, which produces 3-phosphoglyceric acid (PGA) from glycerate.

In addition, the oxygenation of RuBP will change the relative rates of the reductive phase of the Calvin cycle [PGA kinase/glyceraldehyde-3-phosphate dehydrogenase (GAPDH), ATP/NADPH=1] to the regeneration phase (ATP required by the phosphoribulokinase reaction). When both carboxylation and oxygenation operate in parallel, photorespiratory carbon metabolism in the chloroplast will increase the ATP/NADPH requirement from 1.5 to 1.62. However, if photorespiratory ammonia re-assimilation is taken into account, the effect of photorespiration on the required ATP/reductant ratio will be less marked. The effects of photorespiration on ATP/reductant ratios may seem minor, and they are certainly less dramatic than the impact of oxygenation on net carbon fixation. However, the key point is that even relatively small changes in the required ATP/NADPH ratio, such as from 1.5 to 1.6, are predicted to cause rapid de-regulation of chloroplast metabolite pools, particularly if the new demands are not met by appropriate adjustments in production by PET and the ATP synthase. This prediction is a consequence of the high turnover rates of adenylate and reductant pools during photosynthesis.

The balance between CO2 fixation and photorespiration and hence the ATP/reductant ratios that are required in C3 leaves depends on atmospheric CO2 availability. If CO2 fixation and photorespiration alone are considered, then the relative requirement for ATP and reductant (NADPH or two electrons) at 25 °C decreases from a ratio of 1.75 in the absence of carboxylation but with continued oxygenation (e.g. Ci=0 Pa, O2=210 kPa) to ~1.5 in the absence of oxygenation, but with continued carboxylation (Fig. 3). When photorespiration rates are high, as occurs when the stomata are closed, the ATP/reductant (NADPH) requirement will be close to 1.75. Conversely, when the atmospheric CO2 content is enriched to values of <1000 µl l"1, the required ATP and reductant ratios are close to 1.5 (Fig. 3). Changes in the ratio of carboxylation and oxygenation of RuBP also occur in response to temperature changes. In the absence of oxygenation, the ATP/NADPH ratio is at the minimum requirement level of 1.5.

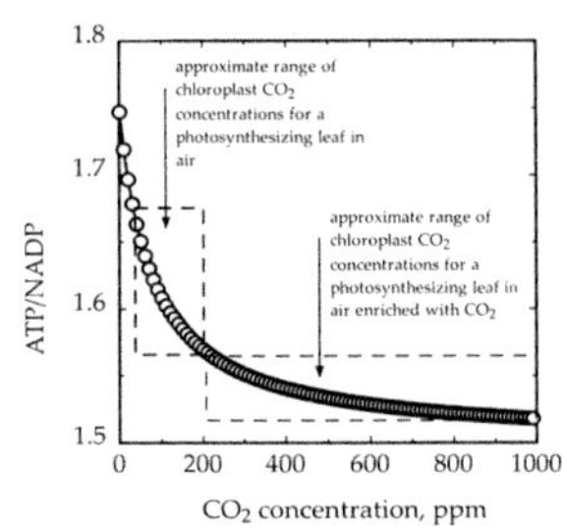

Fig. 3. Relationships between CO_2 concentration at the site of carboxylation and the ratio of demand for ATP and NADPH for the combined activities of the Calvin–Benson and the photosynthetic carbon oxidation cycles in C_3 leaves. Values were calculated using equations and parameters in an atmosphere containing 21% (21 kPa) oxygen. For a leaf in air and with open stomata the CO_2 concentration in the stroma is ~200 ppm (20 Pa). With partial or complete stomatal closure, values can fall to the CO_2 compensation point, <"50 ppm (5 Pa), depending on the temperature and the respiration rate. Increases in atmospheric CO_2 concentrations will result in higher stromal CO_2 concentrations and a lower ATP/NADPH demand.

As linear electron transport translocates three H^+ into the lumen per electron, the formation of two NADPH would be coupled to the synthesis of only 2.6 ATP (12/4.7, assuming the 4.7 H^+ per ATP is correct); that is, a shortfall of 0.4 ATP. Increased rates of oxygenation of the substrate, RuBP, will increase the ATP/NADPH ratio because of the additional requirement of the photorespiratory pathway. Starch synthesis in the chloroplasts and sucrose synthesis in the cytosol also require ATP, and thus are another factor in the chloroplast's budget for ATP and reductant, but one which is more difficult to account for in a quantitative manner. Starch synthesis and degradation occur simultaneously in chloroplasts, the flux in each direction being dependent on the supply of triose phosphate and the regulation of the component enzymes.

PATHWAYS TO BOOST ATP PRODUCTION

Two well known pathways are potentially able to boost ATP production while not themselves supporting easily measurable oxygen evolution. These are (i) cyclic electron transport and (ii) pseudocyclic electron transport, with the latter having other commonly used names (e.g. Mehler reaction, water–water cycle). As the operation of both these pathways is 'hidden' from standard measurements of O_2 exchange, their extent remains subject to debate. Increases in the rate of cyclic electron transport offer a mechanism for increasing the rate of ATP formation relative to the formation of reductant.

In addition to this thylakoid level flexibility, there is also flexibility arising from processes in the stroma. Not all the reductant formed by linear electron transport is associated with processing the products of carboxylation and oxygenation of RuBP, and these other sinks for reductant change the ratio of ATP/reductant demand in the stroma. Under all conditions some reductant is used to reduce O2 to (O2–) in the Mehler reaction.

This is the first step of the water–water cycle. This traditionally incorporates pseudocyclic electron transport, the Mehler–peroxidase reaction, and the ascorbate/glutathione cycle, but it now most probably also includes peroxiredoxin-dependent pathways. Two electrons are required to produce two superoxide molecules from two molecules of O2, supporting the evolution of 0.5 O2 at PSII. Assuming that hydrogen peroxide (H2O2) production occurs only through superoxide dismutation (rather than superoxide reduction), two superoxide (O2·–) radicals undergo a dismutation reaction to produce one molecule each of O2 and H2O2.

Reduction of the H2O2 molecule requires two electrons to regenerate reductants oxidized via ascorbate peroxidase activity, and this also supports the evolution of 0.5 O2 at PSII. The end result is that one O2 is evolved at PSII, two are consumed by the electron transport chain, and one is evolved by SOD. Overall, this process necessarily involves four electrons (i.e. four per O2 evolved from or reduced to water), and neither easily measurable O2 exchange nor net production of reducing power for biosynthetic reactions.

Pseudocyclic electron transport occurs alongside processes that support net O_2 evolution, but its hidden nature means that it is usually detected by indirect methods. In consequence, the literature contains very variable estimates of the proportion of electrons that flow to O_2 in the Mehler reaction. One consequence of O_2 reduction is some contribution to the transthylakoid proton gradient. As well as the protons released into the thylakoid from H_2O oxidation or during PQH_2 oxidation by the cytb6f complex, this mechanism also involves proton consumption in the stroma as O_2 is reduced to water. It will therefore build up DμH+ and allow the formation of more ATP without net reductant formation. It is also expected to increase in activity as the stroma becomes more reduced because the regeneration of electron acceptors (NADP+, oxidized Fd) becomes limiting. Because this is an expected first consequence of a shortfall of ATP relative to reductant, O_2 reduction may be an important self-correcting mechanism to balance ATP production relative to NADPH.

This aspect of $O_2^{\cdot -}$ production and metabolism has become increasingly overlooked in recent years, as the research focus has shifted to the roles of ROS in oxidative stress and/or signalling.

Another mechanism that could operate when the stroma becomes over-reduced is the reduction of oxaloacetate by NADP-malate dehydrogenase (NADP-MDH), which oxidizes NADPH, with the malate formed then being available for export from the chloroplast via the malate shuttle. The export of malate removes reductant from the ATP/reductant balance of the stroma; the malate is re-oxidized in the cytosol to oxaloacetate, which can then exchange for malate via the shuttle. The activity of NADP-MDH is subject to regulation via reduction of regulatory thiol groups by TRX, with increasing reduction of the TRX pool increasing the activation state of the enzyme. Changes in the activity of NADP-MDH have been observed under a range of conditions, such as increasing irradiance and during photosynthetic induction.

The malate valve is therefore considered to be a central regulator of the export of reductant from the stroma. Finally, although most of stromal demand for reductant (as NADPH or reduced Fd) and ATP (estimated to be ~90%) is associated with photosynthetic carbon metabolism, other chloroplast processes also require reductant and/or ATP. Of these perhaps the most important are associated with primary nitrate assimilation and sulphur assimilation, as enzymes such as nitrite reductase and sulphite reductase are localized in the chloroplasts. These reductive processes would result in the net build up of the DμH+, though nitrite reduction is likely to be more significant than the slower reduction of sulphite.

The total net formation of ATP and reductant needs therefore to be viewed at a whole system level; flexibility at the thylakoid level needs to be integrated with flexibility at the stromal and cytosolic levels (any reductant exported from the chloroplast needs, after all, to be used somewhere). Of the various processes that influence the ratio of ATP and (net) reductant formation, the rates of pseudocyclic and cyclic electron transport have been proposed to increase in response to the accumulation of reductant in the stroma.

In addition, activation of NADP-MDH in response to increasing stromal reduction would be expected to increase the export of reductant via the malate valve. So any increase in the degree of reduction of the stroma will increase the rate of reactions that tend to increase the DμH+ with no net increase in stromal reduction. This can compensate for the relatively low ATP/reductant ratio of linear electron transport. Finally, the presence of other metabolic pathways in the stroma other than the predominant pathways associated with CO2 and photorespiration play a role in determining the total demand for ATP and reductant.

Cyclic electron transport is not normally considered as a major flux in C3 leaves. At most, it is probably <10% of the linear flux. In some circumstances, however, such as the induction phase of photosynthesis or in the absence of CO2, cyclic electron transport rates can be much higher. Cyclic electron transport is more important in the bundle sheath cells in C4 leaves, where it may achieve much high rates.

Pseudocyclic electron flow to O2 generally accounts for only ~10% of the linear flux, but higher rates have been reported in plants exposed to stress. It is also assumed that up to ~10% of linear electron transport could be used for processes other than carbon metabolism such as nitrite reduction. Using a model-based approach, it has been estimated that the export of reductant from the chloroplast as malate accounts for ~5% of linear electron transport.

The mechanisms that regulate chloroplast ATP/reductant ratios could also act to down-regulate electron transport processes by decreasing intrathylakoid pH. If metabolic demand for reductant were to be diminished, then the increase in the reduction state of the stromally located electron transport system (Fd, etc.) that would then develop would increase the rates of cyclic electron transport, the Mehler reaction, and export via the malate shuttle. The result of this would be a decrease in the intrathylakoid pH, slowing electron transport and resulting in an increase in ATP/ADP.

An elegant, alternative mechanism developed by Kramer and co-workers suggests that the decrease in intrathylakoid pH can arise as the result of an increased kinetic limitation of the ATPase. This would require a greater pH gradient, and thus a lower intrathylakoid pH, for the same rate of ATP synthesis. Any decrease in intrathylakoid pH would result in down-regulation of the reaction between PQH2 and the cytb6f complex, slowing electron transport of all kinds.

The hierarchy of mechanisms acting in vivo to restrict linear electron transport in response to decreased metabolic demand remains to be demonstrated. Arabidopsis mutants that lack both the PGR5 and chlororespiratory reduction (CRR) proteins have a greatly restricted capacity for cyclic electron transport. PGR5 is believed to lack FQR activity, while the CRR mutants lack NDH activity. These mutants have restricted growth at low irradiances and have much decreased levels of chl in the leaves. The simplest interpretation of these results is that other mechanisms that build up the DμH+ cannot compensate for a shortfall in cyclic electron transport capacity in Arabidopsis.

REGULATION OF LIGHT USE EFFICIENCY AND NPQ IN PSII

A characteristic of PSII is its sensitivity to light-induced damage because of singlet oxygen (1O2) production in the reaction centre of PSII. 1O2 directly damages the PSII D1 reaction centre protein. This vulnerability means that each PSII reaction centre is rebuilt about once every 20–30 min, even under optimal irradiances. The photodamage to PSII requires light and increases when the reaction centre quinone in the PSII reaction centre (QA) is reduced. Limitation of electron transport after PSII will increase the extent of QA reduction. This phenomenon arises with increasing irradiance because of the increasingly limited capacity for electron transport through the cytb6f complex. This effect will be exacerbated if electron transport through cytb6f is restricted by down-regulation.

Singlet oxygen is a major oxidative signal in chloroplasts, 1O2 accumulation leading to substantial re-programming of gene expression. Evidence in support of this role was provided by analysis of the Arabidopsis fluorescent (flu) mutant that accumulates the sensitizing compound, protochlorophyllide, an intermediate in chl biosynthesis in the dark.

When transferred into the light after a period of darkness, the flu mutant produces large amounts of 1O2, leading to major changes in nuclear gene expression. Annulling expression of the EXECUTOR1 and 2 genes, which encode thylakoid proteins, in the flu background suppresses the flu gene response, suggesting a role for these proteins in the retrograde 1O2 signalling. Singlet oxygen selectively activates nuclear genes that are either not responsive or are less responsive to superoxide or H2O2, suggesting that activation by 1O2 might require promoter elements that are different from those used by H2O2.

The production of singlet oxygen and the protection of PSII from damage is minimized under conditions of limiting electron transport capacity through the cytb6f complex by protective mechanisms, the most important of which is called non-photochemical quenching (of chl fluorescence), or NPQ. Sometimes the name is qualified to imply that it is inducible or regulated to distinguish it from the basal or constitutive mechanisms that would operate to dissipate excitation energy in PSII even in the absence of NPQ.

In the absence of stress, energy-dependent quenching (qE) is the dominant component of NPQ that develops at irradiances above light-limiting levels and it increases with irradiance. The formation of qE depends on lumen acidification resulting from electron transport and upon the presence of the PsbS protein of PSII. Lumen acidification results in protonation of the PsbS protein, triggering qE formation. The formation of the xanthophyll pigment, zeaxanthin, is also induced by lumen acidification. The binding of a fraction of the zeaxanthin pool to an as yet unknown protein also facilitates the quenching process induced by lumen acidification. Zeaxanthin accumulation results in enhanced quenching at higher (although still acidic) lumen pH values, consistent with earlier observations.

Much uncertainty remains regarding the precise mechanism of the qE mechanism, and several models have been proposed. While it is clear that pH changes, facilitated by the presence of zeaxanthin and PsbS, result in structural changes in PSII, it is still not certain how these changes result in the formation of quenching centres. The consequence of qE is, however, certain: 1chl* formed in PSII is quenched and thus PSII is protected from damage. NPQ and qE work alongside QA reduction to diminish the quantum yield for PSII electron transport. In this way, the rate of PSII electron transport, i. e. the product of the quantum yield and light absorbed by PSII, is kept in balance with the capacity for electron transport through the rest of the thylakoid electron transport chain. Although qE is the most commonly encountered form of PSII regulation, other mechanisms exist and are important under some circumstances.

ACCLIMATION OF PHOTOSYNTHESIS IN C3 PLANTS TO HIGH CO2 AND ITS ABSENCE IN C4 SPECIES

The responses of different plant species, such as Arabidopsis, soybean, maize, rice, and poplar, to growth with CO2 enrichment has been extensively characterized over the last 30 years. Growth with CO2 enrichment has a positive effect on the plant growth, and yield of C3 species by the stimulation of yield is often much smaller than expected in many crop species. When C3 leaves are exposed to elevated CO2 there is an immediate increase in photosynthesis because of decreased photorespiration, which is negatively affected by an increase in atmospheric CO2/O2 ratios. A stimulation of net photosynthesis can be observed even when plants are grown for long periods at high CO2.

However, photosynthetic capacity is often decreased in plants grown for long periods at elevated CO_2, a process that is known as 'acclimation'. The acclimation of photosynthesis observed in C_3 plants grown for long periods at elevated CO_2 is related to a substantial re-programming of gene expression.

When C_3 plants are grown for long periods under high CO_2, they tend to become carbohydrate rich and CO_2 fixation becomes progressively attenuated by decreases in

Rubisco protein and/or activity. Alterations of gene expression patterns in response to elevated CO_2 are at least in part related to altered sugar signalling because the expression of photosynthetic genes is rapidly inhibited when plants are supplied with exogenous sugars.

Field studies have consistently confirmed that reduced or insufficient sink capacity leads to an increase in leaf carbohydrates and down-regulation of photosynthetic capacity. While genetic factors play an important role in photosynthetic response to high CO_2, the down-regulation of transcripts encoding Rubisco and other proteins associated with CO_2 fixation and an up-regulation of those encoding proteins involved in RuBP regeneration and starch synthesis is a common response to long-term growth with CO_2 enrichment.

The acclimation of photosynthesis to long-term growth with CO_2 enrichment is also related to enhanced nutrient demands, particularly for nitrogen and phosphorus. Nitrogen uptake and assimilation can fail to keep pace with the higher rates of photosynthesis and assimilate production. There is therefore a tendency for the carbon/nitrogen ratios of plant tissues to increase under high CO_2 and the leaves suffer a progressive nitrogen deficit. Furthermore, growth at elevated CO_2 can inhibit shoot nitrate assimilation in C_3 plants if primary nitrate assimilation in the cytosol requires photorespiration for the provision of NADH. Long-term growth with CO_2 enrichment improves nitrogen use efficiency and favours higher rates of dark respiration.

The acclimation of photosynthesis to elevated CO_2 observed in C_3 plants is largely absent from C_4 plants. Experiments conducted in free-air concentration enrichment (FACE) experiments have consistently shown that photosynthesis in C_4 plants such as maize is not changed when plants are grown at elevated CO_2. Moreover, leaf transcriptome patterns are similar in leaves from plants at high CO_2 and those grown in air.

CONTROL OF PHOTOSYNTHETIC GENE EXPRESSION

The expression of nuclear genes encoding photosynthetic proteins is influenced by a plethora of metabolic, developmental, and environmental triggers. However, the unifying central feature of this control is that expression of photosynthetic proteins is highly responsive to chloroplast (retrograde) signals that reflect the metabolic state of the organelle. The components of the photosynthetic electron transport chain are important sensors of light quantity and quality, over-reduction, overoxidation, or imbalances between components that trigger long- and short-term light acclimation processes that serve to optimize light harvesting and CO_2 fixation for growth and development.

Changes in photosynthetic energy metabolism are sensed by oxidative and reductive signalling pathways that regulate gene transcription and post-transcriptional processing). Metabolites and proteins that are exchanged between the chloroplasts, cytosol, and nucleus also make a key contribution to the coordination of gene expression in the nucleus and chloroplast Mitochondria are also important in cellular energy and redox metabolism, and it is therefore not surprising that they influence the transmission of retrograde chloroplast signals to the nucleus.

Light drives photosynthetic electron transport and is the most important factor in the control of photosynthetic gene expression. The abundance of transcripts encoding photosynthetic proteins is generally highest at the end of the night period and at the beginning of the light period.

The diurnal patterns of photosynthetic gene expression are fundamentally controlled by circadian clock regulators with input from photoreceptors such as phytochromes, crytochromes, phototropins, and Zeitlupe-like proteins. Phytochromes mediate the control of plant development by light and influence plant architecture and flowering dormancy. Phototropins control light-induced movement responses that are considered to optimize photosynthesis. Crytochromes are important in regulating plant responses to excess excitation energy responses. Excess excitation energy describes light energy that cannot be used to drive photosynthetic metabolism.

Exposure to high light causes rapid re-programming of gene transcription. This can involve the rapid repression of genes encoding light-harvesting complex (LHC) components as well as PSII and PSI reaction centre subunits. The repression of LHC genes by high light is found in organisms as diverse as barley and Dunaliella. Transcripts encoding proteins involved in photoprotection such as the PSII-S and Early Light-inducible Protein 2 (ELIP2) are enhanced in high light. PSII-S is involved in NPQ and energy dissipation in PSII. ELIP2 is considered to be a sensor and negative regulator of chl biosynthesis.

Responses to changing light intensity involve changes not only in photosynthesis but also in leaf water status and temperature that activate multiple signalling pathways including abscisic acid (ABA), salicylic acid (SA), and other hormone-mediated signal transduction systems. Light-induced changes in stomatal conductance result in a photorespiratory H2O2 burst. H2O2 signalling from the peroxisome contributes to the information on the redox changes in the PET chain, particularly the PQ pool, that influence photosynthetic gene expression. Limitations in CO2 availability that stimulate photorespiratory flux, which occur concomitantly with over-reduction of the PQ pool, trigger programmed cell death responses.

Transcriptome responses to excess excitation energy can thus share common features with responses to infection by biotrophic pathogens with induction of local and systemic defence responses and defence gene expression. High CO2 is able to prevent many of these negative effects of exposure to high light by preventing over-reduction of the PET chain. The level of CO2 can also influence photosystem composition. For example, high light stress under high CO2 led to the specific induction of genes encoding LHC and PSI subunits in Chlamydomonas.

Exposure to UV-B often decreases photosynthetic gene expression. However, in Arabidopsis seedlings, transcripts encoding PSII, PSI, and LHC were increased following exposure to UV-B. The UVB-Resistance 8 (UVR8) protein is considered to be an important UV-B photoreceptor. It functions upstream of the Elongated Hypocotyl 5 (HY5) and Constitutively Photomorphogenic 1 (COP1) transcription factors. COP1 is inactivated by photoreceptor signalling in the light and participates in the inactivation of light-responsive genes in the dark. HY5 is also important in photoreceptor signalling pathways.

Chl biosynthesis precursors (porphyrins) have long been known to play an important role in chloroplast to nucleus retrograde signalling pathways. The genomes uncoupled (gun) mutants, gun2, 3, 4, and 5, are affected in enzymes of porphyrin metabolism in Arabidopsis. The concept that Mg-protoporphyrin IX and its monomethylester are signals in retrograde signalling was challenged by findings showing no correlation between the quantity of Mg-protoporphyrin IX and the expression of chloroplast-associated nuclear genes.

The recent characterization of a new Arabidopsis gun mutant (gun6-1D) suggests a role for haem synthesized by the plastid ferrochelatase I as a regulator of photosynthesis-associated nuclear genes. The ABA-insensitive (ABI) 4 transcription factor has been implicated in the GUN1 (a chloroplast-localized pentatricopeptide repeat protein) pathway. Arabidopsis mutants lacking a æ-carotene desaturase have impaired biosynthesis of carotenoids, which are required for both photoprotection of photosystem function and ABA synthesis. These mutants also show a repression of Chlorophyll A Oxygenase (CH1 or CAO), Protochlorophyllide Oxidoreductase B (PORB), Lhcb1.1, Lhcb1.4, and RbcS expression.

THE CENTRAL ROLE OF THE PLASTOQUINONE POOL

The photosystems have different ranges of light absorption, with maxima at 680 nm for PSII and 700 nm for PSI, respectively. Changes in light quality can therefore lead to imbalances in light capture by the photosystems. It is not surprising, therefore, that the relative abundance of PSII and PSI is regulated by light quality to ensure optimal photosynthesis efficiency. The specific excitation of PSII (PSII-light) leads to the induction of genes encoding PSI proteins and the repression of genes encoding PSII proteins. Similarly, excitation of PSI (PSI-light) induces an inverse pattern of expression.

In early studies, the electron transport inhibitors 3-(3,4-dichlorophenyl)-1,1-dimethylurea (DCMU) and 2,5-dibromo-3- methyl-6-isopropyl-p-benzoquinone (DBMIB), which induce either oxidation or reduction of the PQ pool, respectively, were used extensively to study the effects of the reduction state of the PQ pool on chloroplast to nucleus retrograde signalling pathways. Accumulating evidence demonstrates that the redox state of the PQ pool modulates several independent redox signalling pathways that serve to regulate photosynthetic efficiency by the post-translational modification of existing light-harvesting and reaction centre proteins and by the activation of reaction centre gene transcription.

Enhanced reduction of the PQ pool, which can result, for example, from an imbalance in the rate of excitation of PSII relative to PSI, activates the redox-regulated STN7 protein kinase in the thylakoid membranes. The STN7 kinase phosphorylates the LHCIIs associated with PSII. Phosphorylation of LHCII causes a subpopulation of LHCII to move from PSII to PSI. This process re-distributes absorbed excitation energy so that the relative rates of photochemical conversion and turnover in each reaction centre are balanced. Moreover, STN7 activation provides an input signal for a further kinase signalling network that involves STN8, an orthologue of STN7. The chloroplast sensor kinase (CSK) also plays an important role in the adjustment of PSII and PSI stoichiometries. Unlike STN7, CSK does not regulate LHCII phosphorylation but rather it controls photosystem stoichiometry by the regulation of the transcription of photosynthetic genes encoding photosystem proteins.

A second major target for redox-regulated protein phosphorylation is the chloroplast RNA polymerase complex called the plastid-encoded polymerase (PEP). A Ser/Thr protein kinase, which was originally called the plastid transcription kinase and is now better known as cpCK2 because it is a nuclear-coded plastid-targeted casein kinase 2, is associated with the phosphorylation of PEP and plastid sigma factors. cpCK2 plays a key role in the regulation of the chloroplast phosphoproteome, and is considered to be particularly important in the cross-talk between chloroplast metabolism and gene expression. The plastid sigma factors, which are encoded by six genes (AtSig1–AtSig6) in Arabidopsis thaliana, regulate the transcription of the photosynthetic reaction centre genes such as PsbA, which encodes the PSII D1 reaction centre protein, and PsaA/PsaB, which encode the PSI reaction centre apoproteins.

Glutathione, TRX, and other reductants such as NADPH and Fd are also involved in the regulated expression of PET chain components. Exposure to PSII-light favours increased reduction of the PQ pool and an oxidation of chloroplast TRX, whereas PSI-light promotes the inverse effects of PQ oxidation and TRX reduction. The signalling cascades triggered by reduction of the PQ pool interface with the other signalling pathways triggered by the PET chain and metabolism, particularly the singlet oxygen,

H_2O_2, and glutathione pathways. Redox signalling pathways involving some of these components have been considered at length in other reviews and will therefore not be considered here in detail.

EVIDENCE FOR PHOTOSYNTHETIC CONTROL OF GENE EXPRESSION ASSOCIATED WITH PHOTORESPIRATION

Substantial re-programming of photosynthetic gene expression occurs when C3 plants are grown under conditions where photorespiration is diminished or inhibited. In contrast, in comparisons of leaf transcriptomes of C4 plants grown with CO_2 enrichment or with air, such changes in gene expression patterns are much less apparent. Such observations are consistent with the absence of a high CO_2-dependent stimulation of photosynthesis or growth in C4 plants such as maize. However, growth with CO_2 enrichment alters stomatal opening and conductance in a similar manner in C3 and C4 plants. For example, stomatal conductance values were significantly higher in leaves of maize plants grown under ambient CO_2 conditions compared with those from plants grown at high CO_2. Regardless of the CO_2-dependent effects on whole plant water status, C4 leaf transcriptome patterns are similar in plants unchanged by atmospheric CO_2 availability. Patterns of transcriptional changes occurring following the transition from growth at high CO_2 to air in C3 leaves might therefore be used as an indicator of the influence of photorespiration on the photosynthetic control of gene expression. The published data sets for A. thaliana leaves of plants grown in high CO_2 (3000 μl l"1) and then transferred to ambient conditions (390 μl l-1 CO_2) for 2 d and 4 d in air were used to determine the differential expression of photosynthesis genes. Of the probe sets (22 810) present on the A. thaliana microarray chips, ~10% displayed modified expression under photorespiratory conditions compared with high CO_2, using a t-test calculation with selection of genes with P-values <0.05 and fold changes (in log2) of >0.5 and less than -0.5 for each comparison (either high CO_2/air 2 d or high CO_2/air 4 d). The significantly induced or repressed genes were classified into functional categories by using MAPMAN. A large number of genes involved in signalling, protein synthesis, and degradation or encoding transcription factors were modified following the transition from high CO_2 to photorespiratory conditions.

Similar observations were made in rice and soybean leaves, where a large number of transcripts encoding transcription factors were responsive to atmospheric CO_2 availability. Taken together, available data show that CO_2 availability exerts a much greater major influence on the genes involved in signal transduction and protein metabolism in C3 species than on those encoding proteins with functions in primary processes such as carbon, nitrogen, or lipid metabolism. Of the 104 differentially expressed genes involved in photosynthesis, photorespiration, respiration, and carbon metabolism in A. thaliana, only 31 transcripts encode proteins annotated to be involved in photosynthesis and carbohydrate metabolism (Figs 4, 5).

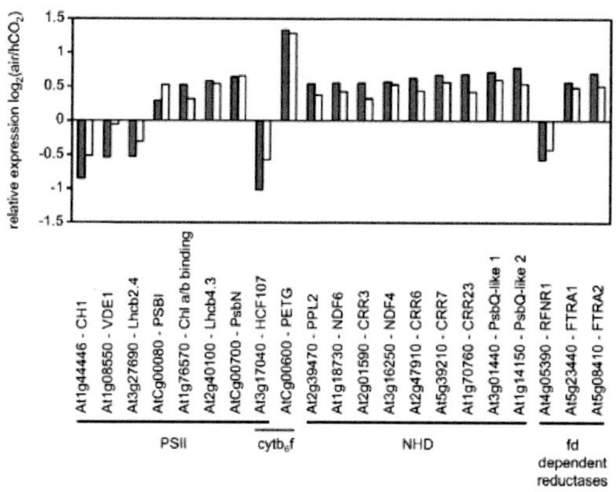

Fig, 4. Transcripts encoding thylakoid electron transport chain components that were differentially expressed 2 d and 4 d after transfer from growth at high CO_2 to air. Transcripts with *P*-values <0.05 and fold changes (log_2) of >0.5 and less than −0.5 are shown. Grey bars, 2 d after transfer; white bars, 4 d after transfer.

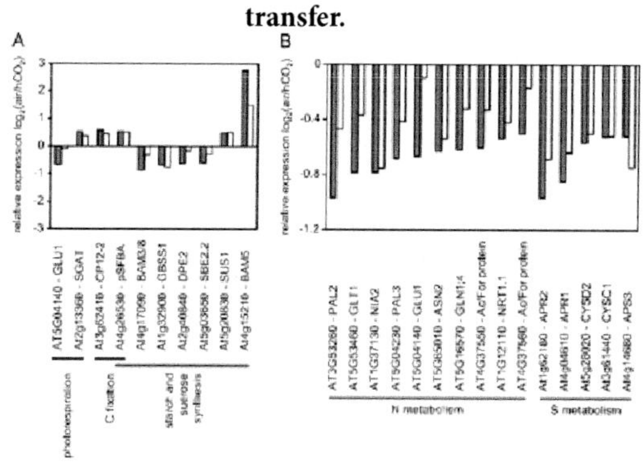

Fig. 5. Transcripts encoding proteins associated with photorespiration, carbon fixation, and sucrose and starch synthesis that were differentially expressed 2 d and 4 d after transfer from growth at high CO2 to air. Transcripts with P-values <0.05 and fold changes (log2) of >0.5 and less than −0.5 are shown. Grey bars, 2 d after transfer; white bars, 4 d after transfer. A, carbon metabolism; B, nitrogen and sulphur metabolism.

Table 2. Arabidopsis thaliana genes that respond to atmospheric CO2 availability

Functional categories	Number of genes Induced	Repressed
Protein metabolism	95	200
Transcription factors	94	151
Signalling	26	135
Stress response	38	68
Photosynthesis, (photo)respiration and carbon metabolism	34	70
Transport	30	68
DNA synthesis and repair	5	74

Development	26	49
Hormone metabolism	30	33
Secondary metabolism	28	20
Nitrogen and sulphur metabolism	9	26
Lipid metabolism	16	17
Cell wall metabolism	12	21
Cell cycle and division	0	32
Redox processes	17	10
Unknown function	315	381

The number of transcripts that were significantly induced or repressed in leaves 2 d and 4 d after transfer from high CO_2 growth conditions to air are listed according to Functional categories.

PHOTORESPIRATION INCREASES TRANSCRIPTS ENCODING CYCLIC ELECTRON FLOW COMPONENTS

The transcriptome comparisons of A. thaliana leaves 2 d and 4 d after the transfer from high CO2 growth conditions showed that transcripts encoding the major PSII or PSI components were unaffected by photorespiration. However, transcripts encoding the PSII reaction centre proteins PsbN and PsbI and LHCB4.3, were increased. In contrast, LHCB2.4, encoding another LHCII protein, was repressed on transfer to air. The location of the proteins encoded by PsbN, PsbI, LHCB2.4, and LHCB4.3 within the PET chain is indicated in Fig. 6. Other transcripts that were decreased 2 d and 4 d after the transfer to air were CH1, a chlorophyllide a oxygenase that converts chlorophyllide a to chlorophyllide b, and HCF107, which is required for the translation of three PSII subunit genes PsbB, PsbT, and PsbH (Fig. 6). Two cytb6f complex genes (PETB and PETD) and transcripts encoding violaxanthin de-epoxidase (VDE1), which is required for zeaxanthin synthesis and therefore involved in NPQ regulation, were also decreased 2 d and 4 d after the transfer to air (Fig. 6).

Simplified scheme for the thylakoid electron transport chain showing the location of proteins encoded by genes whose expression is altered following the transfer from high CO2 to air. White lettering denotes transcripts with P-values <0.05 and fold changes (log2) of >0.5 and less than −0.5. Letters that are underlined indicate that the abundance of transcripts was lower in air relative to high CO2, while the absence of underlining indicates transcripts whose abundance was higher in air than in high CO2.

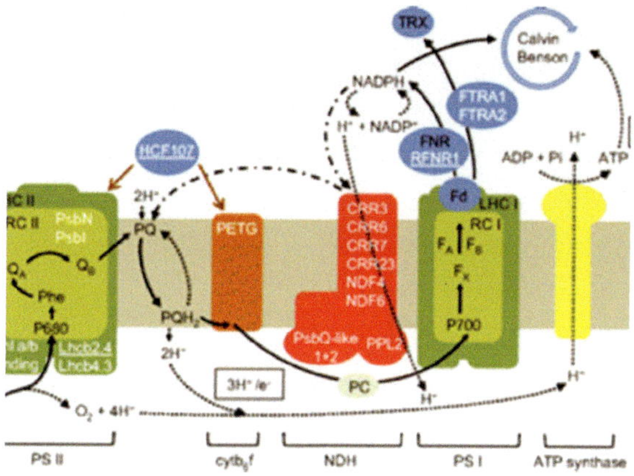

Fig. 6. Simplified scheme for the thylakoid electron transport chain showing the location of proteins encoded by genes whose expression is altered following the transfer from high CO2 to air. White lettering denotes transcripts with P-values <0.05 and fold changes (log2) of >0.5 and less than -0.5. Letters that are underlined indicate that the abundance of transcripts was lower in air relative to high CO2, while the absence of underlining indicates transcripts whose abundance was higher in air than in high CO2. Gene lists for photosynthesis light reactions were assembled from metabolic and signalling pathways illustrated by the KEGG PATHWAY Database

As discussed above, a decrease in atmospheric CO2 availability requires a change in the generation of ATP relative to NADPH by the PET chain in order to accommodate the alterations in chloroplast metabolism associated with enhanced flux through the photorespiration pathways. The higher demand for ATP relative to NADPH in photorespiratory conditions can be met by several mechanisms, such as cyclic or pseudocyclic electron transport. One of the most striking features of the transcriptome signature of Arabidopsis leaves photosynthesizing in air compared with high CO2 is the enhanced abundance of transcripts encoding components involved in cyclic electron flow around PSI (Figs 4, 6). Of the genes encoding photosynthetic components that were induced after transfer to photorespiratory conditions, four are involved in chlororespiration (CRR3, CRR6, CRR7, and CRR23), two are NDH-dependent cyclic electron flow genes (NDF4 and NDF6), two are PsbQ-like proteins, while PPL2 encodes a PsbP-like protein. The enhanced abundance of transcripts encoding components of the NDH complex and associated cyclic electron flow pathways suggests that photosynthetic control of gene expression operates to enhance the expression of cyclic electron flow to meet the increased ATP/NADPH demand of photorespiration. Moreover, the repression of VDE1 might suggest that photosynthetic control of gene expression operates to decrease NPQ in the face of the enhanced demand for protons to make ATP.

Photorespiration decreases transcripts associated with starch synthesis and nitrogen and sulphur metabolism

The second striking feature of the transcriptome signatures of *Arabidopsis* leaves in air compared with high CO_2 is the absence of large changes in transcripts encoding enzymes of carbon fixation or photorespiration. However, transcripts encoding two proteins involved in photosynthetic carbon metabolism were increased in expression in air relative to high CO_2. These are *pSFBA*, which encodes a putative aldolase superfamily protein, and *CP12-2*, which encodes a redox-regulated scaffold protein that forms a complex with GAPDH and phosphoribulokinase. Only six genes encoding proteins involved in starch and sucrose synthesis (Fig. 6 above) were changed in expression 2 d and 4 d after the transfer from high CO_2 to air.

Transcripts encoding two enzymes of starch synthesis, *GBSS1*, which encodes a granule-bound starch synthase, and *SBE2.2*, encoding a starch branching enzyme, were decreased in abundance 2 d and 4 d after transfer to air (Fig. 6). In contrast, a β-amylase gene (*BAM5*), which is involved in starch degradation, and *SUS1* (sucrose synthase 1) were higher under photorespiratory conditions. These results are consistent with decreased starch production when CO_2 availability is low and photorespiration is increased.

PHOTOSYNTHETIC CONTROL OF TRANSCRIPTION FACTORS

A surprisingly large number (245) of transcription factors were differentially expressed in response to photorespiration. The MAR binding filament-like protein 1 (MFP1), which is associated with the formation of thylakoids and nucleoids, was decreased in expression under photorespiratory conditions. Interestingly, of the photorespiration-induced transcription factors, eight belong to the dehydration-responsive element-binding (DREB) subfamily. This finding is consistent with the higher level of stomatal opening, transpiration, and water transport that accompanies the transfer of C3 leaves from high CO2 growth conditions to air. None of the eight photorespiration-induced DREB transcription factors has yet been ascribed roles in the regulation of photosynthetic gene expression.

The apetala 2.4 (RAP2.4) transcription factor has been shown to be induced by cold, drought, and osmotic stress and may have a role in water homeostasis by controlling the expression of aquaporins. C-repeat binding factors 1 and 3 (CBF1 and CBF3) are involved in cold acclimation and they are regulated by a core component of the circadian clock. Other transcription factors that were altered in expression following the transfer from high CO2 growth conditions to air are also involved in circadian regulation.

These changes are consistent with recently described interactions between CO2 and daylength in the control of gene expression. Transcripts encoding the pseudo-response regulator 7 (PRR7) that is an element involved in the feedback loop controlling circadian rhythms are decreased in abundance following the transfer to air. In contrast, circadian 1 (CIR1) is up-regulated by photorespiratory conditions. The expression of CIR1 follows a circadian rhythm, and perturbations in its expression affect flowering time and the circadian regulation of several genes, including Lhcb.

The MYB-related protein 1 (MYR1) gene is a repressor of flowering and was repressed after transfer to air. Conversely, transcripts encoding the nuclear factor YB2 (NF-YB2), which promotes flowering, were increased in abundance following the transfer to photorespiratory conditions. Transcripts encoding the LOB domain 41 (LBD41) transcription factor, which is responsive to the oxygen level, were greatly decreased in abundance following the transfer to photorespiratory conditions.

Hypoxia induces the expression of LBD41; this transcription factor activates the promoter of an anaerobic gene [non-symbiotic haemoglobin 1 (HB1)]. Since the ambient oxygen concentration was maintained constant at air levels in all conditions, it is of note that the expression of LBD41 is highly sensitive to the transfer from non-photorespiratory to photorespiratory conditions. Similarly, the absence of detectable changes in the abundance of transcription factor-encoding genes such as ZAT 10, ANAC036, ANAC090, ORA47, and ERF104 that are associated with changes in redox buffering capacity and H2O2 accumulation (Kerchev et al., 2011) suggests that the photosynthetic control of gene expression that accompanies the transfer from non-photorespiratory to photorespiratory conditions does not involve large changes in ROS or associated signalling pathways.

Conclusions and perspectives

Photosynthetic control operates at multiple levels in order to balance energy supply and energy demand and optimize photosynthetic efficiency, while minimizing potentially harmful back reactions. This balancing act not only involves matching output (assimilated carbon) to input (light and CO_2) within a fluctuating environment (temperature, water status, and other potential stress factors) but it also requires the production of ATP and reductant at appropriate ratios. These ratios are responsive to metabolic factors, requiring flexibility in the relative rates of production. This flexibility is achieved through rapid, short-term modifications in post-translational controls and longer term regulation through effects on gene expression. The classical example of photosynthetic control of gene expression is the altered expression of genes that encode the core apoproteins of the photosystems that accompanies state transitions.

It is rarely applied to transcriptome changes associated with varying metabolic demands for ATP and reductant, such as occur when the rate of photorespiration is varied relative to photosynthesis. The evidence presented here demonstrates that appropriate transcriptional responses are triggered in A. thaliana leaves alongside the rapid short-term adjustments in the photosynthetic control of electron transport flux in the days immediately following the transfer to air from high CO2 growth conditions. Moreover, transcripts associated with cyclic electron transport and NPQ respond in a way that is consistent with the expected alterations in supply and demand of ATP relative to reductant.

The changes in transcriptome signatures that accompany alterations in atmospheric CO2 availability are similar in different C3 species, with the most impact on genes involved in signalling and protein turnover. CO2 enrichment also has an impact on the expression of genes encoding proteins involved in photosynthesis, such as light reaction components that tend to be decreased under CO2 enrichment and enzymes of carbondrate metabolism such as starch synthesis that are increased under these conditions. The change in the transcription patterns of cyclic electron transport genes in A. thaliana points to an important role for cyclic pathways in adjusting ATP/reductant requirements. Thus transient changes in photorespiration that may occur, for example, in response to changes in stomatal conductance could influence the expression of genes encoding cyclic pathway proteins.

C4 grasses consistently exhibit lower maximum stomatal conductance to water than C3 grasses, together with shifts towards smaller stomata at a given density. Carbonic anhydrases have been suggested to function early in the CO2 signalling pathway that controls stomatal movement and hence gas exchange between plants and the atmosphere. The response of specific photosynthesis genes to changes in CO2 concentration is accompanied by modified expression of specific transcription factors.

As previously discussed, the finding that many of the most highly expressed transcription factors that were differentially expressed 2 d and 4 d after the transfer from high CO2 to air encode DREB family members is consistent with altered water relations resulting from increased stomatal opening in air relative to high CO2. Because patterns of CO2-regulated stomatal closure (and hence associated signalling) are similar in C3 and C4 plants, key questions remain concerning why C3 and C4 leaves do not show similar gene expression changes related to altered leaf water status as atmospheric CO2 availability is changed.

A major difference between the two photosynthetic groups is the low level of photorespiration in C4 plants and the absence of any effects of high CO2 on photosynthesis in these species. One possibility is that the altered expression of transcription factors such as the DREB family members observed in the C3 species, A. thaliana, following the transfer to air from high CO2, are related directly to the onset of photorespiration and related photosynthetic control of electron transport rather than plant water status per se.

Thus, changes associated with photorespiration and PET could be a proxy sensor for water status in C3 plants, or at least act as a reinforcing signal that tones the response to changes in this status. This hypothesis requires further study to elucidate relationships of cause and effect.

Considerable literature evidence supports the view that photosynthetic control is embedded not only within the whole cell physiology of the leaf but also in genome biology and control. This is therefore an important point to consider when identifying potential targets for improving photosynthesis through selection or genetic modification. Models describing how metabolism and electron (and proton) transport activities interact are also an important tool in better understanding the integration of metabolism with electron transport and ATP synthesis.

They can be used to estimate, based on parameters such as the H+/e– and H+/ATP ratios, the extent to which linear electron transport alone could meet the demands of metabolism, and what activities of other electron transport processes could compensate for a shortfall in ATP. They can thus establish some boundary conditions for chloroplast function. These models cannot, however, replace a direct experimental analysis of the regulation of electron transport in relation to metabolism.

Abbreviations

- A assimilation
- ABA abscisic acid
- 1chl* singlet excited state of chlorophyll a
- Ci CO2 concentration in the substomatal cavity
- CRR chlororespiratory reduction
- CSK chloroplast sensor kinase
- cytb6f cytochrome b6/f complex
- FACE free-air concentration enrichment
- Fd ferredoxin
- FNR ferredoxin-NADP+ oxidoreductase
- FQR ferredoxin quinone oxidoreductase
- GAPDH glyceraldehyde-3-phosphate dehydrogenase
- H2O2 hydrogen peroxide
- LHC light-harvesting complex
- MDH malate dehydrogenase
- NDH NAD(P)H dehydrogenase
- NPQ non-photochemical quenching of chlorophyll fluorescence

- 1O2 singlet oxygen
- O2·– superoxide
- PAR photosynthetically active radiation
- PEP plastid-encoded RNA polymerase
- PET photosynthetic electron transport
- PGA 3-phosphoglyceric acid
- pmf proton motive force
- PQ plastoquinone
- PQH2 plastoquinol
- PS photosystem
- QA primary stable electron acceptor of photosystem II
- ROS reactive oxygen species
- RuBP ribulose-1,5-bisphosphate
- Rubisco ribulose-1,5-bisphosphate carboxylase/oxygenase
- TRX thioredoxin
- DµH+ transthylakoid proton potential
- DpH transthylakoid pH gradient
- FPSI quantum yield for electron transport by photosystem I
- FPSII quantum yield for electron transport by photosystem II

CHAPTER

EVOLUTION AND DIVERSITY OF GREEN AND LAND PLANTS 8

INTRODUCTION

Plant evolution refers to all the changes water plants had to make to be successful on land. In water, buoyancy meant plants did not have to worry about gravity and their water supply was plentiful. Aquatic plants already possessed chlorophyll a and b, which allowed them to trap light for photosynthesis. However, moving to land created some challenges.

There needed to be a method for plants on land to obtain water from the soil, causing roots to evolve. Plants needed a sturdy form to overcome gravity, so stems evolved. Plants also had to have a way to avoid drying out, so a waxy covering called a cuticle formed. To get carbon dioxide in through the cuticle, plants formed stomata, which are openings in the leaves. Another challenge plants would have on land is protecting the embryo so it would not die out. The answer to this problem was that seeds evolved so their outer covering would protect the embryo of the plant.

The colonization of land by plants was one of the most important events in the earth's history, setting the stage for the greening of the continents, increased oxygenation of the atmosphere, and the provision of food and habitat for the animals and microorganisms that evolved in parallel—and in many cases, diversified in their shadows. They are the foundation of agriculture and the source of timber, fibers, pharmaceuticals, and psychoactives. Fossilized plants, in particular, drive our global economy, being the source of coal, petroleum, and natural gas. And living plants represent huge above- and below-ground (as roots and microalgae) carbon sinks.

Continental colonists who appeared before land plants (embryophytes) included prokaryotic microbes, eukaryotic microalgae, and fungi. The conditions necessary for colonization—sufficient areas of stable land accessible to marine and aquatic organisms, formation of soils, and equable atmospheric and climatic conditions—appear to have been established approximately 540–440 million years ago (MA; Cambrian and Ordovician). By about 470–460 MA (middle Ordovician), plant microfossils first appear in the fossil record, including spores, cuticular fragments, and tracheid cells.

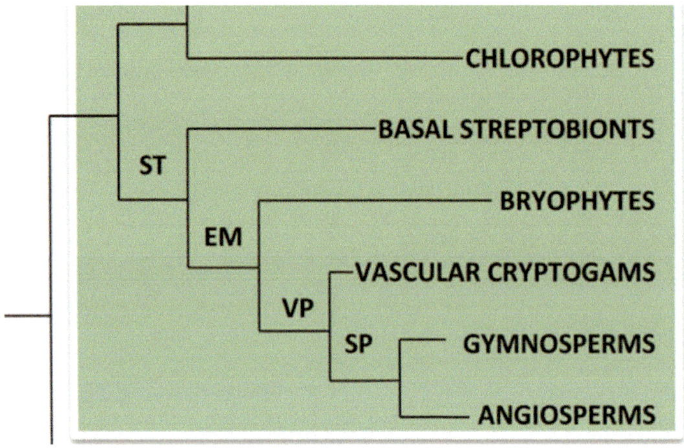

Fig. 1. Major Clades of the Green Plants. This system reflects all of these changes in the taxonomy of the Viridiplantae with two subkingdoms: Chlorobionta and Streptobionta

Early plant micro- and macro-fossils, when considered together, displayed adaptations that facilitated survival on land, including coatings to reduce desiccation, stomata to facilitate gas exchange (in all lineages but liverworts), and specialized cells such as tracheids for the movement of water and nutrients (precursors of a vascular system). By about 410 MA (late Silurian to early Devonian), vegetation comprising small plants with a well-developed vascular system was establishing itself on multiple continents. From about 390-360 MA (middle to late Devonian), the number of species increased rapidly, and by 360 MA the major extant lineages of land plants were established.

Major innovations in plant form began to appear, including the capacity to attain great height and size, and in the late Devonian period, the evolution of seeds. The seed provided protection during periods of dormancy and facilitated survival at greater distances from water. These events were followed during the Cretaceous/ Cenozoic by one of the most spectacular radiations of any terrestrial clade: the rise of the angiosperms, which in the 21st century constitute about 90 percent of all land plant diversity.

GREEN PLANTS

Green plants are those which include all organisms commonly known as green algae and land plants, including live-words, mosses, ferns and other non-seed plants and seed plants. Green plants are termed as a broad assemblage of photosynthetic organisms that all contain chlorophylls a and b, store their photosynthetic products as starch inside the double membrane-bounded chloroplasts in which it is produced and have cell walls made of cellulose.

In this group are several thousand species of what is classically considered green algae plus several hundred thousand land plants.

Several apomorphies unite the Viridiplantae. One possible novelty for this group is a cellulosic cell wall. Cellulose, like starch, is a polysaccharide, but one in which the glucose sugar units are bonded in the beta-l,4 position (=3-1,4-g1ucopyranoside). This slight change in chemical bond position results in a very different molecule. Cellulose is secreted outside the plasma membrane as micro scopic fiber-like units called microfibrils that are further intertwined into larger fibril units, forming a supportive meshwork. The function of cellulose is to impart rigidity to the cells, acting as a sort of cellular exoskeleton. The evolu tion of a cellulosic cell wall was a preamble to the further evolution of more complex types of growth, particularly of self-supporting shoot systems. It is not clear if a cellulosic cell wall constitutes an apomorphy for the Viridiplantae alone, as it may have evolved much earlier, constituting an apomorphy for the Viridiplantae plus one or more other groups; in any case, its adaptive significance seems clear.

Perhaps the primary apomorphy for the Viridiplantae is a specialized type of chloroplast. Chloroplasts are one of the major defining charac teristics of traditionally defined "plants"; their adaptive sig nificance as organelles functioning in photosynthesis, the conversion -of light energy to chemical energy, is unques tioned. Chloroplasts in the Viridiplantae, the green plants, differ from those of most other organisms, such as the red and brown "algae," in (1) containing chlorophyll b in addition to chloro phyll a, the former of which acts as an accessory pigment in light capture; (2) having thylakoids, the chlorophyll-containing membranes, that are stacked into grana, which are pancakelike aggregations; and (3) manufacturing as a storage product true starch, a polymer of glucose sugar units (= polysaccharide) in which the glucose molecules are chemically bonded in the alpha-1,4 position (ci-l,4-glucopyranoside).

Thus, all green plants, from filamentous green "algae" in a pond or tide pooi to giant sequoia or Eucalyptus trees have this same type of chloroplast. Recent data imply that chioroplasts found in the green plants today were modified from those that evolved via endosymbiosis, the intracellular cohabitation of an independently living, uni cellular prokaryote inside a eukaryotic cell.

EMBRYOPHYTA- LAND PLANTS

Embryophyta is a major grouping of plants, sometimes known as "land plants," that includes both the non-vascular bryophytes (mosses, hornworts, and liverworts) and the vascular land plants, which are those so familiar with their vascular system and true roots, leaves, and stems, such as the ferns, flowering plants, conifers, and ginkgos.

Embryophytes are characterized by an alternation of generations life cycle, apical cell growth, cuticle, antherida (male gametophyte organs), and archegonia (female gametophyte organs. Embryophytes are distinguished from the mostly aquatic algae, which do not develop embryos, nor have true roots, stems, or leaves, whereas the embryophytes do form embryos, and have differentiated stems and leaves, and in the case of the vascular plants, true roots.

The origin of embryophytes, as these multicellular plants arose and conquered the land, was a pivotal event in the history of life on earth. Without embryophytes, there would be no animals or humans surviving on the land portion of our planet. The provide food, habitat, energy, oxygen, protection, and numerous vital other functions for the world's creatures. Human beings also benefit from the aesthetic beauty, medicines, and innumerable products derived from these diverse plants.

Embryophytes are the most familiar group of plants. They include trees, flowers, ferns, mosses, and various other green land plants. All are complex multicellular eukaryotes with specialized reproductive organs. With very few exceptions, embryophytes obtain their energy through photosynthesis (that is, by absorbing light); and they synthesize their food from carbon dioxide.

Traditionally, plants were divided into the two groups of embryophytes (Embryobionta), which do develop embryos, and thallophytes (subkingdom Thallobionta), which do not develop embryos and which historically included both algae and fungi. However, fungi are no longer considered plants, and are placed in their own Kingdom. Note that most members of the thallophyte grouping do undergo alternation of generations, with two alternating generations, but all embryophytes undergo such alternation of generations.

Embryophyta may be distinguished from chlorophyll-using multicellular algae by having sterile tissue within the reproductive organs. Furthermore, embryophytes are primarily adapted for life on land, although some are aquatic (which some assume to be secondarily evolved). Accordingly, they are often called land plants or terrestrial plants.

The following are the synapomorphies of the embryophytes: a life cycle with an alternation of generations, apical cell growth (meristem-like growth organization), antheridia, archegonia, and a cuticle (outer covering used to control water loss on land).

On a microscopic level, embryophyte cells remain very similar to those of green algae. They are eukaryotic, with a cell wall composed of cellulose and plastids surrounded by two membranes. These usually take the form of chloroplasts, which conduct photosynthesis and store food in the form of starch, and characteristically are pigmented with chlorophylls a and b, generally giving them a bright green color. Embryophytes also generally have an enlarged central vacuole or tonoplast, which maintains cell turgor and keeps the plant rigid. They lack flagella and centrioles except in certain gametes.

Another evolutionary innovation for the land plants was the antheridium. The antheridium is a type of specialized gametangium of the haploid (n) gametophyte, one that contains the sperm-producing cells. It is distin guished from similar structures in the Viridiplantae in being surrounded by a layer of sterile cells, the antheridial wall. The evolution of the surrounding layer of sterile wall cells, which is often called a sterile "jacket" layer, was probably adaptive in protecting the developing sperm cells from desic cation. In all of the nonseed land plants, the sperm cells are released from the antheridium into the external environment and must swim to the egg in a thin film of water.

Thus, a wet environment is needed for fertilization to be effected in the nonseed plants, a vestige of their aquatic ancestry. Members of the Charales also have a structure termed an antheridium, which has an outer layer of sterile cells. However, because of its differing anatomy, the Charales antheridium may not be homologous with that of the land plants, and thus may have evolved independently.

Another land plant innovation was the evolution of the archegonium, a specialized female gametangium. The archegonium consists of an outer layer of sterile cells, termed the venter, that immediately surround the egg, plus others that extend outward as a tube-like neck. The archegonium is stalked in some taxa; in others the egg is rather deeply embedded in the parent gametophyte. The egg cell is located inside and at the base of the archegonium. Immediately above the egg is a second cell, called the ventral canal cell, and above this and within the neck region there may be several neck canal cells. The archegonium may have several adaptive functions.

It may serve to protect the developing egg. It may also function in fertilization. Before fertilization occurs, the neck canal cells and ventral canal cell break down and are secreted from the terminal pore of the neck itself; the chemical compounds released function as a attractant, acting as a homing device for the swimming sperm. Sperm cells enter the neck of the archegonium and fertilize the egg cell to form a diploid (2n) zygote. In addition to effecting fertilization, the archegonium serves as a site for embryo/sporophyte development and the establishment of a nutritional dependence of the sporophyte upon gametophytic tissue.

The land plants share other possible apomorphies: the presence of various ultrastructural modifications of the sperm cells, flavonoid chemical compounds, and a proliferation of heat shock proteins. These are not discussed here.

Subgroups: Bryophytes and vascular plants

There are two major groupings within Embryophyta.

Bryophytes, or nonvascular land plants, includes the mosses (division Bryophyta), hornworts (division Anthocerotophyta), and liverworts (division Marchantiophyta). Originally, the three groups were brought together as the three classes or three phyla within the division Bryophyta. However, since the three groups of bryophytes form a paraphyletic group, they now are placed in three separate divisions. They are grouped together as bryophytes because of their similarity as non-vascular land plants. Algae are also non-vascular, but are not "land plants." Note that bryophytes do require water to propagate, and thus live in water or moist habitats.

Like the vascular plants, bryophytes do have differentiated stems, and although these are generally only a few millimeters tall, they do provide mechanical support. They also have leaves, although these typically are one cell thick and lack veins. However, they lack true roots, with their filamentous rhizomes having a primary function of mechanical attachment rather than extracting soil nutrients (Palaeos 2008).

Vascular plants comprise the other group of embryophytes. These are really the true land plants, distinguished by a vascular system, and also characterized by true stems, leaves, and roots. Vascular plants have specialized tissues for conducting water, with water transport occurring either in xylem or phloem. The xylem carries water and inorganic solutes upward toward the leaves from the roots, while phloem carries organic solutes throughout the plant. Vascular plants include the familiar ferns, clubmosses, horsetails, flowering plants (angiosperms), and conifers, and other gymnosperms. Scientific names for this group include Tracheophyta and Tracheobionta, but neither is very widely used.

EVOLUTION AND CLASSIFICATION

The higher-level classification of plants varies considerably. Some authors have restricted the kingdom Plantae to include only embryophytes, others have given them various names and ranks. The groups listed here are often considered divisions or phyla, but have also been treated as classes, and they are occasionally compressed into as few as two divisions. Some classifications, indeed, consider the term Embryophyta at the superphylum (superdivision) level, and include land plants and some Charophyceae in a subkingdom named Streptophyta.

Embryophytes are considered to have developed from complex green algae (Chlorophyta) during the Paleozoic era. The Charales, or stoneworts, appear to be the best living illustration of that developmental step. These alga-like plants undergo an alternation between haploid and diploid generations (respectively called gametophytes and sporophytes). In the first embryophytes, however, the sporophytes became very different in structure and function, remaining small and dependent on the parent for their entire brief life. Such plants are informally called bryophytes. They include three surviving groups:

- Bryophyta (mosses)
- Anthocerotophyta (hornworts)
- Marchantiophyta (liverworts)

All of the above bryophytes are relatively small and are usually confined to moist environments, relying on water to disperse their spores. Other plants, better adapted to terrestrial conditions, appeared during the Silurian period. During the Devonian period, they diversified and spread to many different land environments, becoming the vascular plants or tracheophytes.

Tracheophyta have vascular tissues or tracheids, which transport water throughout the body, and an outer layer or cuticle that resists drying out. In most vascular plants, the sporophyte is the dominant individual, and develops true leaves, stems, and roots, while the gametophyte remains very small.

Many vascular plants, however, still disperse using spores. They include two extant groups:

- Lycopodiophyta (clubmosses)
- Pteridophyta (ferns, whisk ferns, and horsetails)

Other groups, which first appeared towards the end of the Paleozoic era, reproduce using desiccation-resistant capsules called seeds. These groups are accordingly called spermatophytes or seed plants. In these forms, the gametophyte is completely reduced, taking the form of single-celled pollen and ova, while the sporophyte begins its life enclosed within the seed. Some seed plants may even survive in extremely arid conditions, unlike their more water-bound precursors. The seed plants include the following extant groups:

- Cycadophyta (cycads)
- Ginkgophyta (ginkgo)
- Pinophyta (conifers)
- Gnetophyta (gnetae)
- Magnoliophyta (flowering plants)

The first four groups are referred to as gymnosperms, since the embryonic sporophyte is not enclosed until after pollination. In contrast, among the flowering plants or angiosperms, the pollen has to grow a tube to penetrate the seed coat. Angiosperms were the last major group of plants to appear, developing from gymnosperms during the Jurassic period, and then spreading rapidly during the Cretaceous. They are the predominant group of plants in most terrestrial biomes today.

CHAPTER

CELL DIVISION AND GENETICS 9

INTRODUCTION

In plants, the presence of cell walls implies that tissue forms a mechanical continuum where cells are attached to their neighbours. The position of new cell walls is thus of importance since it will influence cell shape, tissue topology, patterning, and to some extent the overall organ morphology. The positioning of new cell walls in plants has retained the attention of scientists for more than a century, as their non-random orientation has been noticed quite early, several rules explaining their positions have been proposed. In 1863 Hofmeister proposed that new cell walls formed perpendicular to the main direction of growth of the mother cell.

In relation to Hofmeister's rule, Sachs proposed that new cell walls were inserted at a 90° angle from the mother cell wall. Then in 1886, Errera proposed that most planes of plant cell divisions followed the shortest possible path (i.e. minimal surface area), halving the cell into two daughter cells of equal volume in a way similar to soap film formation that tends to optimize the energy by minimising their surface area. More than 100 years later Flanders *et al.* added another rule on top of these: the new cell wall inserts in the pre-existing one to avoid the formation of four-way junctions.

More recently, Besson and Dumais revised Errera's rule by including a probabilistic component: the path followed by the new wall is selected from several alternative paths with an increased probability toward the shortest one. These rules are based on cell geometry or growth but some authors have recently proposed that the positioning of new cell walls maintains the topological properties of the tissue such as each cell having an average of six neighbours or homogeneity of the cells network.

However, since topology is an emergent property of the division pattern, both geometry-based and topology-based rules predict new cell walls to be (or very close to) the shortest possible one going through the mother cell centroid, halving the cell in two equal volumes. But there are some cases where the selected path deviates significantly from this rule by following a path longer than the shortest one, creating an asymmetry in the volume of the two daughter cells or not passing through the mother cell centroid. Here we present the essential subcellular structures required for the positioning of the new cell wall then we provide an overview of situations where the division path deviates from the shortest one going through the cell centroid.

NEW CELL WALL AND DIVISION ORIENTATION ESTABLISHMENT

The definition of the division path and the formation of the new cell wall is a multistep process initiated by the end of interphase, here we will briefly present these steps before discussing the potential keystones for the definition of division orientation. The first events relating to the definition of the new cell wall position occur at the end of interphase when the cell nucleus often moves to a new position; central for divisions that pass through the centroid and away from the cell centroid for those that do not. The movement of the nucleus is associated with the formation of cytoplasmic strands containing actin and microtubules (MTs) radiating from the nucleus, the nuclear membrane having the ability to nucleate MTs. The Pre Prophase Band (PPB) of microtubules is formed soon after nuclear displacement and consists of highly ordered microtubules.

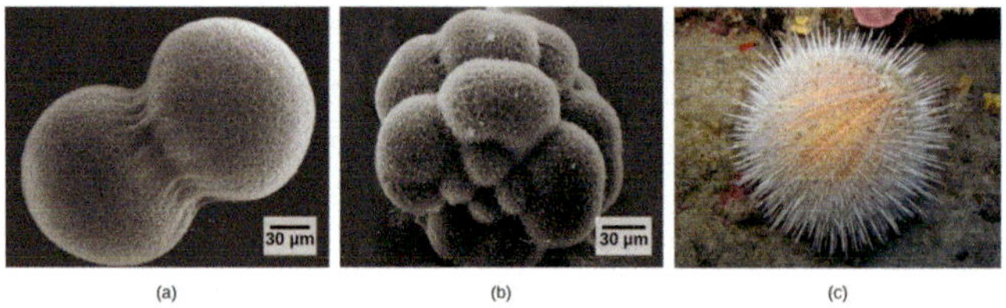

Fig. 1 A sea urchin begins life as a single cell that (a) divides to form two cells, visible by scanning electron microscopy. After four rounds of cell division, (b) there are 16 cells, as seen in this SEM image, (c) mature sea urchin

The cortical division zone (CDZ) that marks the future site of cell plate insertion is formed concomitantly to the PPB and is characterised by a local change in plasma membrane composition and enrichment in a specific type of kinase and microtubule-associated proteins. Changes in plasma membrane composition and associated proteins during cell division are reviewed in detail in. The PPB narrows then coalesce during prophase, the mitotic spindle forms during prophase, and metaphase segregate chromosomes in anaphase.

In telophase, vesicles from golgi and trans-golgi networks fuse at the cell equator to form the cell plate. These vesicles are transported by a dense network of microtubules and actin: the phragmoplast, they carry both plasma membrane material, cell wall material, and proteins responsible for cell wall material biosynthesis. The new cell plate is formed centrifugally following the movement of the phragmoplast, it then fuses with the existing cell wall and the plasma membrane at the division site.

All division rules imply a mechanism to identify the shortest path, measure the potential angle of cell plate insertion, or evaluate the direction of tensile stress. This mechanism could rely on the physical properties of the structures involved in cell division or on the modulation of these properties by additional molecular actors. One important property of plant structures is the level of tension or compression they support. Tension and compression are forces bearing directional information. The main direction of tensile stress within single cells can align with the shortest path across the cell. For instance, the shortest path predictions made by Besson and Dumais can be recapitulated by a simple inflated elastic membrane model predicting the direction of tensile stress. At the subcellular level, MTs have been shown to align with the maximal direction of tensile stress and could be tension sensors per se. During interphase, a dense network of intertwined MT is present at the cell cortex often referenced as Cortical Microtubules array (CMT), its principal direction is influenced by the tensile and compressive force.

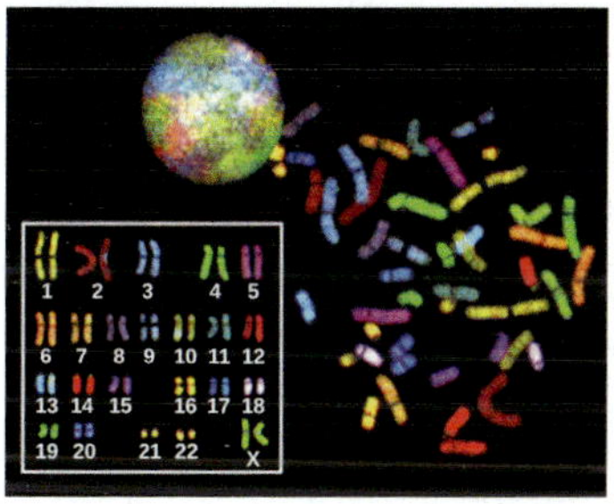

Fig. 2 There are 23 pairs of homologous chromosomes in a female human somatic cell. These chromosomes are viewed within the nucleus (top), removed from a cell in mitosis (right), and arranged according to length (left) in an arrangement called a karyotype. In this image, the chromosomes were exposed to fluorescent stains to distinguish them.

Several structures involved in cell division are made of MTs and as we will now discuss, some of them are under tension or align with the principal direction of tensile stress. For instance, cytoplasmic strands associated with nuclear displacement are under tension, and like a soap film in a rigid frame, they tend to reach the cell edge at 90° following the shortest path. Another structure aligning with tension is the PPB whose orientation often follows the main orientation of cortical microtubules in interphase. PPB position is accurately predicted by the soap-film area minimization model run on real 3D segmented cell templates, in turn, the PPB position predicts the position of the future division plane.

However, PPB is not present in all plant species and its function has recently been re-evaluated. PPB formation is regulated by the Tonneau1 (TON1)-TON1 Recruiting Motif (TRM)-Protein Phosphatase 2A (PP2A) (TTP) protein complex. Several mutants affected in this complex do not form a PPB and have cell division orientation defects. Defective expression of the regulatory subunit of PP2A (fass mutant) severely disrupts division plane placement in Arabidopsis embryo.

TON1 defectives alleles also exhibit strong cell division orientation defects. However, both fass and ton1 have defects in other microtubules structures such as the interphasic cortical microtubules array, and have strong developmental defects (ton1 mutant is named after 'tonneau' the French word for a barrel, as the plantlet looks like a tiny barrel), making the relationships between lack of PPB and cell division defects complex to interpret. Recently, a trm678 triple mutant has been isolated, this line does not form PPB and does not seems to have other microtubules defects.

On the contrary of other PPB defectives lines, the mean orientation of division in trm678 roots is similar to wild type although more variable, indicating that, at least in roots, the PPB would ensure the robustness of cell division orientation rather than determining the orientation. Another main MT-based structure involved in cell division is the spindle.

It is formed perpendicular to the axis of the PPB during prophase but its position does not seem to be the determining factor of division plane position since centrifugation experiments have shown that displaced spindles can trackback to the initial position of the PPB. In addition, rotations ending in the alignment of the spindle or even developing cell plates with the PPB position are frequent during guard mother cell division in Allium.

All these studies indicate that the division orientation and thus the location of cell plate insertion are defined prior to spindle formation by structures under tension or aligned with the principal direction of tension.

After its specification, the division site has to be maintained to ensure the proper guidance of the phragmoplast, here we provide a brief description of the few known actors involved in this process, a more detailed review of these actors is provided in. Most of the actors identified to have a role in phragmoplast guidance are located at the CDZ underlying the PPB and remain there after PPB disassembly. For instance, the presence of Phragmoplast Orienting Kinesin (POK)1 and POK2 at the CDZ are essential for phragmoplast guidance, their absence in a pok1pok2 mutant causes a mismatch between phragmoplast and PPB orientation leading to severe division orientation defects as the cell plate fuses outside of the CDZ.

The positioning of POK1 at the CDZ partially depends on Pleckstrin Homology GAPs (PHGAPS) proteins, mispositioning of POK1 in phgaps1phgaps2 double mutant also give rise to division orientation defects. The presence of POK1 at the CDZ is required to maintain other important actors at the CDZ after the PPB disassembly. This is the case for Tangled1 (TAN1) whose location at the CDZ is not maintained after PPB disassembly in pok1pok2. TAN1 is a microtubule-associated protein whose loss of function in the maize line tan1 results in strong division orientation defects due to a mismatch between phragmoplast and PPB orientation. It has been proposed that the interaction between TAN1 at the division site and microtubules at the leading edge of the phragmoplast would maintain its trajectory.

In addition to the CDZ, POK2 is also present in the midzone of the expanding phragmoplast and interacts with microtubule cross-linker MAP65-3/PLEIADE, this interaction would stabilize the phragmoplast and favour its rapid expansion toward the division site. The microtubule severing activity of KATANIN is essential for proper orientation of the spindle and guidance of the phragmoplast to the division site. Strong katanin mutant alleles show defects in these processes resulting in cell division plane defects. In addition, a line with a mutation in CORD4, a protein recruiting katanin to the phragmoplast, shows a delayed progression of the phragmoplast to the division site.

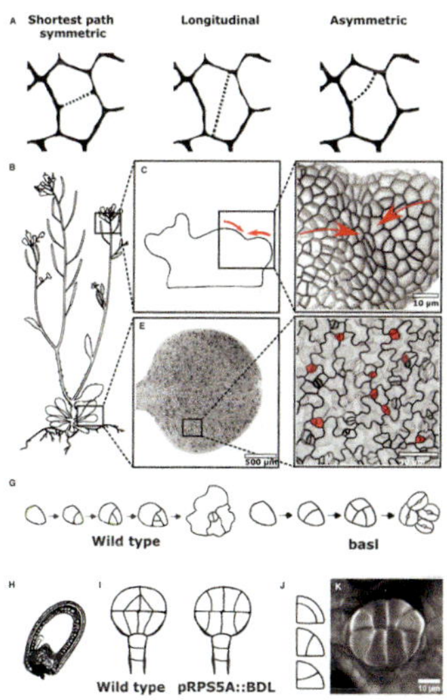

Fig. 3. Deviations from the shortest symmetric path (A) Schematic representations of different cell divisions: shortest path symmetric, longitudinal symmetric, and shortest path asymmetric. (B) Drawing of Arabidopsis thaliana with highlights of some locations where deviations from the shortest path are found. (C) Illustration of the Shoot Apical Meristem (SAM) where differential growth associated with organogenesis generate compressive forces (red arrows). (D) Confocal image of the region of the SAM under compression exhibiting longitudinal divisions (as in (c), red arrows represents compressive forces). (E) Confocal image of a cotyledon. (F) Close up of the cotyledon in (E) showing asymmetric divisions in the stomatal lineage (highlighted in red). (G) BASL polar distribution (in green) is required to make the divisions asymmetric. In a basl mutant divisions of the stomatal lineage are symmetric and lead to the formation of stomatal clusters. (H) Arabidopsis thaliana seed. (I) In wild type embryos the presence of an auxin response is associated with asymmetric divisions. In transgenic lines expressing a negative regulator of auxin signaling (pRPS5A::bdl), these embryonic divisions become symmetric and are reminiscent of the conformation adopted by soap film in quadrant. (J) Schematic representation of the spontaneous formation of soap film in a quadrant. (K) Glandular trichome of Pilea glaucophylla

After the initial definition of the future division plane by the PPB orientation, the spatial information is retained by some of the above-mentioned proteins, but the precise mechanism by which the leading edge of the phragmoplast grows in the precise direction of the division site remains unclear. Although the guidance of the phragmoplast to the division site is an important part of the division orientation process, the actors modulating the division path are likely to act on the early steps of the cell division process prior to spindle formation. Indeed, in a wild type context, no new division has been reported not to match the position of the division site. In the next section, we will describe the situations where the selected division path is not the shortest one going through the cell centre halving the cell in two equal volumes as well as the factors associated with the definition of these alternative division paths.

INTRODUCTION TO REPRODUCTION AT THE CELLULAR LEVEL

The individual sexually reproducing organism—including humans—begins life as a fertilized egg, or zygote. Trillions of cell divisions subsequently occur in a controlled manner to produce a complex, multicellular human. In other words, that original single cell was the ancestor of every other cell in the body. Once a human individual is fully grown, cell reproduction is still necessary to repair or regenerate tissues. For example, new blood and skin cells are constantly being produced. All multicellular organisms use cell division for growth, and in most cases, the maintenance and repair of cells and tissues. Single-celled organisms use cell division as their method of reproduction.

The Genome

The continuity of life from one cell to another has its foundation in the reproduction of cells by way of the cell cycle. The cell cycle is an orderly sequence of events in the life of a cell from the division of a single parent cell to produce two new daughter cells, to the subsequent division of those daughter cells. The mechanisms involved in the cell cycle are highly conserved across eukaryotes. Organisms as diverse as protists, plants, and animals employ similar steps.

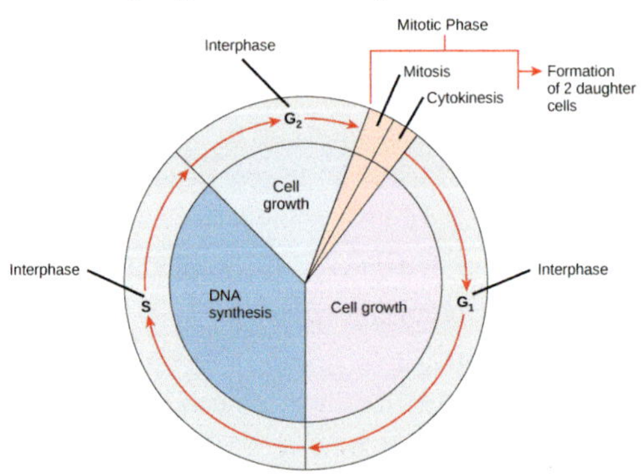

Fig. 4 A cell moves through a series of phases in an orderly manner. During interphase, G1 involves cell growth and protein synthesis, the S phase involves DNA replication and the replication of the centrosome, and G2 involves further growth and protein synthesis. The mitotic phase follows interphase. Mitosis is

nuclear division during which duplicated chromosomes are segregated and distributed into daughter nuclei. Usually the cell will divide after mitosis in a process called cytokinesis in which the cytoplasm is divided and two daughter cells are formed.

Genomic DNA

Before discussing the steps a cell undertakes to replicate, a deeper understanding of the structure and function of a cell's genetic information is necessary. A cell's complete complement of DNA is called its genome. In prokaryotes, the genome is composed of a single, double-stranded DNA molecule in the form of a loop or circle. The region in the cell containing this genetic material is called a nucleoid. Some prokaryotes also have smaller loops of DNA called plasmids that are not essential for normal growth.

In eukaryotes, the genome comprises several double-stranded, linear DNA molecules (Fig. 2) bound with proteins to form complexes called chromosomes. Each species of eukaryote has a characteristic number of chromosomes in the nuclei of its cells. Human body cells (somatic cells) have 46 chromosomes. A somatic cell contains two matched sets of chromosomes, a configuration known as diploid. The letter n is used to represent a single set of chromosomes; therefore a diploid organism is designated 2n. Human cells that contain one set of 23 chromosomes are called gametes, or sex cells; these eggs and sperm are designated n, or haploid.

The matched pairs of chromosomes in a diploid organism are called homologous chromosomes. Homologous chromosomes are the same length and have specific nucleotide segments called genes in exactly the same location, or locus. Genes, the functional units of chromosomes, determine specific characteristics by coding for specific proteins. Traits are the different forms of a characteristic. For example, the shape of earlobes is a characteristic with traits of free or attached.

Each copy of the homologous pair of chromosomes originates from a different parent; therefore, the copies of each of the genes themselves may not be identical. The variation of individuals within a species is caused by the specific combination of the genes inherited from both parents. For example, there are three possible gene sequences on the human chromosome that codes for blood type: sequence A, sequence B, and sequence O.

Because all diploid human cells have two copies of the chromosome that determines blood type, the blood type (the trait) is determined by which two versions of the marker gene are inherited. It is possible to have two copies of the same gene sequence, one on each homologous chromosome (for example, AA, BB, or OO), or two different sequences, such as AB.

Minor variations in traits such as those for blood type, eye color, and height contribute to the natural variation found within a species. The sex chromosomes, X and Y, are the single exception to the rule of homologous chromosomes; other than a small amount of homology that is necessary to reliably produce gametes, the genes found on the X and Y chromosomes are not the same.

Summary

Prokaryotes have a single loop chromosome, whereas eukaryotes have multiple, linear chromosomes surrounded by a nuclear membrane. Human somatic cells have 46 chromosomes consisting of two sets of 22 homologous chromosomes and a pair of nonhomologous sex chromosomes. This is the 2n, or diploid, state. Human gametes have 23 chromosomes or one complete set of chromosomes. This is the n, or haploid, state. Genes are segments of DNA that code for a specific protein or RNA molecule. An organism's traits are determined in large part by the genes inherited from each parent, but also by the environment that they experience. Genes are expressed as characteristics of the organism and each characteristic may have different variants called traits that are caused by differences in the DNA sequence for a gene.

VARIATIONS FROM THE SHORTEST SYMMETRIC PATH

Although most divisions seem to follow the shortest path, there are a number of examples where this is not the case (Fig. 3A). Divisions parallel to the longest axis of the cell, referred to as longitudinal divisions hereafter, are frequent during leaf development in maize. However, they tend to occur in nearly isotropic cells, making the selected path not so distant from the shortest path]. Besides this situation that can be interpreted as an extreme case of the Besson and Dumais rule, some deviations from the shortest path cannot be explained by cell geometry-based rules alone.

For instance in Shoot Apical Meristems (SAM), cells in the boundary region that separates organ primordia from the rest of the meristem almost systematically divide longitudinally. These divisions occur perpendicular to the main direction of compression generated by the differential growth between organ primordia and SAM (Fig. 3B–D). This is in accordance with the classical experiments using compression or wounding on dividing plant tissues that have repeatedly reported divisions to occur perpendicular to the direction of compression or parallel to the main direction of tensile stress.

More recently, cell laser ablation experiments further confirmed the influence of mechanical stress on the orientation of cell division. In Arabidopsis SAM, cell ablation leads to the formation of a circumferential pattern of mechanical stress as visualized by the alignment of cortical microtubules and cell division occurring parallel to the main direction of tensile stress . The local or global mechanical stress pattern influences the orientation of cell division but they are not the only factor affecting cell divisions.

Beside longitudinal divisions, asymmetric divisions represent another type of deviation from the shortest symmetric path. Asymmetric cell divisions are found in distinct contexts during plant development, for instance within the early steps of the stomatal lineage in Arabidopsis, during the formation of subsidiary cells in monocotyledons or the first steps of lateral roots development. In Arabidopsis, the very first division giving rise to the suspensor and embryo lineages or divisions leading to the formation of protodermal layers between the 8 and 16 cells stage of the embryo are asymmetric.

In the case of the division to form the protoderm the division plane still passes through the centroid of the cell but does not produce daughter cells of equal volume. The division plane is a possible surface area minimisation, however, there is evidence that other factors influence its placement, perhaps to ensure its selection over other minimal surfaces. Polarly distributed proteins and auxin signalling are the two main factors commonly associated with asymmetric divisions in various developmental contexts.

In the amplifying divisions of the stomata lineage, the nucleus and division plane are displaced from the centroid of the cell. The polar distribution of Breaking of Asymmetry in the Stomatal Lineage (BASL) and its interacting members BREVIX RADIX (BRX) or BRX-LIKE (BRXL) is essential to ensure the asymmetry of divisions in the stomatal lineage. BASL is present in a crescent in meristemoids, and its polarity switches following division to generate the characteristic spiral pattern seen in stomatal lineages. BASL triggers nucleus migration away from its location prior to cell division and then towards its location following division with the new cell wall forming away from the cell edge where BASL is enriched. Defective expression of BASL or BRX and BRXL family members leads to symmetric divisions in the stomatal lineage.

In Maize, the polar distribution of PANGLOSS receptors like kinase (PAN1 and PAN2) in the subsidiary mother cell is essential for setting the asymmetry of the division which produces the subsidiary cells. The importance of protein polarity for asymmetric cell division could be wider than these specific contexts, indeed recently identified polar proteins of the SOSEKI family are widely expressed in plants and ectopic expression of their members leads to the formation of asymmetric divisions in various organs. In addition to polar proteins, auxin signalling has been shown to be associated with the formation of asymmetric divisions.

Auxin is involved in many aspects of plant development from early embryogenesis to the late developmental stages of lateral organs. In many of these developmental processes, divisions associated with auxin signalling are asymmetric. For instance, while high auxin signalling is associated with the asymmetric divisions of the stomatal lineage leading to the guard mother cell formation, the last symmetric division giving rise to the guard cells is preceded by a decrease in auxin responses. Another developmental process where auxin is associated with asymmetric division is the formation of lateral roots (LR). LR are initiated from the pericycle by a pair of anticlinal asymmetric divisions taking place after a local auxin response maxima.

Pericycle divisions can be triggered by laser ablation of endodermal cells, but only concomitant auxin application can recapitulate the anticlinal asymmetric division. This relationship between auxin and asymmetric divisions is further illustrated by genetic approaches in the embryo. Defective alleles of Auxin Response Factor 5 are widely affected in auxin responses and presents a shift from asymmetric to symmetric division during embryonic ground tissue initiation\. Consistently, ectopic expression of the stabilized negative regulator of auxin responses bodenlos/iaa12 directs symmetric divisions instead of asymmetric at the 8 to 16 cell transition. These symmetric divisions are reminiscent of the conformation adopted by soap films in quadrant and the resulting pattern is frequently observed in other globular structures such as glandular trichomes.

This similar pattern between constrained soap film, glandular trichomes, and auxin response deficient embryos suggests that auxin is able to shift divisions from symmetric to asymmetric. Interestingly auxin triggered asymmetric divisions seem to require local tuning of microtubule stability both in embryonic and lateral root development. Indeed the auxin triggered asymmetric division of pericycle cells requires functional microtubule severing protein KATANIN and pericycle divisions are misplaced in katanin mutants.

In addition, a study combining modelling and imaging demonstrated that cell divisions in the embryo could be predicted from the observed CMT orientation. In turn, the CMT orientation prior to division could be predicted by cell geometry with the addition of a cell-edge triggered catastrophe and microtubule stabilization at a selected cell-face. The authors proposed that auxin is the factor responsible for tuning the stability of microtubules at the faces of the cells.

Auxin, like mechanical stress and polarity, is a factor that can impact the division plane orientation, but the processes involving these factors are highly intertwined as illustrated by the influence of mechanical stress on both polarly distributed proteins BRXL or auxin transporter PIN FORMED 1 (PIN1) and microtubules dynamics. While microtubule dynamics are not required for mediating mechanical-stress triggered changes in BRXL or PIN1 polarity, BASL and BRX effects on pre-division nuclear migration and division asymmetry do require functional microtubule dynamics. Microtubules appear to be central in the integration of multiple signals in the process of division path selection.

THE CELL CYCLE

The cell cycle is an ordered series of events involving cell growth and cell division that produces two new daughter cells. Cells on the path to cell division proceed through a series of precisely timed and carefully regulated stages of growth, DNA replication, and division that produce two genetically identical cells. The cell cycle has two major phases: interphase and the mitotic phase (Fig. 4). During interphase, the cell grows and DNA is replicated. During the mitotic phase, the replicated DNA and cytoplasmic contents are separated and the cell divides.

Interphase

During interphase, the cell undergoes normal processes while also preparing for cell division. For a cell to move from interphase to the mitotic phase, many internal and external conditions must be met. The three stages of interphase are called G_1, S, and G_2.

G_1 Phase

The first stage of interphase is called the G_1 phase, or first gap, because little change is visible. However, during the G_1 stage, the cell is quite active at the biochemical level. The cell is accumulating the building blocks of chromosomal DNA and the associated proteins, as well as accumulating enough energy reserves to complete the task of replicating each chromosome in the nucleus.

S Phase

Throughout interphase, nuclear DNA remains in a semi-condensed chromatin configuration. In the S phase (synthesis phase), **DNA replication** results in the formation of two identical copies of each chromosome—sister chromatids—that are firmly attached at the centromere region. At this stage, each chromosome is made of two sister chromatids and is a duplicated chromosome. The centrosome is duplicated during the S phase. The two centrosomes will give rise to the mitotic spindle, the apparatus that orchestrates the movement of chromosomes during mitosis. The centrosome consists of a pair of rod-like centrioles at right angles to each other. Centrioles help organize cell division. Centrioles are not present in the centrosomes of many eukaryotic species, such as plants and most fungi.

G_2 Phase

In the G_2 phase, or second gap, the cell replenishes its energy stores and synthesizes the proteins necessary for chromosome manipulation. Some cell organelles are duplicated, and the cytoskeleton is dismantled to provide resources for the mitotic spindle. There may be additional cell growth during G_2. The final preparations for the mitotic phase must be completed before the cell is able to enter the first stage of mitosis.

The Mitotic Phase

To make two daughter cells, the contents of the nucleus and the cytoplasm must be divided. The mitotic phase is a multistep process during which the duplicated chromosomes are aligned, separated, and moved to opposite poles of the cell, and then the cell is divided into **two new identical daughter cells**. The first portion of the mitotic phase, mitosis, is composed of five stages, which accomplish nuclear division. The second portion of the mitotic phase, called cytokinesis, is the physical separation of the cytoplasmic components into two daughter cells.

MITOSIS

Mitosis is divided into a series of phases—prophase, prometaphase, metaphase, anaphase, and telophase—that result in the division of the cell nucleus.

Which of the following is the correct order of events in mitosis?

- Sister chromatids line up at the metaphase plate. The kinetochore becomes attached to the mitotic spindle. The nucleus re-forms and the cell divides. The sister chromatids separate.
- The kinetochore becomes attached to the mitotic spindle. The sister chromatids separate. Sister chromatids line up at the metaphase plate. The nucleus re-forms and the cell divides.
- The kinetochore becomes attached to metaphase plate. Sister chromatids line up at the metaphase plate. The kinetochore breaks down and the sister chromatids separate. The nucleus re-forms and the cell divides.
- The kinetochore becomes attached to the mitotic spindle. Sister chromatids line up at the metaphase plate. The kinetochore breaks apart and the sister chromatids separate. The nucleus re-forms and the cell divides.

During prophase, the "first phase," several events must occur to provide access to the chromosomes in the nucleus. The nuclear envelope starts to break into small vesicles, and the Golgi apparatus and endoplasmic reticulum fragment and disperse to the periphery of the cell. The nucleolus disappears. The centrosomes begin to move to opposite poles of the cell. The microtubules that form the basis of the mitotic spindle extend between the centrosomes, pushing them farther apart as the microtubule fibers lengthen. The sister chromatids begin to coil more tightly and become visible under a light microscope.

During prometaphase, many processes that were begun in prophase continue to advance and culminate in the formation of a connection between the chromosomes and cytoskeleton. The remnants of the nuclear envelope disappear. The mitotic spindle continues to develop as more microtubules assemble and stretch across the length of the former nuclear area. Chromosomes become more condensed and visually discrete. Each sister chromatid attaches to spindle microtubules at the centromere via a protein complex called the kinetochore.

During metaphase, all of the chromosomes are aligned in a plane called the metaphase plate, or the equatorial plane, midway between the two poles of the cell. The sister chromatids are still tightly attached to each other. At this time, the chromosomes are maximally condensed.

During anaphase, the sister chromatids at the equatorial plane are split apart at the centromere. Each chromatid, now called a chromosome, is pulled rapidly toward the centrosome to which its microtubule was attached. The cell becomes visibly elongated as the non-kinetochore microtubules slide against each other at the metaphase plate where they overlap.

During telophase, all of the events that set up the duplicated chromosomes for mitosis during the first three phases are reversed. The chromosomes reach the opposite poles and begin to decondense (unravel). The mitotic spindles are broken down into monomers that will be used to assemble cytoskeleton components for each daughter cell. Nuclear envelopes form around chromosomes.

CYTOKINESIS

Cytokinesis is the second part of the mitotic phase during which cell division is completed by the physical separation of the cytoplasmic components into two daughter cells. Although the stages of mitosis are similar for most eukaryotes, the process of cytokinesis is quite different for eukaryotes that have cell walls, such as plant cells.

In cells such as animal cells that lack cell walls, cytokinesis begins following the onset of anaphase. A contractile ring composed of actin filaments forms just inside the plasma membrane at the former metaphase plate. The actin filaments pull the equator of the cell inward, forming a fissure. This fissure, or "crack," is called the cleavage furrow. The furrow deepens as the actin ring contracts, and eventually the membrane and cell are cleaved in two (Fig. 5).

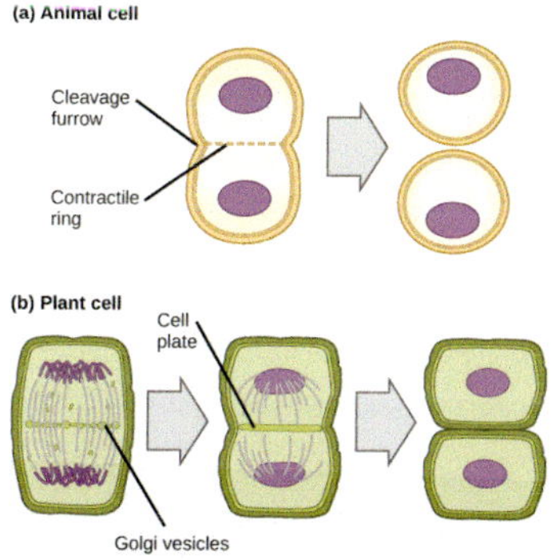

Fig. 5 In part (a), a cleavage furrow forms at the former metaphase plate in the animal cell. The plasma membrane is drawn in by a ring of actin fibers contracting just inside the membrane. The cleavage furrow deepens until the cells are pinched in two. In part (b), Golgi vesicles coalesce at the former metaphase plate in a plant cell. The vesicles fuse and form the cell plate. The cell plate grows from the center toward the cell walls. New cell walls are made from the vesicle contents.

In plant cells, a cleavage furrow is not possible because of the rigid cell walls surrounding the plasma membrane. A new cell wall must form between the daughter cells. During interphase, the Golgi apparatus accumulates enzymes, structural proteins, and glucose molecules prior to breaking up into vesicles and dispersing throughout the dividing cell.

During telophase, these Golgi vesicles move on microtubules to collect at the metaphase plate. There, the vesicles fuse from the center toward the cell walls; this structure is called a cell plate. As more vesicles fuse, the cell plate enlarges until it merges with the cell wall at the periphery of the cell. Enzymes use the glucose that has accumulated between the membrane layers to build a new cell wall of cellulose. The Golgi membranes become the plasma membrane on either side of the new cell wall (Fig. 5).

G_0 Phase

Not all cells adhere to the classic cell-cycle pattern in which a newly formed daughter cell immediately enters interphase, closely followed by the mitotic phase. Cells in the G_0 phase are **not actively preparing to divide**. The cell is in a quiescent (inactive) stage, having exited the cell cycle. Some cells enter G_0 temporarily until an external signal triggers the onset of G_1. Other cells that never or rarely divide, such as mature cardiac muscle and nerve cells, remain in G_0 permanently (Fig. 6).

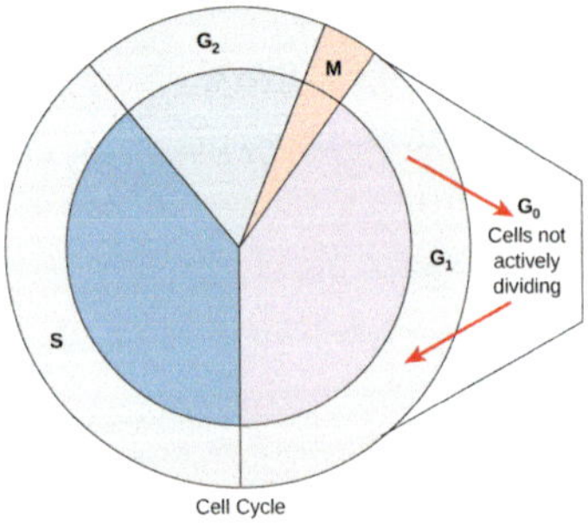

Fig. 6 Cells that are not actively preparing to divide enter an alternate phase called G0. In some cases, this is a temporary condition until triggered to enter G1. In other cases, the cell will remain in G0 permanently.

CONTROL OF THE CELL CYCLE

The length of the cell cycle is highly variable even within the cells of an individual organism. In humans, the frequency of cell turnover ranges from a few hours in early embryonic development to an average of two to five days for epithelial cells, or to an entire human lifetime spent in G0 by specialized cells such as cortical neurons or cardiac muscle cells. There is also variation in the time that a cell spends in each phase of the cell cycle. When fast-dividing mammalian cells are grown in culture (outside the body under optimal growing conditions), the length of the cycle is approximately 24 hours. In rapidly dividing human cells with a 24-hour cell cycle, the G1 phase lasts approximately 11 hours. The timing of events in the cell cycle is controlled by mechanisms that are both internal and external to the cell.

REGULATION AT INTERNAL CHECKPOINTS

It is essential that daughter cells be exact duplicates of the parent cell. Mistakes in the duplication or distribution of the chromosomes lead to mutations that may be passed forward to every new cell produced from the abnormal cell. To prevent a compromised cell from continuing to divide, there are internal control mechanisms that operate at three main cell cycle checkpoints at which the cell cycle can be stopped until conditions are favorable. These checkpoints occur near the end of G_1, at the G_2–M transition, and during metaphase (Fig.7).

The G_1 Checkpoint

The G_1 checkpoint determines whether all conditions are favorable for cell division to proceed. The G_1 checkpoint, also called the restriction point, is the point at which the cell irreversibly commits to the cell-division process. In addition to adequate reserves and cell size, there is a check for damage to the genomic DNA at the G_1 checkpoint. A cell that does not meet all the requirements will not be released into the S phase.

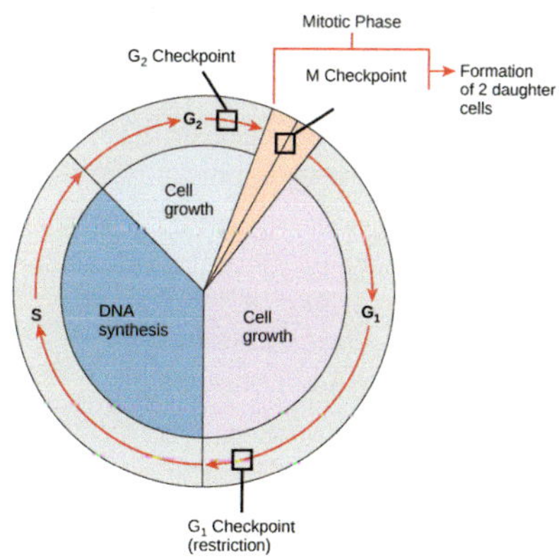

Fig. 7 The cell cycle is controlled at three checkpoints. Integrity of the DNA is assessed at the G1 checkpoint. Proper chromosome duplication is assessed at the G2 checkpoint. Attachment of each kinetochore to a spindle fiber is assessed at the M checkpoint.

The G_2 Checkpoint

The G_2 checkpoint bars the entry to the mitotic phase if certain conditions are not met. As in the G_1 checkpoint, cell size and protein reserves are assessed. However, the most important role of the G_2 checkpoint is to ensure that all of the chromosomes have been replicated and that the replicated DNA is not damaged.

The M Checkpoint

The M checkpoint occurs near the end of the metaphase stage of mitosis. The M checkpoint is also known as the spindle checkpoint because it determines if all the sister chromatids are correctly attached to the spindle microtubules.

Because the separation of the sister chromatids during anaphase is an irreversible step, the cycle will not proceed until the kinetochores of each pair of sister chromatids are firmly anchored to spindle fibers arising from opposite poles of the cell.

Summary

The cell cycle is an orderly sequence of events. Cells on the path to cell division proceed through a series of precisely timed and carefully regulated stages. In eukaryotes, the cell cycle consists of a long preparatory period, called interphase. Interphase is divided into G_1, S, and G_2 phases.

Mitosis consists of five stages: prophase, prometaphase, metaphase, anaphase, and telophase. Mitosis is usually accompanied by cytokinesis, during which the cytoplasmic components of the daughter cells are separated either by an actin ring (animal cells) or by cell plate formation (plant cells).

Each step of the cell cycle is monitored by internal controls called checkpoints. There are three major checkpoints in the cell cycle: one near the end of G_1, a second at the G_2–M transition, and the third during metaphase.

Abbreviations

- BASL Breaking of Asymmetry in the Stomatal Lineage
- BRX Brevix Radix
- BRXL Brx-Like
- CDZ Cortical Division Zone
- LR Lateral Roots
- MTs Microtubules
- PAN1 Pangloss1
- PAN2 Pangloss2
- PHGAPS Pleckstrin Homology Gaps
- PIN1 Pin Formed 1
- POK1/2Phragmoplast Orienting Kinesin1/2
- PP2A Protein Phosphatase 2a
- PPB Pre-Prophase Band of microtubules
- SAM Shoot Apical Meristem
- TAN1 Tangled1

- TON1 Tonneau1
- TRM Ton1 Recruiting Motif
- TTP Ton1-Trm-PP2A

CHAPTER

USE OF ENERGY FROM SUNLIGHT BY PHOTOSYNTHESIS

10

When plants undergo photosynthesis, they make use of light energy from the sun. The chlorophyll a pigment present in leaves, traps light energy. Hence, plants utilize sun energy to transform carbon dioxide and water into glucose (a sugar). This glucose is then utilized by plants as their fuel to synthesize starch and cellulose and to perform other essential life processes.

Photosynthesis is an endothermic process, wherein light energy is converted to chemical energy with the use of chlorophyll. This pigment absorbs blue and red light. Oxygen is released as a byproduct of photosynthesis.

The process by which green plants and certain other organisms transform light energy into chemical energy. During photosynthesis in green plants, light energy is captured and used to convert water, carbon dioxide, and minerals into oxygen and energy-rich organic compounds.

It would be impossible to overestimate the importance of photosynthesis in the maintenance of life on Earth. If photosynthesis ceased, there would soon be little food or other organic matter on Earth. Most organisms would disappear, and in time Earth's atmosphere would become nearly devoid of gaseous oxygen. The only organisms able to exist under such conditions would be the chemosynthetic bacteria, which can utilize the chemical energy of certain inorganic compounds and thus are not dependent on the conversion of light energy.

Energy produced by photosynthesis carried out by plants millions of years ago is responsible for the fossil fuels (i.e., coal, oil, and gas) that power industrial society. In past ages, green plants and small organisms that fed on plants increased faster than they were consumed, and their remains were deposited in Earth's crust by sedimentation and other geological processes. There, protected from oxidation, these organic remains were slowly converted to fossil fuels. These fuels not only provide much of the energy used in factories, homes, and transportation but also serve as the raw material for plastics and other synthetic products. Unfortunately, modern civilization is using up in a few centuries the excess of photosynthetic production accumulated over millions of years. Consequently, the carbon dioxide that has been removed from the air to make carbohydrates in photosynthesis over millions of years is being returned at an incredibly rapid rate. The carbon dioxide concentration in Earth's atmosphere is rising the fastest it ever has in Earth's history, and this phenomenon is expected to have major implications on Earth's climate.

Plants are called autotrophs because they can use energy from light to synthesize, or make, their own food source. Many people believe they are "feeding" a plant when they put it in soil, water it, or place it outside in the Sun, but none of these things are considered food. Rather, plants use sunlight, water, and the gases in the air to make glucose, which is a form of sugar that plants need to survive. This process is called photosynthesis and is performed by all plants, algae, and even some microorganisms. To perform photosynthesis, plants need three things: carbon dioxide, water, and sunlight. Just like you, plants need to take in gases in order to live. Animals take in gases through a process called respiration. During the respiration process, animals inhale all of the gases in the atmosphere, but the only gas that is retained and not immediately exhaled is oxygen. Plants, however, take in and use carbon dioxide gas for photosynthesis. Carbon dioxide enters through tiny holes in a plant's leaves, flowers, branches, stems, and roots. Plants also require water to make their food. Depending on the environment, a plant's access to water will vary. For example, desert plants, like a cactus, have less available water than a lilypad in a pond, but every photosynthetic organism has some sort of adaptation, or special structure, designed to collect water. For most plants, roots are responsible for absorbing water.

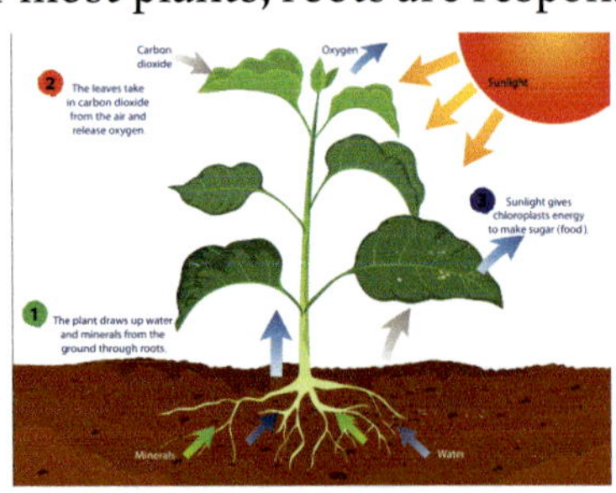

By taking in water (H_2O) through the roots, carbon dioxide (CO_2) from the air, and light energy from the Sun, plants can perform photosynthesis to make glucose (sugars) and oxygen (O_2)

The last requirement for photosynthesis is an important one because it provides the energy to make sugar. How does a plant take carbon dioxide and water molecules and make a food molecule? The Sun! The energy from light causes a chemical reaction that breaks down the molecules of carbon dioxide and water and reorganizes them to make the sugar (glucose) and oxygen gas. After the sugar is produced, it is then broken down by the mitochondria into energy that can be used for growth and repair. The oxygen that is produced is released from the same tiny holes through which the carbon dioxide entered. Even the oxygen that is released serves another purpose. Other organisms, such as animals, use oxygen to aid in their survival.

If we were to write a formula for photosynthesis, it would look like this:

$6CO_2 + 6H_2O$ + Light energy '! $C_6H_{12}O_6$ (sugar) + $6O_2$

The whole process of photosynthesis is a transfer of energy from the Sun to a plant. In each sugar molecule created, there is a little bit of the energy from the Sun, which the plant can either use or store for later.

Imagine a pea plant. If that pea plant is forming new pods, it requires a large amount of sugar energy to grow larger. This is similar to how you eat food to grow taller and stronger. But rather than going to the store and buying groceries, the pea plant will use sunlight to obtain the energy to build sugar. When the pea pods are fully grown, the plant may no longer need as much sugar and will store it in its cells.

A hungry rabbit comes along and decides to eat some of the plant, which provides the energy that allows the rabbit to hop back to its home. Where did the rabbit's energy come from? Consider the process of photosynthesis. With the help of carbon dioxide and water, the pea pod used the energy from sunlight to construct the sugar molecules. When the rabbit ate the pea pod, it indirectly received energy from sunlight, which was stored in the sugar molecules in the plant.

TYPES OF PHOTOSYNTHETIC PROCESSES

There are two types of photosynthetic processes: oxygenic photosynthesis and anoxygenic photosynthesis. They both follow very similar principles, but oxygenic photosynthesis is the most common and is seen in plants, algae and cyanobacteria.

During oxygenic photosynthesis, light energy transfers electrons from water (H_2O) taken up by plant roots to CO_2 to produce carbohydrates. In this transfer, the CO_2 is "reduced," or receives electrons, and the water is "oxidized," or loses electrons. Oxygen is produced along with carbohydrates.

Oxygenic photosynthesis functions as a counterbalance to respiration by taking in the CO_2 produced by all breathing organisms and reintroducing oxygen to the atmosphere.

Anoxygenic photosynthesis, meanwhile, uses electron donors that are not water and do not produce oxygen, according to "Anoxygenic Photosynthetic Bacteria" by LibreTexts. The process typically occurs in bacteria such as green sulfur bacteria and phototrophic purple bacteria.

THE PHOTOSYNTHESIS EQUATION

Though both types of photosynthesis are complex, multistep affairs, the overall process can be neatly summarized as a chemical equation.

The oxygenic photosynthesis equation is:

$6CO_2 + 12H_2O + \text{Light Energy} \rightarrow C_6H_{12}O_6 + 6O_2 + 6H_2O$

Here, six molecules of carbon dioxide (CO_2) combine with 12 molecules of water (H_2O) using light energy. The end result is the formation of a single carbohydrate molecule ($C_6H_{12}O_6$, or glucose) along with six molecules each of oxygen and water.]

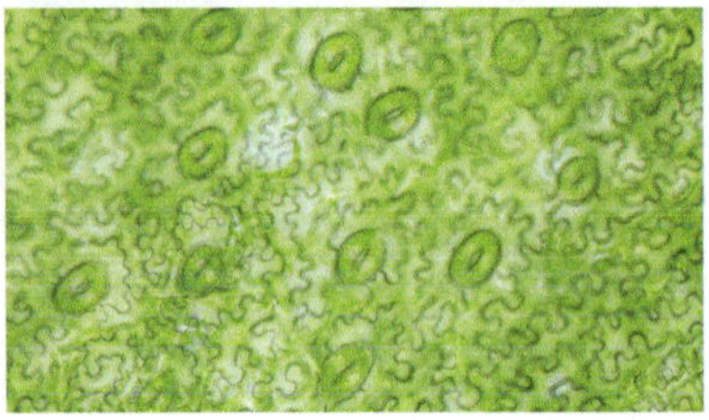

Fig. 2. Stomata are the gatekeepers of the leaf, allowing gas exchange between the leaf and surrounding air

Similarly, the various anoxygenic photosynthesis reactions can be represented as a single generalized formula:

$CO_2 + 2H_2A + \text{Light Energy} \rightarrow [CH_2O] + 2A + H_2O$

The letter A in the equation is a variable, and H_2A represents the potential electron donor. For example, "A" may represent sulfur in the electron donor hydrogen sulfide (H_2S).

How is carbon dioxide and oxygen exchanged?

Plants absorb CO_2 from the surrounding air and release water and oxygen via microscopic pores on their leaves called stomata. Stomata are the gatekeepers of gas exchange between the inside of plants and the external environment.

When stomata open, they let in CO_2; however, while open, the stomata release oxygen and let water vapor escape. In a bid to reduce the amount of water lost, stomata close, but that means the plant can no longer gain CO_2 for photosynthesis. This tradeoff between CO_2 gain and water loss is a particular problem for plants growing in hot, dry environments.

How do plants absorb sunlight for photosynthesis?

Plants contain special pigments that absorb the light energy needed for photosynthesis.

Chlorophyll absorbs red and blue light to use in photosynthesis and reflects green light. Chlorophyll is a large molecule and takes a lot of resources to make; as such, it breaks down towards the end of the leaf's life, and most of the pigment's nitrogen (one of the building blocks of chlorophyll) is resorbed back into the plant, according to Harvard University's The Harvard Forest. When leaves lose their chlorophyll in the fall, other leaf pigments such as carotenoids and anthocyanins begin to show their true colors. While carotenoids primarily absorb blue light and reflect yellow, anthocyanins absorb blue-green light and reflect red light.

Fig. 3. Plants need energy from sunlight for photosynthesis to occur

Pigment molecules are associated with proteins, which allow them the flexibility to move toward light and toward one another. A large collection of 100 to 5,000 pigment molecules constitutes an "antenna," according to an article by Wim Vermaas , a professor at Arizona State University. These structures effectively capture light energy from the sun, in the form of photons.

Where in the plant does photosynthesis take place?

Photosynthesis occurs in chloroplasts, a type of plastid (an organelle with a membrane) that contains chlorophyll and is primarily found in plant leaves. Double-membraned plastids in plants and algae are known as primary plastids, while the multiple-membraned variety found in plankton are called secondary plastids.

Chloroplasts are similar to mitochondria, the energy centers of cells, in that they have their own genome, or collection of genes, contained within circular DNA. These genes encode proteins that are essential to the organelle and to photosynthesis.

Inside chloroplasts are plate-shaped structures called thylakoids that are responsible for harvesting photons of light for photosynthesis. The thylakoids are stacked on top of each other in columns known as grana. In between the grana is the stroma — a fluid containing enzymes, molecules and ions, where sugar formation takes place.

Ultimately, light energy must be transferred to a pigment-protein complex that can convert it to chemical energy, in the form of electrons. In plants, light energy is transferred to chlorophyll pigments. The conversion to chemical energy is accomplished when a chlorophyll pigment expels an electron, which can then move on to an appropriate recipient.

The pigments and proteins that convert light energy to chemical energy and begin the process of electron transfer are known as reaction centers.

The reactions of plant photosynthesis are divided into two major stages: those that require the presence of sunlight (light-dependent reactions) and those that do not (light-independent reactions). Both types of reactions take place in chloroplasts:

light-dependent reactions in the thylakoid and light-independent reactions in the stroma.

LIGHT-DEPENDENT REACTIONS

When a plant absorbs solar energy it first needs to convert it into chemical energy. When a photon of light hits the reaction center, a pigment molecule such as chlorophyll releases an electron.

The released electron manages to escape by traveling through an electron transport chain, which generates the energy needed to produce ATP (adenosine triphosphate, a source of chemical energy for cells) and NADPH - both of which are required in the next stage of photosynthesis in the Calvin cycle. The "electron hole" in the original chlorophyll pigment is filled by taking an electron from water. This splitting of water molecules releases oxygen into the atmosphere.

LIGHT-INDEPENDENT REACTIONS: THE CALVIN CYCLE

The Calvin cycle uses energy stored from the light-dependent reactions to fix CO_2 into sugars needed for plant growth. According to the Khan Academy, these reactions take place in the stroma of the chloroplasts and are not directly driven by light - hence their name "light-independent reactions." However, they are still related to light as the Calvin cycle is fuelled by ATP and NADPH (both from the previously mentioned light-dependent reactions).

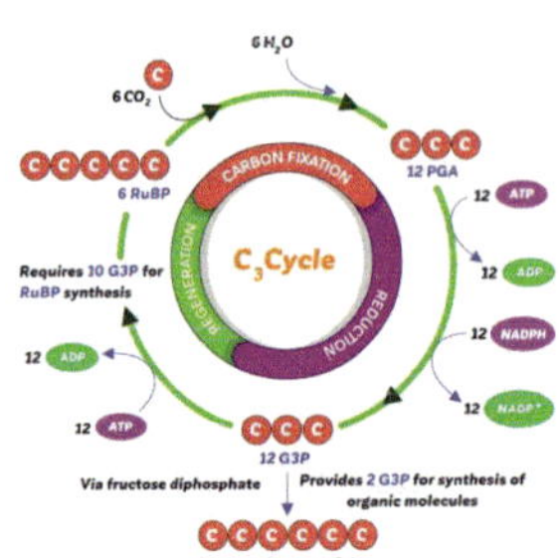

Fig. 4. Photosynthesis involves a process called the Calvin cycle to use energy stored from the light-dependent reactions to fix CO_2 into sugars needed for plant growth

Firstly, CO_2 combines with ribulose-1,5-bisphosphate (RuBP) which is a five-carbon acceptor, according to the Khan Academy. Next, it splits into two molecules of a three-carbon compound — 3-phosphoglyceric acid (3-PGA). The reaction is catalyzed by an enzyme called RuBP carboxylase/oxygenase, also known as rubisco.

The second stage of the Calvin cycle involves converting 3-PGA into a three-carbon sugar called glyceraldehyde-3-phosphate (G3P) — the process uses ATP and NADPH. Finally, while some G3P molecules are used to make glucose, others are recycled back to make RuBP, which is used in the first step to accept CO_2. For every one molecule of G3P that makes glucose, five molecules are recycled to generate three RuBP acceptor molecules.

PHOTORESPIRATION

Photorespiration is a particularly big problem when plants have their stomata closed to conserve water and are therefore not taking in any more CO2. Rubisco has no other choice but to fix oxygen instead, which in turn lowers the photosynthetic efficiency of the plant. This means that less plant food (sugars) will be produced, which could result in a slowdown of growth and therefore smaller plants.

This is a big problem for agriculture, as smaller plants mean a smaller harvest. There are mounting pressures on the agricultural industry to increase plant productivity to feed our ever-expanding global population. Scientists are constantly looking for ways to increase photosynthetic efficiency and reduce the occurrence of wasteful photorespiration.

Types Of Photosynthesis

There are three main types of photosynthetic pathways: C_3, C_4 and CAM. They all produce sugars from CO_2 using the Calvin cycle, but each pathway is slightly different.

C3 photosynthesis

Most plants use C_3 photosynthesis, according to the photosynthesis research project Realizing Increased Photosynthetic Efficiency (RIPE) , including cereals (wheat and rice), cotton, potatoes and soybeans. C_3 photosynthesis is named for the three-carbon compound called 3-phosphoglyceric acid (3-PGA) that it uses during the Calvin cycle. 3-PGA is produced when rubisco fixes CO_2, forming the three-carbon compound.

C4 photosynthesis

Plants such as maize and sugarcane use C_4 photosynthesis. This process uses a four-carbon compound intermediate (called oxaloacetate) which is converted to malate , according to Biology Online. Malate is then transported into the bundle sheath where it breaks down and released CO_2, which is then fixed by rubisco and made into sugars in the Calvin cycle (just like C_3 photosynthesis). C_4 plants are better adapted to hot, dry environments and can continue to fix carbon even when their stomata are closed (as they have a clever storage solution), which reduces their risk of photorespiration, according to Biology Online.

CAM photosynthesis

Crassulacean acid metabolism (CAM) is found in plants adapted to very hot and dry environments, such as cacti and pineapples, according to the educational website Khan Academy. When stomata open to take in CO_2, they risk losing water to the external environment. Because of this, plants in very arid and hot environments have adapted. One adaptation is CAM, whereby plants open stomata at night (when temperatures are lower and water loss is less of a risk). According to the Khan Academy, CO_2 enters the plants via the stomata and is fixed into oxaloacetate and converted into malate or another organic acid (like in the C_4 pathway). The CO_2 is then available for light-dependent reactions in the daytime, and stomata close, reducing the risk of water loss.

How photosynthesis could combat climate change

Photosynthetic organisms are a possible means to generate clean-burning fuels such as hydrogen. A research group at the University of Turku in Finland tapped into the ability of green algae to produce hydrogen. Green algae can produce hydrogen for a few seconds if they are first exposed to dark, anaerobic (oxygen-free) conditions and then exposed to light. The researchers devised a way to extend green algae's hydrogen production for up to three days, as reported in their 2018 study published in the journal Energy & Environmental Science.

Scientists have also made advances in the field of artificial photosynthesis. For instance, a group of researchers from the University of California, Berkeley, developed an artificial system to capture CO_2 using nanowires, or wires that are a few billionths of a meter in diameter. The wires feed into a system of microbes that reduce CO_2 into fuels or polymers by using energy from sunlight.

In 2016, members of this same group published a study in the journal Science that described another artificial photosynthetic system in which specially engineered bacteria were used to create liquid fuels using sunlight, water and CO_2. In general, plants are only able to harness about one percent of solar energy and use it to produce organic compounds during photosynthesis. In contrast, the researchers' artificial system was able to harness 10% of solar energy to produce organic compounds.

Scientists found that this bacteria is particularly good at concentrating CO_2 in its cells, which helps stop rubisco from accidentally binding to oxygen. By understanding how the bacteria achieve this, scientists hope to incorporate the mechanism into plants to help boost photosynthetic efficiency and reduce the risk of photorespiration.

CHAPTER

NITROGEN AND SULFUR METBOLISM 11

INTRODUCTION

Sulfur is a crucial microelement for plant nutrition, required by the biosynthesis of sulfolipids, antioxidants, cofactors, secondary metabolites, and amino acids that are strictly or conditionally essential for human nutrition (cysteine and methionine). Also, grain S content is essential to form protein disulfide bridges and dictates flour processing properties (such as dough extensibility) in bread, pasta or biscuit industry. It is often assumed that S assimilation positively correlates with photosynthesis, because sulfate incorporation is stimulated by sugars in plant roots and inhibited under non-physiological conditions where CO_2 is omitted from the atmosphere (CO_2-free air). Nevertheless, the relationship with photosynthesis has never been tested experimentally and thus it is presently uncertain as to whether S assimilation can be impacted by environmental conditions that affect photosynthetic metabolism.

As mentioned above, solving the question of a possible correlation between photosynthesis and S metabolism in the long term is critically important because CO_2 mole fraction in Earth's atmosphere inexorably increases and favours plant photosynthesis, and also current metabolic engineering aims to increase photosynthesis by suppressing photorespiration. In the past 25 years, it has been found that plants cultivated in a CO_2-enriched atmosphere contain less elemental including in grains, thereby affecting flour quality.

Accordingly, cultivation under CO_2-enriched conditions has been found to alter the content in S-containing antioxidant (glutathione). Therefore, present atmospheric changes seem to be detrimental to S metabolism. This problem may worsen in the near future because of limited available S in the environment, due to decreased pollution-driven SO_2 emissions and lower utilization of S-containing fertilisers (such as superphosphate). However, possible mechanisms explaining the lower S content when plants are grown at high CO_2 have not been studied.

The question of a possible correlation between photosynthesis and S assimilation in the short term is also important since in the field, photosynthesis can vary considerably depending on environmental conditions, and thus plant S nutrition could be affected accordingly. Furthermore, S metabolism is involved in electron consumption and redox metabolism in illuminated leaves and thus can affect leaf photosynthetic capacity.

In principle, changes in S metabolism when photosynthesis varies could stem from metabolic interactions with photorespiration. The key enzymatic activity of photorespiratory metabolism is the conversion of glycine to serine. In this biochemically complicated reaction, a glycine molecule is cleaved thereby liberating CO_2 and ammonium (NH^+_4), and a one-carbon unit is fixed onto another glycine molecule to form serine. This reaction involves tetrahydrofolate (H_4F) as a cofactor to transfer the one-carbon unit. One-carbon (C_1) metabolism is thus directly involved in photorespiration (due to H_4F requirement) and this is visible in, e.g., mutants affected in formyl tetrahydrofolate deformylase (which regenerates H_4F) that have growth defects and accumulate glycine to very high levels under photorespiratory conditions.

Also, C_1 metabolism is essential not only for photorespiration but also for many one-carbon requiring reactions of metabolism, including methionine synthesis. The involvement of N-containing compounds (ammonium, glycine, serine) in photorespiration explains why photorespiration is intimately linked to enzyme activities of N assimilation and how it may stimulate nitrate reduction. There is no such direct relationship with sulfur but presumably, photorespiration may stimulate S assimilation because serine and one-carbon units are the building blocks used to synthesize S-containing amino acids methionine and cysteine.

Experiments with glycine leaf feeding in the dark or manipulation of glutathione synthesis (via enzyme overexpression) have suggested that glycine provision by photorespiration might be of importance for glutathione synthesis and thus S metabolism. Also, under salt stress, the increase in glutathione content has been proposed to be due to augmented photorespiration rates, which increase metabolic availability in glycine and serine.

Furthermore, the provision in redox power (in the form of NADH or via the malate valve, for nitrate and sulfate reduction), and aspartate (for methionine synthesis) may in principle benefit from photorespiration. When photorespiration is low, N and S assimilation is believed to be downregulated in leaves because there is an increase in nitrate reduction in roots resulting in a generally lower N/C ratio in leaves and higher xylem transport of organic N from roots to shoots and furthermore, C_3-C_4 *Flaveria* interspecies grafts suggested that in C_4 plants (where photorespiration is suppressed), glutathione is preferentially synthesized in roots.

Conversely, the response to S-deficiency includes an increase in photorespiratory transcripts and metabolites and an increase in the photosynthetic CO_2 compensation point, suggesting that photorespiration participates in S homeostasis.

N/SULFUR COMPOUNDS

The L-asparagine analog, L-3-cyanoalanine, is synthesized from L-cysteine + HCN.

An interesting group of cysteine derivatives found in garlic and onions, *Allium cepa*, *A. sativum* and *A. ursinum*, are **alliin** and derivatives. Alliin and propenylallin are hydrolyzed (cleaved) to allylsulfenic acid, propenyl sulfenic acid and pyruvic acid + NH_3 by alliin lyase when garlic or onion tissues are damaged. Allylsulfenic acid spontaneously condenses to the disulfides, allicin and to dialylsulfide. Allicin and dialylsulfide have strong antimicrobial activity. Propenyl sulfenic acid spontaneously rearranges to *syn*-propanethial S-oxide. Allicin and syn-propanethial S-oxide have strong feeding deterrent activity toward herbivores such as insects. *syn*-Propanethial S-oxide is a lachrymator - the reason why slicing onions causes tears to form (Fig.1):

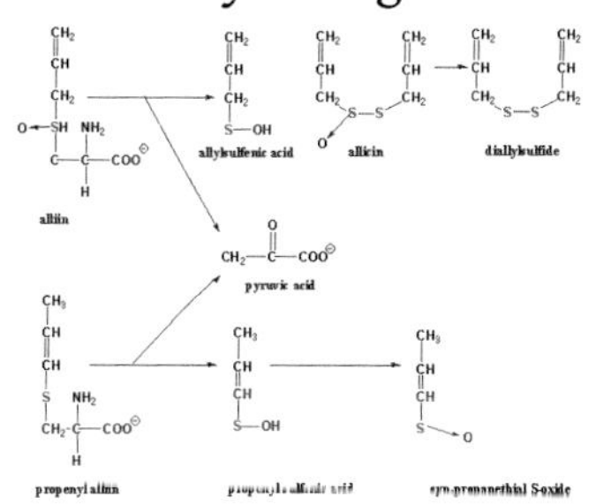

Fig. 1.

Certain plants in the dicot order Capparales such as in the Cruciferae accumulate thioglucosides known as glucosinolates in vacuoles that are similar to cyanogens in many aspects. All plant cells that sequester glucosinolates in vacuoles also have thioglucoside glucohydrolases known as myrosinase stored in places other than vacuoles. Upon tissue damage or wounding or large increases in membrane permeability, myrosinase and glucosinolates come together releasing the pungent, repellent and microbicidal compound, isothiocyanate.

The distinctive, pungent flavor and odor of mustard, radishes and horseradish is due to isothiocyanates. Glucosinolates are synthesized from various protein and non-protein amino acids, cysteine and glucose in a manner very similar to the biosynthesis of cyanogenic glucosides except for the addition of sulfur from cysteine. The precursor amino acid is converted to an aldoximine and to thiohydroximic acid after addition of a sulfate group. A thioglucoside, desulfo-glucosinolate, is formed at the next step with addition of glucose from UDP-glucose. Finally another sulfur group is added in the form of sulfate by transfer from phosphoadenosine-phosphosulfate (PAPS) forming a glucosinolate (Fig. 2):

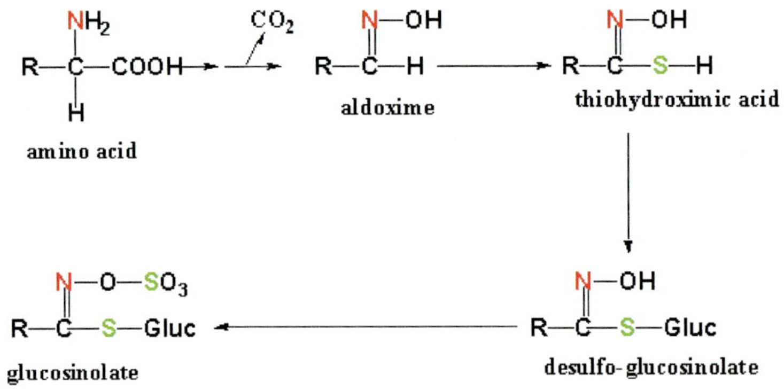

Fig. 2.

PLANT NITRATE ASSIMILATION

The most common form of N plants acquire from the soil is nitrate. Nitrate is transported into plant cells by nitrate transporters (Fig. 3). Three uptake systems are responsible for uptake of nitrate into the roots: a constitutive high affinity uptake system, an inducible high affinity system an a low affinity system. However, further transporters are necessary to facilitate xylem loading and distribution of nitrate throughout the plant as well as its storage in the vacuoles. Two major classes of nitrate transporters exist in plants, the NRT1/PRT family responsible for low affinity nitrate uptake as well as amino acid and peptide transport, which contains 53 genes in Arabidopsis, and NRT2 family of high affinity nitrate transporters composed of seven members in this species. Genes of both classes have been found throughout the plant kingdom, including C_4 plant.

Nitrate is reduced to ammonium in two spatially separated steps. In the cytosol, nitrate reductase (NR) transfers electrons from NADH to nitrate to form nitrite. Nitrite is transported into plastids and reduced to ammonium by ferredoxin dependent nitrite reductase. Ammonium is assimilated into organic compounds by glutamine synthetase (GS) which uses glutamate as the ammonium acceptor. GS is coupled with glutamate synthase (GOGAT) which transfers the amino group of glutamine to 2-oxoglutarate to form two molecules of glutamate, in a GS/GOGAT cycle. The cycle thus uses one molecule of 2-oxoglutarate to assimilate one NH 3 molecule and export one molecule of the amino acid glutamate (Fig. 3). Glutamate is the source of reduced N for the synthesis of other amino acids, with the first step in this route being usually transamination with oxaloacetate to form aspartate catalyzed by aspartate aminotransferase.

Ammonium in plants originates not only from nitrate reduction. It is present as a nutrient in the soil and can be transported into plants by ammonium transporters. In fact, ammonium is the preferred source of inorganic N for many plant species and optimal growth is often achieved only when both nitrate and ammonium are present. Ammonium is also produced by photorespiration, in the mitochondrial glycine decarboxylation reaction. An efficient assimilation and re-assimilation of ammonium thus must occur in all compartments to prevent its accumulation to toxic levels. Indeed, plastidic and cytosolic isoforms of GS exist in all plant species and the plastidic GS is dual targeted also to mitochondria. GOGAT is present in plants in multiple forms as well, the major ferredoxin dependent enzyme is found predominantly in leaves, whereas in non-photosynthetic cells NADH–GOGAT is the prevalent isoform.

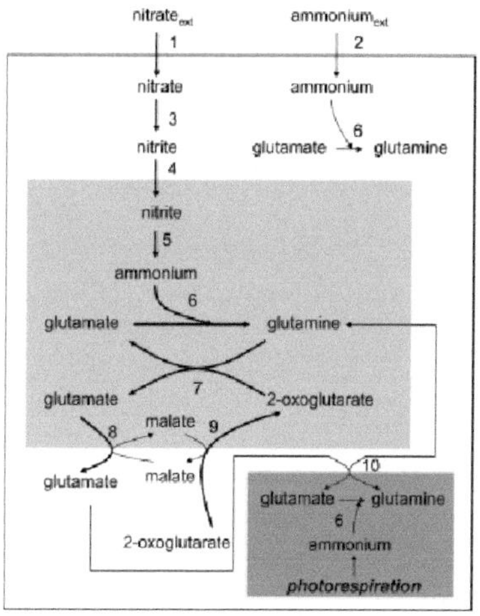

Fig. 3. Schematic representation of plant nitrate assimilation. Dark shaded rectangle represents mitochondria, light shaded one denotes plastid. Enzymes are symbolized by numbers: 1 – nitrate transporter, 2 – ammonium transporter, 3 – nitrate reductase, 4 – nitrite transporter, 5 – nitrite reductase, 6 – glutamine synthetase, 7 – glutamate synthase, 8 – plastidic glutamate-malate translocator, 9 – plastidic 2-oxoglutarate- malate translocator, 10 – mitochondrial glutamate–glutamine translocator. The major pathway of nitrate assimilation is printed bold.

Both forms of GOGAT are, however, localized to plastids. Another enzyme, the glutamate dehydrogenase, which in vitro synthesizes glutamate from ammonium and 2-oxoglutarate, has been long implicated to participate in ammonium assimilation. Recent findings however suggest that the major function of this enzyme is the provision of carbohydrate skeletons for carbon metabolism from amino acids during protein degradation.

REGULATION OF NITRATE ASSIMILATION

Nitrate and ammonium assimilation is strongly regulated by the N demand of the plant and N supply and is closely connected to carbon metabolism. The components of the pathway undergo a coordinated regulation which often occurs on multiple levels. So, nitrate uptake is induced in the presence of nitrate and feedback inhibited by amino acids and ammonium. It is regulated by light and CO 2 concentration via the availability of carbohydrates. Accordingly, nitrate uptake was demonstrated to be under diurnal regulation with maximum activity during day and minimum activity at night.

The decrease of uptake at night can be reversed by feeding sucrose. The coordination between N demand and N supply in control of nitrate uptake was proposed to be achieved by internal cycling of amino acids and/or cytokinins between the roots and the shoots. Ammonium transport seems to be regulated in a similar manner to nitrate transport: it undergoes a day night rhythm and the reduction of uptake in night can be prevented by sucrose treatment and is inhibited by glutamine. However, in contrast to nitrate transport ammonium uptake seems to be regulated primarily by local and not by systemic signals.

NR is a well studied enzyme which undergoes a complex regulation. NR expression is controlled by circadian clock and it is also induced by light and by sugars in the dark. The level of NR transcript is modulated by CO_2 availability; it decreases when CO_2 assimilation is diminished, e.g. due to water stress and increases upon exposure of plants to elevated CO_2. The rapid response of NR activity to environmental stimuli is, however, caused mostly by a rapid post-translational regulation. Upon sudden decrease in photosynthesis NR is rapidly and reversibly phosphorylated.

The phosphorylation of NR enables its binding to 14-3-3 proteins, which seems to be the actual mechanism of inactivation and might be a signal for NR degradation. The rapid response seems to be necessary to prevent accumulation of toxic levels of nitrite. NR is present in leaves and roots with different distribution of the activity among these organs in different plant species. Even a relatively low activity, however, is important for correct C/N balance as revealed by experiments with tobacco lacking root NR.

A central role in controlling nitrate and ammonium assimilation is occupied by GS. In contrast to NR and nitrite reductase, which are usually encoded by one to two genes, GS and GOGAT are encoded by small multigene families of three to six members. This accounts for a great versatility of regulation with some genes expressed in tissue specific manner and at certain developmental stages while some being universal.

The organel-targeted GS2 is especially important for reassimilation of the photorespiratory NH_3 and is, therefore, highly expressed in green tissues and inducible by light. Cytosolic GS1 often fulfils specialized roles, such as in nodules to assimilate the ammonium produced by nitrogen fixation in the bacteroids or in tissues involved in transport of reduced N. GS is regulated transcriptionally but undergoes also post- transcriptional and post-translational regulation. Both GS forms are subjected to phosphorylation and bind 14-3-3 proteins. Cytosolic GS is increased during senescence to facilitate the remobilisation of N. GS is also often associated with improved yield of crop plants by quantitative genetics. Indeed, overexpression of pine cytosolic GS in poplar resulted in significantly enhanced growth whereas disruptions of specific GS1 genes in maize led to reduced kernel size and/or number.

Despite the importance of N assimilation surprisingly little is known about the molecular mechanisms of the regulation of N acquisition and metabolism. Nitrate transporter NRT1;1 seems to play an important role in N sensing. Nitrate has clearly a role as a signal since microarray analysis of Arabidopsis mutants in nitrate reductase revealed that expression of 595 genes including several transcription factors responded to nitrate treatment.

Different amino acids act as feedback inhibitors of N assimilation, whereas glutamine seems to be most important in regulation of nitrate and ammonium uptake, glutamate, cysteine, and asparagine inhibit nitrate reduction. Sugars are clearly involved in the regulation of N assimilation, both as inductors and repressors, when their content is low, however the mechanisms of their action are not understood. A special role in regulation of N assimilation, specifically in signalling N-status of the plant, is played by the phytohormones cytokinins. Cytokinins are produced in the root at sufficient N supply and after transport into the shoots induce expression of genes of N assimilation. They are, however, also transported from the shoot to the root and affect nitrate uptake.

NITRATE ASSIMILATION IN C_4 PLANTS

Most of our knowledge on regulation of N metabolism has been derived from C_3 plants and no differences in organization and regulation of nitrate assimilation between C_3 and C_4 plants were described. However, C_4 plants differ significantly in the compartmentalization of N metabolism and also in N use efficiency. It has long been known that in terms of growth rate C_4 grasses respond better to applied N than C3 grasses. The C4 species clearly exhibited a higher N use efficiency, expressed as biomass per unit N in plant. The greater N use efficiency seems to be connected with a lower content of Rubisco in the leaves due to the CO_2 concentrating mechanism.

The intracellular localization of N assimilation in maize follows the pattern of C_3 plants, i.e. cytosolic localization of NR and the presence of nitrite reductase in chloroplast. Because of the observed differences in distribution of carbon metabolism in C_4 plants and in structures of MS and BSC chloroplasts the question of intercellular localization of nitrate assimilation has frequently been addressed in the early years of C_4 photosynthesis research. By enzyme activity assays researchers found that NR is prevalently present in MC of three C_4 species with different types of BSC chloroplasts: maize, Sorghum sudanense, and Gomphrena globosa. NADH dependent glutamate dehydrogenase, which at that time was believed to be the only ammonium assimilating enzyme, was localized in BSC. Nitrite reductase activity has been found in MC of maize and S. sudanense, but in BSC of G. globosa which seems to imply that it is associated with the presence of chloroplasts with grana.

Also in Eleusine coracana NR and nitrite reductase occurred predominantly in MC. In a more systematic approach analysing five C_4 species of all three metabolic subtypes, i.e. NADP-malic enzyme, NAD-malic enzyme, and phoshoenolpyruvate carboxykinase, NR, nitrite reductase, GS, and GOGAT were coordinately localized in MC, except Panicum maximum, where nitrite reductase was present both in MC and BSC. On the other hand, in another study with maize, GS and GOGAT have been found in both cell types but predominantly in BSC. Similarly, in Digitaria sanguinalis NR and nitrite reductase were found exclusively in MC, GS was equally distributed between MC and BSC, and GOGAT was predominantly localized to BSC.

Since all the localization studies were based on distribution of enzyme activities in differentially homogenized leaves, it was difficult to conclude on exclusivity of the intercellular distribution of the N assimilation pathway and its components. Importantly, therefore, a clear evidence for an exclusive localization of maize NR in the cytosol of MC was obtained by immunogold labelling. Similarly, the ambiguities in distribution of ammonium assimilation were clarified by researchers who showed that both cytosolic and plastidic GS are present in both cell types, whereas ferredoxin dependent GOGAT is almost completely confined to BSC chloroplasts.

An interesting feature of C_4 plants is a relatively high cytosolic GS activity. Whereas in most C_3 species cytosolic GS1 accounts for less than 30% of total leaf GS activity, in most C_4 plants analyzed to date GS1 activity is higher or equal to the plastidic GS2. Analyzing Panicum species of different type of photosynthesis researchers showed that the presence of GS1 activity in leaf correlated with C_4 photosynthesis, with the C_3-C_4 intermediate P. milliodes possessing foliar GS1 activity intermediate between the C_3 and C_4 species. The ratio in accumulation of the two GS isoforms also differs in between the two cell types of C_4 plants. GS1 seems to be more abundant in MC, whereas both isoform are present at the same level in BSC. Clearly, due to substantially reduced and BSC confined photorespiration much less ammonia is produced in C_4 leaves than in C_3 leaves and therefore the demand for its rapid refixation by GS is lower. The MC localization of nitrate reduction implies an increased need for transport of reduced N between MC and BSC, which might explain the need for higher cytosolic GS. The evolutionary advantage of the spatial distribution of nitrate assimilation in C_4 plants is, however, not known. Interestingly, assimilation of another essential mineral nutrient, sulfate, is also distributed in cell-specific manner in C_4 plant species.

SULFATE ASSIMILATION

Sulfur is essential for all living organisms as a constituent of the amino acids cysteine and methionine, many coenzymes such as iron sulfur centers, thiamine, lipoic acid, etc., and various other compounds of primary and secondary metabolism. In most of these compounds sulfur is present in the reduced (-2) form of organic sulfide. The most common form of sulfur in nature is, however, oxidized as inorganic sulfate (+6). Plants, algae, and many microorganisms are able to directly utilize sulfate, reduce it and incorporate into bioorganic molecules, which are the form of sulfur accessible to animals and other organisms. The pathway of sulfate assimilation is thus an essential component of plant primary metabolism.

Plant Sulfate Assimilation

The pathway of sulfate assimilation in plants has been completely resolved only relatively recently and has lately been subjected to several comprehensive reviews. Sulfate uptake into plant cells is facilitated by sulfate transporters (Fig. 4). Also sulfate transporters form a large multigene family with 12–15 members differing in affinity to sulfate and tissue distribution. Because it is a very stable and inert compound, before reduction sulfate has to be activated to adenosine 5'-phosphosulfate (APS) by adenylation catalyzed by ATP sulfurylase (ATPS). APS is reduced to sulfite by APS reductase (APR). Sulfite is further reduced to sulfide by ferredoxin dependent sulfite reductase. Sulfide is than incorporated into the amino acid skeleton of O-acetylserine (OAS), which is synthesized by acetylation of serine by serine acetyltransferase, to form cysteine in a reaction catalyzed by O-acetylserine (thiol)lyase (Fig. 4). Cysteine is the source of reduced sulfur for synthesis of methionine, iron sulfur clusters, and other compounds.

Sulfate assimilation is confined to plastids, however, some reactions occur also in other compartments. The cysteine synthesis, e.g., proceeds in all three compartments capable of protein synthesis, i.e. plastids, cytosol, and mitochondria.

ATPS activity was detected both in plastids and in the cytosol. On the other hand the enzymes involved in reductive steps of the pathway, APR and sulfite reductase, are localized exclusively in plastids. Total foliar ATPS and APR activity decrease with the leaf age in Arabidopsis, however, the cytosolic and plastidic ATPS are regulated differently. Whereas the plastidic activity declines with time, the cytosolic is increasing with leaf age. This indicates different roles of ATPS in the two compartments: a provision of APS for sulfate reduction for biosynthetic processes required for growth in plastids and involvement in synthesis of secondary compounds in the cytosol.

The reduction of sulfate occurs predominantly in leaves, and reduced sulfur compounds are distributed to sink tissues via phloem. It appears, however, that most tissues are capable of sulfate reduction, including roots and developing seeds. Indeed, available microarray data in the Genevestigator database reveals the presence of mRNA for the genes of sulfate assimilation, such as APR or sulfite reductase, in all Arabidopsis organs, including flowers and siliques. These findings were corroborated using promoter::GUS fusions which clearly showed activity of

APR and ATPS promoters in all tissues of Arabidopsis.

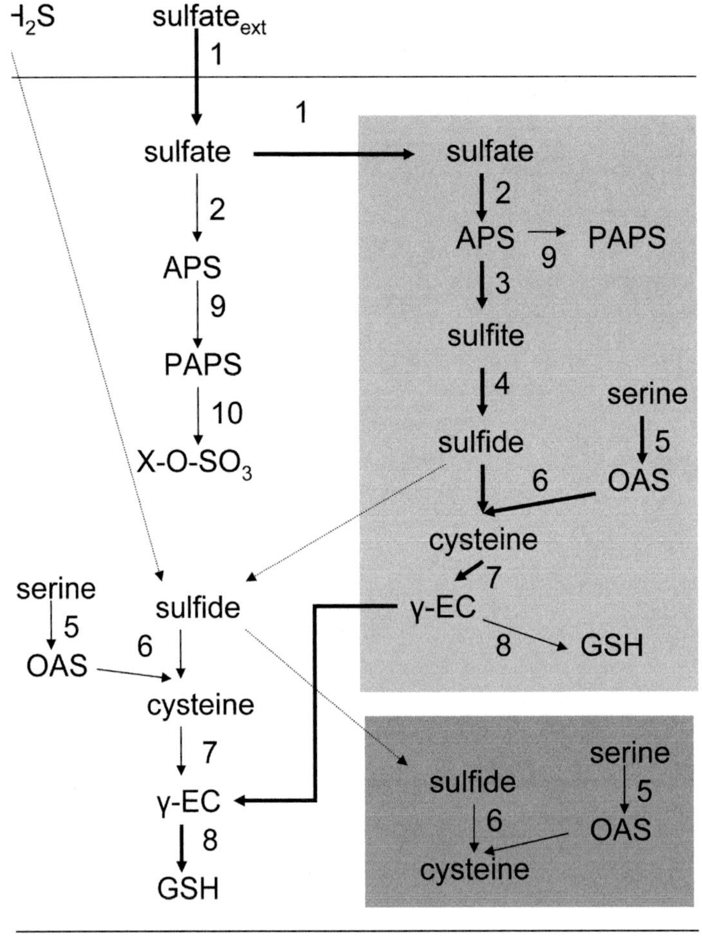

Fig. 4. Schematic representation of plant sulfate assimilation. Dark shaded rectangle represents mitochondria, light shaded one denotes plastid. Enzymes are symbolized by numbers: 1 – sulfate transporter, 2 – ATP sulfurylase, 3 – APS reductase, 4 – sulfite reductase, 5 – serine acetyltransferase, 6 – *O*-acetylserine (thiol)lyase, 7 – gglutamylcysteine synthetase, 8 – glutathione synthetase, 9 – APS kinase, 10 – sulfotransferase. The major pathway of sulfate assimilation and glutathione synthesis (in Arabidopsis) is printed bold.

However, whether the sulfate reduction rate in these cells is sufficient to cover their needs for reduced sulfur instead of relying on long distance transport of organic sulfur compounds, such as glutathione or S-methylmethionine, remains to be seen. However, there is a group of plants that lacks the ability to reduce sulfate in a great portion of their cells, the monocot C_4 plants.

Regulation of Sulfate Assimilation

The sulfate assimilation pathway is extensively regulated in a demand-driven manner to prevent accumulation of toxic intermediate and to provide optimal rate of cysteine production. Thus, when demand for reduced sulfur is increased due to enhanced protein synthesis or increased turnover of the sulfur containing tripeptide glutathione the flux through the pathway is increased. On the other hand, when plants are exposed to reduced sulfur in the atmosphere or rhizosphere or when there is not sufficient supply of the amino acid acceptor due to carbon or nitrogen deficiency, the pathway is downregulated.

Although regulation of all components of the sulfate assimilation has been described, control flux analysis revealed that APR and sulfate uptake possess the highest control over the pathway. In line with these observations, APR activity and mRNA accumulation undergoes a diurnal rhythm with a maximum during light and minimum at night. The high activity coincides with the highest flux through the pathway.

Despite a comprehensive knowledge on the physiological responses of sulfate assimilation to various environmental stimuli, very little is known about the molecular mechanisms of regulation and the signals involved. Several compounds have been proposed as molecular signals in regulation of the pathway. OAS accumulates during sulfur deficiency and when supplied exogenously induces mRNA accumulation of many genes of sulfate uptake and assimilation. Indeed, system biology approaches revealed a correlation of OAS content with transcript accumulation of many genes during a sulfur starvation response and large set of genes regulated in the same way by sulfate starvation and OAS treatment.

However, not all genes induced by sulfur deficiency are regulated by OAS and the timing of OAS accumulation seems to fall behind the induction of sulfate uptake by sulfur starvation in potato, so that OAS is probably not the only signal in the sulfur starvation response. Glutathione (GSH) may also represent the signal of sulfur status of the plant, as depletion of GSH by treatment with an inhibitor of its synthesis, buthionine sulfoximine (BSO), leads to upregulation of APR, whereas GSH itself inhibits sulfate uptake and reduction.

ATPS and APR activity are reduced also in plants treated with cysteine, however, the feedback inhibition is alleviated when simultaneously BSO blocks the synthesis of GSH from the additional cysteine. In addition, several phytohormones have been shown to modulate expression or activity of various components of sulfate assimilation, such as jasmonate, or abscisic acid. Only very recently, however, the first transcription factor and cis element responsible for regulating genes for sulfate transporter by sulfur starvation have been identified.

Sulfate Assimilation in C_4 Plants In a search for further metabolic processes spatially distributed in C_4 plants it was soon discovered that in maize 75–100% of total leaf ATPS activity is confined to BSC. These findings were extended to 17 other C4 species of all three C_4 subtypes, where 95-100% of total leaf ATPS was localized in BSC chloroplasts. Also APR was found almost exclusively in BSC of maize, while the activities of sulfite reductase and OAS(thiol)lyase could be measured at comparable levels in MC and BSC.

In attempts to decipher the mechanism of the spatial distribution of these enzymes, northern analysis of BSC and MC specific RNA revealed that mRNA levels for ATPS, APR, and sulfite reductase were detected in BSC only, whereas the mRNA for OAS(thiol) lyase was found in both MC and BSC. The cell-specific localization of enzymes of sulfate assimilation in maize seems, therefore, to be regulated on the transcriptional level, at least under standard growth conditions. However, the situation might be different under stress. Maize is especially sensitive to chilling which induces a strong oxidative stress characterized by production of reactive oxygen species (ROS).

In maize plants subjected to chilling APR activity and mRNA level were greatly increased in BSC, and mRNA but not enzyme activity was also detectable in MC. This indicates an additional post-transcriptional mechanism to ensure the BSC specific localization of sulfate assimilation in maize.

The exclusive localization of ATPS and APR in BSC of maize means that an efficient transport of reduced sulfur compounds from BSC to MC must exist. MC are capable of cysteine synthesis, therefore, the transport form of reduced sulfur could be sulfide, cysteine, methionine or glutathione. Feeding of isolated bundle sheath strands from maize with [^{35}S]sulfate resulted in secretion of labelled cysteine into the nutrient solution. It seems therefore, that cysteine (or its oxidized form cystine) is the most probable transport metabolite for reduced sulfur (Fig. 4). Interestingly, [^{35}S] sulfate feeding experiments also imply that glutathione synthesis was predominantly localized in MC.

The biological significance of the BSC specific localization of sulfate assimilation in C_4 plants is not obvious and, similarly, it is not clear whether this localization is a pre-requisite or a consequence of C_4 photosynthesis. To answer this question we addressed the distribution of ATPS and APR in Flaveria species with different types of photosynthesis. The dicot genus Flaveria (Flaveriinae-Asteraceae) is an excellent model to study the evolution of C_4 photosynthesis because, beside C_3 and C_4 species, it comprises a relatively large number of C_3–C_4 intermediates. A continuous gradation in the physiology and biochemistry of C_4 photosynthesis can be found among Flaveria species. Indeed, we showed previously that the C_3–C_4 intermediate Flaveria species are true evolutionary intermediates in the path from C_3 to C_4 photosynthesis, based on phylogenetic analysis of the H-protein subunit of glycine decarboxylase.

The aim of the study with Flaveria was to compare the intercellular distribution of sulfate assimilation in C_3, C_3-C_4, and C_4 species. We expected a BSC localization of APR and ATPS in C4 Flaveria species and a ubiquitous distribution in C_3 species. The distribution of the two enzymes in the intermediate ones would be an excellent indication for the evolutionary sequence of the processes leading to the BSC localization of sulfate assimilation. Surprisingly however, northern analysis of cell-specific RNA and in situ hybridization revealed that in the C_4 species F. trinervia mRNA for ATPS and APR were present at comparable levels in both MC and BSC. Immunogold electron microscopy confirmed the presence of APR protein in chloroplasts of both cell types. Consequently, the localization of assimilatory sulfate reduction in BSC cannot be a general C_4 trait.

How can these findings be explained taking into account the results of researchers who showed the exclusive BSC localization of ATPS in 17 C_4 species? Whereas the 17 species analyzed before were monocots, *F. trinervia* and *F. australasica* were the first C_4 dicots where the subcellular localization of the pathway was addressed. It has to be concluded that sulfate assimilation is exclusively localized in BSC only in C_4 monocots and this distribution is thus neither a pre-requisite nor a consequence of C_4 photosynthesis.

However, the BSC localization of ATPS in maize and other C_4 plants analyzed by researchers is not a general trait of monocots either, since in wheat, a C_3 monocot, ATPS and APR are present in all cell types. Whether C_4 Flaveria species are exceptions or the rule for C4 dicots remains to be established, however, it is evident that the previously generally accepted link between C_4 photosynthesis and BSC localization of sulfate assimilation is no longer valid.

The investigations of sulfate assimilation in Flaveria resulted in an additional interesting finding. The APR activity and levels of thiols were significantly higher in leaves of C_4-like and C_4 species than in those of C_3 and C_3–C_4 species. APR, cysteine and GSH content correlated with the degree of development of C_4 photosynthesis expressed by CO_2 compensation points. The actual foliar concentration of GSH is highly dependent on environmental conditions and, therefore, varies significantly among different plant species but also within single species.

Since the Flaveria species were grown at identical controlled conditions, it seems that the clear tendency to towards higher APR activity and GSH content with increasing C_4 photosynthesis might be a result of the adaptation to different habitats. C_4 photosynthesis is especially advantageous in dry and warm conditions. The higher GSH contents in C_4 Flaveria might thus be a mechanism to cope with increased oxidative stress caused by such environmental conditions. According to the demand-driven regulation, APR activity would have to be elevated to supply sufficient cysteine for the GSH synthesis.

GLUTATHIONE SYNTHESIS AND REDUCTION

Among sulfur containing metabolites the tripeptide glutathione (γ–glutamyl-cysteinyl-glycine) plays the most versatile role being essential for plant stress defense, redox regulation and signaling, control of cell cycle, and as a storage and transport form of reduced sulfur. GSH is involved in plant defense against ROS as a reductant of dehydroascorbate in the glutathione–ascorbate cycle. It is indispensable for protection against heavy metals as a substrate for synthesis of phytochelatins, which chelate the metals and enable their sequestration into the vacuoles.

Conjugation of xenobiotics with GSH is the first step in their detoxification and a molecular basis of resistance to some herbicides. Not surprisingly, GSH accumulates to high levels reaching concentration of several mM. Due to its role in stress defense and specifically in the response to chilling, GSH synthesis and its regulation was often addressed in C_4 plants.

REGULATION OF GSH SYNTHESIS

GSH is synthesized from its constituent amino acids in two ATP dependent steps. Firstly a γ-glutamylcysteine synthetase (γ-ECS) synthesizes γ-glutamylcysteine (γ-EC) from glutamate and cysteine. Subsequently, glycine is added to the g-EC by glutathione synthetase (GSHS). GSH synthesis is an essential process, since disruption of g ECS gene by T-DNA insertion is embryolethal. The rate of GSH synthesis is primarily controlled by the availability of the constituent amino acids, with the regulation of γ-ECS playing an additional substantial role.

Most of our knowledge on this regulation is derived from studies with poplar overexpressing bacterial genes for γ-ECS and GSHS. γ-ECS and GSHS are induced at conditions of high demand for GSH, such as after exposure to heavy metals or herbicide safeners. γ-ECS is more important for the control of GSH synthesis because GSH content was increased only in poplars overexpressing γ-ECS and not GSHS. The enzyme seems to undergo a complex regulation on different levels. Heavy metals induce mRNA accumulation of γ-ECS (and GSHS), but a post-transcriptional regulation has also been described. On the other hand, the enzyme is feedback inhibited by GSH, which causes a reversible conformational change due to a reduction of an internal disulfide bond.

The γ-ECS regulation thus seems to follow the same demand-driven manner as described for the components of sulfate assimilation. Indeed, since at most physiological situations cysteine limits GSH synthesis γ-ECS is often regulated in the same way as APR or other enzymes of sulfate assimilation. At some conditions however, e.g. during night or at non-photorespiratory conditions, synthesis of GSH is controlled by availability of glycine.

During its function in the glutathione–ascorbate cycle GSH is oxidized. The oxidized form of glutathione, GSSG, is reduced by the action of glutathione reductase (GR) which utilizes the electrons from NADPH. GR maintains the reducing environment in cells so that GSSG normally forms no more than 5% of total glutathione. Increased ratio of GSSG to total GSH is thus indicative of oxidative stress.

Localization of GSH and GSH Synthesis

GSH is present in all cellular compartments; however, the quantitative distribution is not resolved yet. Currently, two approaches are being used to quantify GSH on the subcellular level. The confocal laser scanning microscopy method makes use of the fluorescence of the GSH conjugate with monochlorobimane. However, since the conjugation is catalyzed enzymatically by GSH transferase, the method can directly measure only cytosolic GSH, whereas the organellar pools can be estimated by HPLC after cell disruption and additional chemical labeling of the remaining GSH by monobromobimane. The second method is based on immunohistochemistry with antibodies against GSH. This approach shows the highest density of labeling in mitochondria and the lowest in plastids. Unfortunately, none of the methods have been used for localization of GSH in C_4 plants.

GSH synthesis has been shown to take place in the cytosol and plastids. However, recently it was revealed that the two steps of GSH biosynthesis may be spatially separated, at least in some plant species. In Arabidopsis and Brassica juncea γ-ECS seems to be exclusively localized in the plastids whereas GSHS is present in both plastids and cytosol. The γ-EC may thus not only be a precursor of GSH but also play important roles in transport of reduced sulfur from plastids to the cytosol and possibly in signaling of the redox status of the chloroplast. This conclusion is supported by the identification of the regulator of APX2 (rax1) mutant in Arabidopsis. This mutant constitutively expresses cytosolic ascorbate peroxidase, which is normally inducible by photooxidative stress, and contains only approximately 50% of normal foliar GSH levels. In rax1 thus the chloroplast– cytosol signaling is clearly disturbed. The rax1 phenotype is caused by an R(229)-K substitution in γ-ECS protein. Several other Arabidopsis mutants are associated with γ-ECS and low GSH levels, the cadmium sensitive cad2, root meristemless rml1, and phytoalexin-deficient pad2, but none of them displays the same effect on the chloroplast–cytosol signaling.

On the other hand, γ-ECS and GSHS were detected by immunolocalization in both chloroplasts and cytosol of maize. Also, expression of the bacterial gene for γ-ECS both in plastids and in the cytosol led to an increase in foliar GSH content in poplar. Interestingly, the increase in GSH content in poplar plants overexpressing γ-ECS in the cytosol seems to be confined to this compartment which implies a limited exchange of GSH between cytosol and plastids. GR is localized in cytosol, plastids, and mitochondria. Interestingly, a single gene encodes both plastidic and mitochondrial isoform of GR due to a presence of a dual-specificity targeting peptide.

GSH is present and can be synthesized in all plant organs. Obviously, in many occasions cell specific differences in GSH levels have been found. In maize roots, GSH levels form a clear gradient from the highest at root tip to the lowest in the mature portion of the root. When investigated at the cellular level by confocal laser scanning microscopy approximately twofold higher cytosolic GSH concentration was measured in atrichoblasts than in trichoblasts. In addition high GSH levels were detected in root cap while markedly lower GSH was found in quiescent centers. On the other hand, in poplar leaves analyzed by the same approach surprisingly uniform concentration of cytosolic GSH has been detected.

Also in leaves, however, certain cell types distinctly differ, as very high GSH concentration has been found in leaf trichoms. Consequently, γ-ECS, GSHS, and genes for enzymes of cysteine synthesis were highly expressed in these cells. These cell specific differences in GSH levels are probably caused by differential rates of GSH synthesis and indicate that GSH transport between cells may not be efficient enough to balance such differences.

STRUCTURAL BIOLOGY OF PLANT SULFUR METABOLISM: FROM SULFATE TO GLUTATHIONE

For plants, the acquisition and use of sulfur and sulfur-containing molecules is essential for cellular processes. Not only is sulfur incorporated into amino acids (i.e. cysteine and methionine) and proteins, but reduced sulfur is used for the biosynthesis of a variety of metabolites that play critical roles in plants, ranging from vitamins/cofactors to protection against biotic and abiotic stresses. The enzymes involved in sulfur assimilation, cysteine biosynthesis, and glutathione biosynthesis (Fig. 5) are also attractive targets for engineering nutrient and sulfur-containing amino acid content in a variety of food crops, for enhancing stress responses, and for improving plants as tools for environmental remediation.

Metabolically, after a plant acquires sulfate from the environment, it must be activated to a useful form for further purposes. The three enzymes of the sulfur assimilation pathway [ATP sulfurylase, adenosine-5'-phosphosulfate (APS) reductase, and sulfite reductase] sequentially form a high-energy molecule (i.e. APS) and reduce this molecule to release sulfite, which is finally converted into sulfide (Fig. 5). Sulfide can then be used for the synthesis of cysteine, which requires serine acetyltransferase and O-acetylserine sulfhydrylase [also known as O-acetylserine(thiol)lyase and cysteine synthase] (Fig. 5). Subsequent incorporation of cysteine into the tripeptide glutathione, a major cellular redox buffer, requires glutamate-cysteine ligase (also known as g-glutamylcysteine synthetase) and glutathione synthetase (Fig. 5).

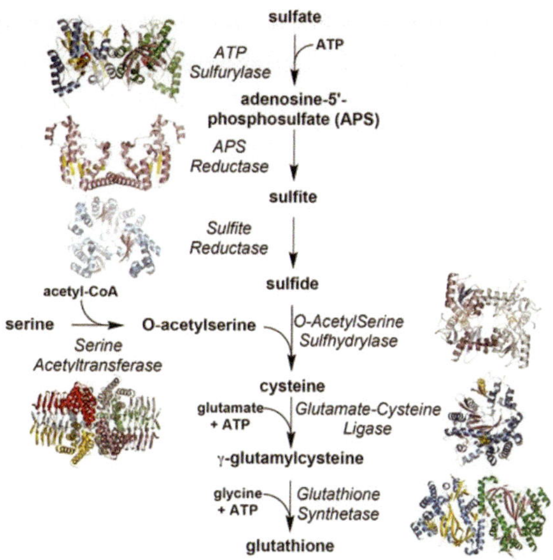

Fig. 5. Overview of plant sulfur metabolism. Metabolites (bold) and enzymes (italic) involved in the sulfur assimilation (ATP sulfurylase, PDB: 4MAF; APS reductase, PDB: 4BWV; sulfite reductase, PDB: 5H8V), cysteine synthesis (serine acetyltransferase, PDB: 4N69; *O*acetylserine sulfhydrylase, PDB: 1Z7Y), and glutathione synthesis (glutamate-cysteine ligase, PDB: 2GWD; glutathione synthetase, PDB: 5OES) are shown. Ribbon diagrams for enzymes involved at each step are shown.

This section covers work aimed at understanding the molecular function of the enzymes in the core pathways of plant sulfur metabolism through the use of structural biology. To date, the X-ray crystal structures of all the plant enzymes in the pathways from sulfur assimilation to glutathione synthesis have been determined. This structural information combined with genetic, genomic, and biochemical studies elaborates the molecular mechanisms and biochemical regulation of every step in how sulfur is assimilated and used by plants to create molecules that support a variety of cellular processes.

Sulfur assimilation: ATP sulfurylase

ATP sulfurylase (ATP:sulfate adenylyl transferase; EC 2.7.7.4) catalyzes the first (and energetically committed) step in the plant sulfur assimilation pathway by forming APS and inorganic pyrophosphate (PPi) from sulfate and ATP (Fig. 5). In plants, multigene families encode ATP sulfurylase isoforms, which are localized to different organelles. Of these isoforms, those targeted to the chloroplast are critical for reductive sulfate assimilation in plants. In the model plant Arabidopsis thaliana, expression of the plastid forms of ATP sulfurylase are partly controlled by the sulfur deprivation-inducible miRNA miR395. Moreover, natural variation in expression levels of ATP sulfurylase isoforms also contributes to modulating sulfate levels in leaves.

The first X-ray crystal structure of a plant ATP sulfurylase identified key amino acids in the reaction mechanism underlying formation of the high-energy mixed anhydride bond in APS and provided insights on the overall structure of this dimeric enzyme and its architectural relationships to homologs from other organisms. The three-dimensional structure of ATP sulfurylase from soybean in complex with APS revealed a dimer, with each monomer consisting of two mixed α/β domains (Fig. 6a). This structure corresponds to a partial product complex with an APS molecule bound in the C-terminal domain of each monomer. The binding of APS defines the location of the active site, which includes Phe245, Val247, Gln246, Arg248, Asn249, His252, His255, Leu258, Met322, His333, Gly348, Arg349, Ala352, and Ala390, as an enclosed cavity with a channel to the surface positioned above the APS α-phosphate moiety (Fig. 6b, c).

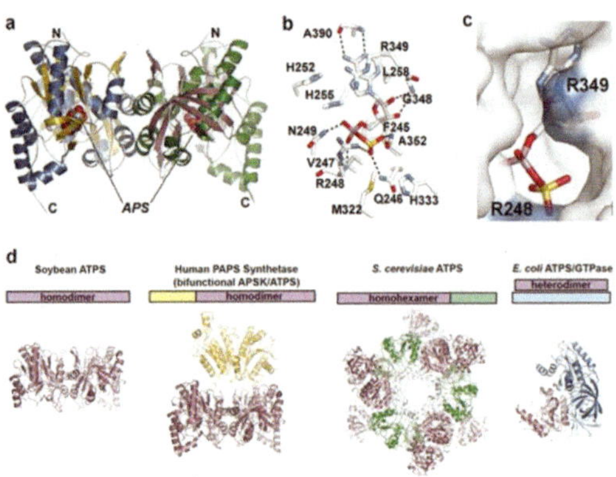

Fig. 6. Plant ATP sulfurylase. (a) Overall structure of the soybean ATP sulfurylase (PDB: 4MAF) dimer. Secondary structures in each monomer are differentially colored. The spacefilling model shows the position of bound APS in each monomer. (b) Soybean ATP sulfurylase active site. Dotted lines represent contacts between active site residues and APS. (c) Molecular surface view of the active site. The view is oriented down the putative pyrophosphate-binding site. Surfaces corresponding to Arg248 and Arg349 are colored (blue). APS is shown as a stick model. (d) Comparison of the structural organization of ATP sulfurylase from selected organisms. The domain architecture of the plant ATP sulfurylase, PAPS synthetase, yeast ATP sulfurylase, and the *E. coli* ATP sulfurylase are shown at the top of the panel. The three-dimensional structures of a homodimeric plant ATP sulfurylase from soybean (PDB: 4MAF), the dimeric bifunctional PAPS synthetase from human (PDB: 1XNJ), the homohexameric ATP sulfurylase from *S. cerevisiae* (PDB: 1G8F), and the heterodimeric ATP sulfurylase, which includes a regulatory GTPase domain, from *E. coli* (PDB: 1ZUN) are shown as ribbon diagrams in the lower part of the panel. The domain structure of each protein is colored by function as follows: ATP sulfurylase, rose; APS kinase, yellow; APS kinase-like, green; and GTPase, blue.

Previous sequence comparisons indicate that His252 and His255 of soybean ATP sulfurylase are part of the HXXH motif, which is conserved across the nucleotidylyl transferase and type I amino-acyl transferase enzyme families, and were proposed to be involved in cleavage of the α/β-phosphate bond during formation of APS. Based on the soybean ATP sulfurylase structure, site-directed mutagenesis of either histidine led to production of insoluble protein for the H252Q, H252A, H255N, H333N, and H333A mutants, which is consistent with a structural and not catalytic role for each histidine.

Interestingly, the H255A mutant showed a 15-fold decrease in turnover rate and suggests the loss of a hydrogen bond interaction in the reaction mechanism, but does not prevent catalysis. Thus, the structure and biochemical analysis of the soybean enzyme suggested a different reaction mechanism.

Modeling of the forward (ATP- and sulfate-bound) and reverse (APS- and PP_i-bound) reaction complexes, along with the crystal structure and results of site-directed mutagenesis studies, implicates Asn249 and Arg349 as having key roles in the chemical mechanism. For APS formation, interactions with Arg248, Asn249, His255, Arg349, and the phosphate groups of ATP distort the nucleotide into a U-shape. This bending opens the ATP α-phosphate group for nucleophilic attack by the sulfate group, which leads to formation of APS and PP_i release. In the reverse reaction (i.e. ATP synthesis), Asn249, His255, and Arg349 bind PP_i to allow for nucleophilic attack on the α-phosphate of APS. The active site facilitates a substitution mechanism in either the forward or reverse reaction by relying on charged reactive groups in close contact with substrates to stabilize a pentavalent transition state during the reaction.

Although ATP sulfurylase is found in multiple organisms, the structural organization of the enzyme widely differs in both domain architecture and oligomerization state. The homodimer observed in the three-dimensional structure of the soybean ATP sulfurylase is just the catalytic core, which is retained across the homologs from organisms across the tree-of-life, without any accessory domains (Fig. 6d). In microbes, such as *Escherichia coli* and *Pseudomonas syringae*, ATP sulfurylases are heterodimeric proteins with a core catalytic domain and a regulatory GTPase subunit.

Yet, in other bacteria and some invertebrates, ATP sulfurylase can be found as either a homodimeric enzyme (like the plant form) or as a bi-functional ATP sulfurylase/APS kinase. Humans also share the bi-functional ATP sulfurylase/APS kinase domain structure, which provides for synthesis of 3'-phosphoadenosine-5'-phosphosulfate (PAPS) as a sulfo-group donor molecule. In contrast, the ATP sulfurylases from *Saccharomyces cerevisiae* and *Penicillium chrysogenum* function as homohexamers, with each monomer consisting of an N-terminal ATP sulfurylase domain and a C-terminal APS kinase-like domain, which provides allosteric feedback regulation.

It remains unclear if these structural variations are related to the evolution of regulatory needs in sulfur assimilation for each organism; however, genomic studies of sulfur assimilation pathway genes in the major eukaryotic and prokaryotic lineages show various origins for these enzymes and that diverse events, including gene fissions and fusions, gene loss, and lateral gene transfers, yielded a diverse range of molecular architectures for these enzymes.

Sulfur assimilation: APS reductase

In plants, the activation of sulfate by ATP sulfurylase into APS is followed by sulfate reduction to sulfite (SO_3^{2-}). This is in contrast to prokaryotes and other eukaryotes that phosphorylate APS to PAPS via APS kinase and then use PAPS either to generate sulfate or as a sulfo-donor molecule in downstream reactions. In plants, APS kinase also forms PAPS, which is used in a variety of pathways such as glucosinolate biosynthesis, and is redox regulated potentially to control the metabolic fate of APS toward either sulfur assimilation or sulfo-donor reactions. Metabolically, the reduction of APS is a critical determinant of flux through the sulfur assimilatory pathway in plants.

Plant APS reductases (adenylyl-sulfate reductase; EC 1.8.99.2) catalyze the reduction of APS to sulfite and AMP (Fig. 7a). As with ATP sulfurylase, plants typically encode multiple isoforms of APS reductase. Initial cloning of the plant APS reductase genes revealed an N-terminal region sharing sequence similarity with microbial PAPS reductases and a C-terminal region encoding a glutathione-dependent glutaredoxin-like domain.

Later studies also identified an iron–sulfur cluster and suggest that this feature is structural and not necessarily catalytic. Similar work on the APS reductase from *Enteromorpha intestinalis*, a marine green alga, revealed a similar domain structure with one disulfide bond serving as a potential electron acceptor from glutathione and a second disulfide providing a catalytically critical structural feature. In the overall APS reductase reaction sequence, formation of an S-sulfo-cysteine is required for release of AMP and sulfite from APS (Fig. 7b).

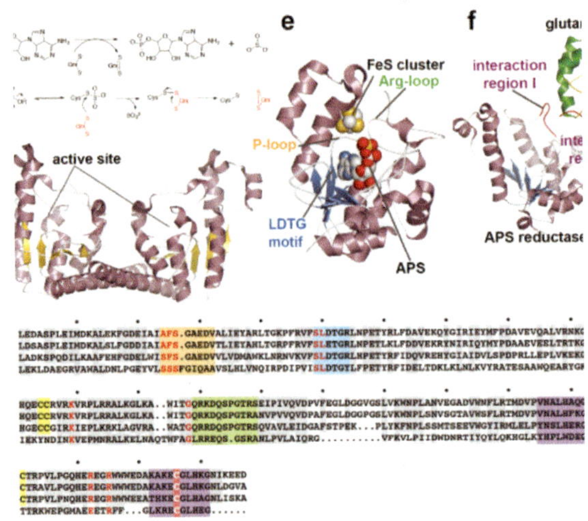

Fig. 7. Plant APS reductase.

To date, the three-dimensional structure of a canonical plant APS reductase remains to be reported; however, the C-terminal domain of Arabidopsis APS reductase 1, which shares homology with thioredoxin yet possesses glutaredoxin activity, has been crystallized. In addition, the protein crystal structure of an atypical APS reductase from the moss *Physcomitrella patens* provides molecular details that can be extrapolated to canonical APS reductases (Fig. 7c).

The moss APR-B protein was originally described as an APS reductase that lacked the iron–sulfur cluster found in homologs from other plants. The X-ray crystal structure of moss APR-B (Stevensen *et al.*, 2013) was the first for a plant sulfonucleotide reductase and showed the $\alpha/\beta/\alpha$-sandwich fold observed in the PAPS reductases from *E. coli* and *S. cerevisiae* and the APS reductase from *P. aeruginosa* (Fig. 7c).

Interestingly, a different oligomeric structure of the moss APR-B protein compared with the other sulfonucleotide reductases was observed. This alteration occurs because of an elongated central α-helix, which orients each monomer outward relative to the microbial sulfonucleotide reductase structures. Comparison of the moss APR-B structure with the other reported structures also revealed a highly conserved active site and a sulfonucleotide-binding site. Thus, key features are similar across the family of sulfonucleotide reductases from plants and microbes.

Sequence analysis of the Arabidopsis APS reductase compared with the *P. aeruginosa* APS reductase and the *E. coli* PAPS reductase shows a high level of amino acid sequence similarity in their catalytic domains and identifies features that provide insight on the biochemical function of the plant enzyme and its possible structure (Fig. 7d, e). For example, each enzyme retains critical motifs surrounding the active site, as follows: a phosphate binding- or P-loop, which provides interactions with the sulfonucleotide; the four amino acid 'LDGT'-motif, which is part of a β-strand that structures the active site with multiple van der Waals contacts; and an arginine-rich loop that encloses the charged sulfo-phospho groups of the bound nucleotide. Moreover, in the microbial and plant APS reductases containing an iron–sulfur (4Fe–4S) cluster, the four cysteine residues that form the cluster are invariant. Structure–function studies of bacterial sulfonucleotides indicate that the iron–sulfur cluster is a major contributor to the evolution of catalytic efficiency and substrate recognition in APS reductases; however, the molecular details of how the cluster directly contributes to catalysis are unclear.

During the catalytic cycle of APS reductase (Fig. 7b), an invariant cysteine residue, which is found in a flexible loop, must react with bound APS to form a covalent *S*-sulfo-cysteine intermediate, which leads to release of AMP. In the plant APS reductase, the C-terminal glutaredoxin-like domain must then resolve this intermediate to produce sulfite for further metabolism. Next, regeneration of the glutaredoxin-like domain requires glutathione. The X-ray crystal structure of the interaction complex between *E. coli* PAPS reductase and thioredoxin, as reported by researchers, provides a template for how the APS reductase and glutaredoxin-like domains of the plant enzyme may interact to complete the catalytic cycle (Fig. 7f).

The *E. coli* PAPS reductase–thioredoxin complex was generated using an inactive thioredoxin and revealed two regions of the reductase that provided protein–protein interactions with the thioredoxin (Fig. 7d, f). The resulting structure also contained the covalent S-sulfo-cysteine intermediate formed by the catalytic PAPS reductase cysteine, which is in the second interaction domain, bound to the thioredoxin active site. Both protein–protein interaction regions are well conserved in the plant APS reductases and may provide contacts with the C-terminal glutaredoxin-like domain during the catalytic cycle.

As previously noted, significant structural rearrangements probably accompany movement of the catalytic cysteine into the proximity of the mixed anhydride bond of APS bound in the active site to generate the S-sulfo-cysteine intermediate. Once formed, the loop containing the modified cysteine needs to next be positioned for presentation to the glutaredoxin-like domain to allow for sulfite release and subsequent glutathione-dependent regeneration of the glutaredoxin active site for additional rounds of catalysis. Presently, the details of how the reduction of APS to AMP and sulfite is orchestrated at the molecular level need to be established.

Summary

In combination with genomic, biochemical, and cellular studies, structural biology has contributed to understanding the molecular function of enzymes involved in plant sulfur metabolism. Multiple studies have established the chemistry underlying activation of sulfur and its subsequent incorporation into cysteine and glutathione, along with other diverse metabolites, and revealed the biochemical regulatory mechanisms for control of these core metabolic steps in plant.

CHAPTER

RESPIRATION IN PLANTS 12

Respiration is a chain of chemical reactions that enables all living entities to synthesize energy required to sustain.

It is a biochemical process wherein air moves between the external environment and the tissues and cells of the species. In respiration, inhalation of oxygen and exhalation of carbon dioxide gas takes place. As an entity acquires energy through oxidising nutrients and hence liberating wastes, it is referred to as a metabolic process.

Let us have a look at the respiration in plants notes provided here to know about the process of respiration, and the different types of respiration that occur in plants.

DO PLANTS BREATHE?

Yes, like animals and humans, plants also breathe.

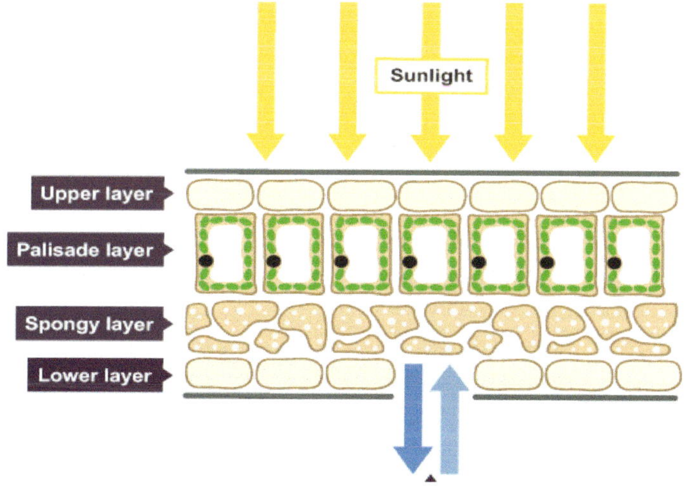

Fig. 1. A cross-section through a leaf showing its main parts

Plants do require oxygen to respire, the process in return gives out carbon dioxide. Unlike humans and animals, plants do not possess any specialized structures for exchange of gases, however, they do possess stomata (found in leaves) and lenticels (found in stems) actively involved in the gaseous exchange. Leaves, stems and plant roots respire at a low pace compared to humans and animals.

Breathing is different from respiration. Both animals and humans breathe, which is a step involved in respiration. Plants take part in respiration all through their life as the plant cell needs the energy to survive, however, plants breathe differently, through a process known as Cellular respiration.

In this process of cellular respiration, plants generate glucose molecules through photosynthesis by capturing energy from sunlight and converting it into glucose. Several live experiments demonstrate the breathing of plants. All plants respire to provide energy for their cells to be active or alive.

Plants are adapted to carry out the process of respiration and photosynthesis. Gases such as oxygen and carbon dioxide can travel through tiny holes underneath the leaf called stomata. Carbon dioxide diffuses into the cells found within the leaf to help carry out the process of photosynthesis, whilst oxygen is released from these cells as a product of photosynthesis.

Respiration is essential for growth and maintenance of all plant tissues, and plays an important role in the carbon balance of individual cells, whole-plants and ecosystems, as well as in the global carbon cycle. Through the processes of respiration, solar energy conserved during photosynthesis and stored as chemical energy in organic molecules is released in a regulated manner for the production of ATP, the universal currency of biological energy transformations, and reducing power (e.g. NADH and NADPH). A quantitatively important by-product of respiration is CO_2 and, therefore, plant and ecosystem respiration play a major role in the global carbon cycle.

Terrestrial ecosystems exchange about 120 Gt of carbon per year with the atmosphere, through the processes of photosynthesis and respiration. Roughly, half of the CO_2 assimilated annually through photosynthesis is released back to the atmosphere by plant respiration. Because terrestrial biosphere-atmosphere fluxes of CO_2 far outweigh anthropogenic inputs of CO_2 to the atmosphere, a small change in terrestrial respiration could have a significant impact on the annual increment in atmospheric [CO_2]. A large body of literature has indicated that plant respiration is reduced in plants grown at high [CO_2]. For example, it was estimated that the observed 15–20% reduction in plant tissue respiration caused by doubling current atmospheric [CO_2] could increase the net sink capacity of global ecosystems by 3·4 Gt of carbon per year, and thus offset an equivalent amount of carbon from anthropogenic CO_2 emissions. Therefore, not only are gross changes in respiration important for large-scale carbon balance issues, changes in specific rates of respiration can have significant impact on basic plant biology such as growth, biomass partitioning or nutrient uptake.

Scaling the effects of an increase in atmospheric [CO_2] on plant respiration at the biochemical level to the whole ecosystem is difficult for at least two important reasons: (1) the fine and coarse control points of respiratory pathways in tissues and whole plants are not well known; and (2) it is unclear how respiration rates actively or passively adjust to the effects that elevated [CO_2] may have both upstream (e.g. on substrate availability) and downstream (e.g. on energy demand) of the carbon oxidation pathways. In addition, accurate measurements of respiratory rates as CO_2 evolution have been proven difficult with current techniques, presenting yet another limitation to scaling up process-based measurements of respiration rates to organism, ecosystem or global levels. In this document, we re-evaluate the theory on respiration responses to elevated [CO_2] in view of recent studies that have provided new insights on the effects of a short- and long-term change in atmospheric [CO_2] on plant respiration.

COMPARING PHOTOSYNTHESIS AND RESPIRATION IN PLANTS

Photosynthesis:

Carbon dioxide + water -> glucose + oxygen

Respiration:

Oxygen + glucose -> water + carbon dioxide

The gas carbon dioxide is needed for photosynthesis to take place in plants and is also produced when plants respire. Oxygen is a product of photosynthesis - it is needed as a reactant during aerobic respiration within a plant. This means that the reactants and products of photosynthesis and aerobic respiration are opposites.

THE PROCESS OF RESPIRATION IN PLANTS

During respiration, in different plant parts, significantly less exchange of gas takes place. Hence, each part nourishes and fulfils its own energy requirements.

Consequently, leaves, stems and roots of plants separately exchange gases. Leaves possess stomata – tiny pores, for gaseous exchange. The oxygen consumed via stomata is used up by cells in the leaves to disintegrate glucose into water and carbon dioxide (Fig. 2).

DIRECT AND INDIRECT EFFECTS OF [CO_2] ON PLANT-SPECIFIC RESPIRATION RATES

Amthor (1991) described two different effects of [CO_2] on plant respiration rates that can be distinguished experimentally: a direct effect, in which the rate of respiration of mitochondria or tissues could be rapidly and reversibly reduced following a rapid increase in [CO_2], and an indirect effect in which high [CO_2] changes the rate of respiration in plant tissues compared with the rate seen for those plants grown in normal ambient [CO_2] when both are tested at a common background [CO_2]. Most of the studies investigating the effects of a change in [CO_2] on respiration rates have described magnitude (from non-significant to 60%) and direction (either stimulation, inhibition or no effect) of these two effects, with little progress on their underlying mechanisms. Confounding measurement artefacts are an added complication for establishing the presence and magnitude of direct and indirect effects of [CO_2] on respiration.

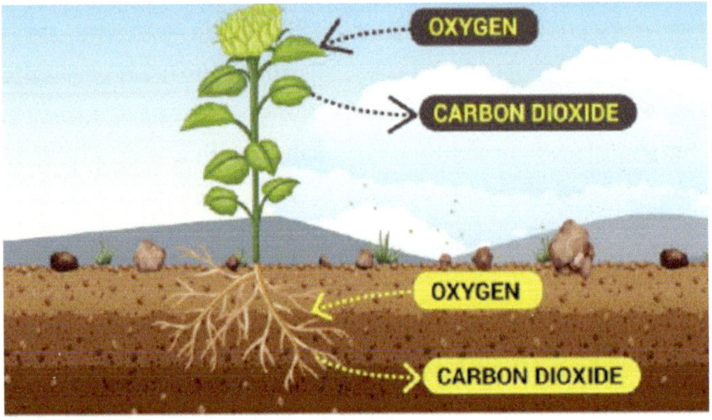

Fig. 2. Respiration in Plants

Direct effects

Previously, a rapid short-term doubling of current atmospheric CO_2 levels was reported to inhibit respiration of mitochondria and plant tissues by 15–20%. Although direct effects of [CO_2] on respiratory enzymes have been reported, the magnitude of the direct inhibition of intact tissue respiration by elevated [CO_2] has now been shown to be explained by measurement artefacts, diminishing the impact that such reductions in respiration rate may have on plant growth and the global carbon cycle.

Direct effects on enzymes

Doubling current levels in [CO_2] (here we review the effects of only two- or three-fold increases in atmospheric CO_2) have been shown to inhibit the oxygen uptake of isolated mitochondria and the activity of mitochondrial enzymes under some conditions. Twice current atmospheric [CO_2] reduces the activity of cytochrome *c* oxidase and succinate dehydrogenase in isolated mitochondria from cotyledon and roots of *Glycine max*. Mitochondrial oxygen uptake may show either no response or up to 15% inhibition by a rapid increase in [CO_2], depending on the electron donor used in the assay, indicating that the response of mitochondrial enzymes to high [CO_2] depends on the cell's metabolism. In scaling the direct effects of high [CO_2]

on respiratory enzymes to intact tissues two factors need to be considered: (1) cytochrome oxidase exerts up to 50% of the total respiratory control at the level of the mitochondrial electron transport chain; and (2) the competitive nature of the cytochrome and alternative pathways of mitochondrial respiration. The activity of the alternative pathway of respiration could increase upon doubling of the [CO_2], masking a direct CO_2 inhibition of the cytochrome pathway.

The oxygen isotope technique allows for the distinction of direct effects of [CO_2] on oxygen uptake activity by the cytochrome and the alternative oxidases. Figure 1 shows that mitochondrial electron transport activity is affected when CO_2 (a mild inhibitor) restricts the normal electron flow through one of the pathways (Cyt pathway – vcyt). Under conditions of low alternative pathway activity, inhibition of the cytochrome pathway (18%) by doubling the ambient [CO_2] is compensated for by a similar increase in alternative pathway activity (valt), resulting in no significant reduction in overall oxygen uptake of the isolated mitochondria. These results show that enhanced activity of the alternative pathway upon an increase in [CO_2] provides a mechanism by which the direct [CO_2] inhibition of the cytochrome oxidase may not be seen in the total oxygen uptake under conditions of high adenylate control (low ADP). Researchers argued that direct effects of CO_2 on respiration exceeding >10% inhibition are likely due to factors other than inhibition of mitochondrial enzymes. Due to confounding measurement artefacts, rates of tissue respiration have been reported to decrease >10% when the atmospheric [CO_2] doubles.

Direct effects on intact tissues

Recent evidence indicates that, in studying direct effects of CO_2 on respiratory CO_2 efflux of intact tissues, investigators should first be concerned about measurement artefacts. Researchers showed that gas CO_2 exchange measurement errors could augment and explain the magnitude of the direct effect. New studies made in this context have shown that respiration rates are little or not at all inhibited by a doubling of atmospheric [CO_2]. Indeed, researchers confirmed that measurement artefacts due to leakage in CO_2-exchange systems could be as large as the previously reported direct inhibitory effects.

They also found that leaks through the aerial spaces of homobaric leaves showed a significant apparent inhibition of CO_2 efflux that was not due to an inhibition of respiration by elevated [CO_2]. Therefore, considerations of the impact direct effects of [CO_2] on plant respiration rates may have on the global carbon cycle were overstated. Corrections for instrument leaks can be applied in most cases. Applying some corrections for gas exchange leaks, researchers reported a significant reduction in respiration after a rapid increase in [CO_2].

Moreover, simultaneously switching the air isotopic composition from $^{12}CO_2$ to pure $^{13}CO_2$ when the [CO_2] was changed, showed that the efflux of $^{12}CO_2$ (which represents net respiratory CO_2 evolution originating from respiratory substrates minus the CO_2 being re-fixed by carboxylases) was reduced in leaves of C_3 plants, but not C_4, when the [CO_2] increased. These recent results illustrate that not all reports of direct inhibition of respiration by elevated [CO_2] have yet been reconciled with each other, but they do not show a clear proof that the direct effect.

An alternate explanation for the apparent reduction in respiration rates upon an increase in [CO_2] is an increase in dark CO_2 fixation catalysed by phosphoenolpyruvate carboxylase (PEPC), resulting in an apparent reduction of CO_2 efflux. However, in *Rumex* and *Glycine* leaves, researchers found an effect of elevated [CO_2] on CO_2 efflux in only a minority of experiments, with no effect of elevated [CO_2] on the O_2 uptake of the same leaves, indicating at most a small effect of elevated [CO_2] on PEPC activity or respiration rates.

Similar results were found for a variety of species, involving 600 measurements, in which [CO_2] increases did not alter the CO_2 and O_2 leaf exchanges in the dark.

It seems that direct effects of [CO_2] on mitochondrial enzymes may have no consequence on the specific respiratory rate of whole tissues. Because mitochondrial respiratory enzymes are generally 'in excess' to the levels required to support normal tissue respiratory activity under most growth conditions, a minor inhibition of enzymatic activity by [CO_2] will not translate to the overall tissue respiration rate.

A small increase in enzyme levels in plants exposed to elevated [CO_2] could be enough to compensate for the direct effects of CO_2 on mitochondrial enzyme activity, as it has been seen in leaves of some plants grown at high [CO_2].

Another factor to consider is that increases in alternative pathway activity upon inhibition of cytochrome oxidase by a change in [CO_2] will result in unaltered dark respiratory rates. However, an increase in the non-phosphorylating activity of the alternative pathway reduces the efficiency with which energy from oxidized substrates supports growth and maintenance of plants. Some studies have reported that the growth of plants exposed to elevated [CO_2] only during the night-time is altered.

Such altered growth patterns have been attributed, in part, to effects of [CO_2] on plant respiration, including increased activity of the non-phosphorylating alternative pathway. These CO_2 effects are considered next.

Overall, our categorical approach attempts to provide a field-independent frame of reference for discussing, modelling, interpreting and altering respiration rates in plants. Where necessary, we distinguish between rates of CO_2 evolution and rates of O_2 consumption (e.g. during day respiration), but otherwise we treat patterns arising from both measurements as revealing highly related properties of plant respiration.

PRINCIPLE 1 – PLANT RESPIRATION PERFORMS THREE DISTINCT FUNCTIONS

It is well understood that in aerobic organisms respiration functions to provide the cell with chemical energy in the form of ATP. However, in plant cells, respiration fulfils two other additional purposes which are greatly enhanced in comparison with most animal cells: production of C precursors and redox balancing. The production of biosynthetic precursors (often referred to as C skeletons) is enhanced because plants produce most of their organic molecules (sugars, amino acids, lipids, etc.) from inorganic materials (CO_2, NO_3^-, etc.). Many biosynthetic pathways begin by branching off from primary C metabolism, often at the location of a metabolite within glycolysis or the citric acid cycle (TCA cycle). Therefore, parts of the same respiratory pathways that oxidize C for energy generation must also meet the demands for C skeletons imposed by biosynthesis.

The balance of cellular redox status has received less attention, but its importance and influence on respiration are the focuses of this section. Central measures of cellular redox status are the ratios of $NADH/NAD^+$ and $NADPH/NADP^+$ as well as the production of reactive oxygen species (ROS). The mitochondrial electron transport chain (mETC) is a unique and essential component of cellular redox balancing because it can act as a flexible sink to reoxidize NAD(P)H and other forms of reductant. As explained later, this ability is a great advantage to plant cells because it allows them to prioritize biosynthesis over energetic efficiency.

These three respiratory functions (ATP production, C skeleton production and redox balancing) can overlap simultaneously, yet they are distinct because they require different modes of operation (flux modes) of the respiratory pathways. The need to successfully coordinate these three functions is a recurring explanation for the characteristic biochemical organization and regulation of plant respiration, as discussed in Principles 2 and 4.

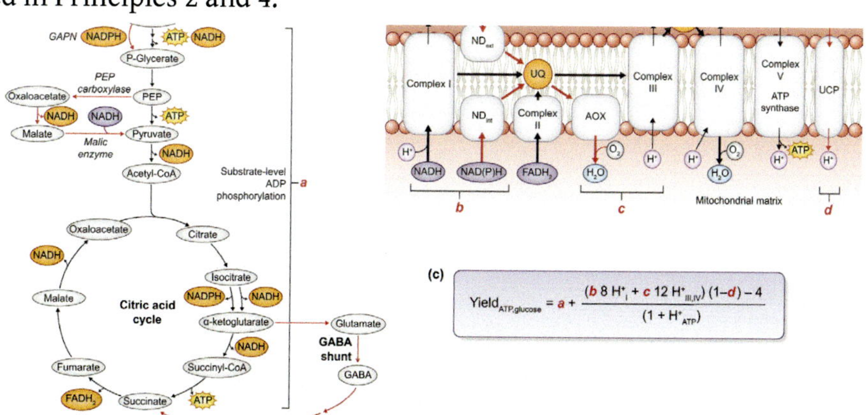

Fig. 3 The yield of chemical energy (ATP) captured throughout plant respiration is variable (a) It is possible to circumvent every ATP-generating metabolic reaction within glycolysis and the TCA cycle.

Note that the production of reductant (e.g. NADH or NADPH) cannot be circumvented, as it is needed to balance electrons in the chemical reactions; it is the coupling of energy-releasing (exergonic) reactions to ATP synthesis that is optional. (b) The mitochondrial electron transport chain (mETC) displays highly variable ATP yield. ND_{ext} and ND_{int} represent alternative nonproton pumping NAD(P)H dehydrogenases on the outer and inner surfaces of the mitochondrial inner membrane, respectively, which bypass complex I. Alternative oxidase (AOX) is a nonenergy-conserving terminal oxidase which bypasses proton pumping complexes III and IV. Members of the mitochondrial uncoupling protein (UCP) family may act to dissipate the electrochemical gradient as in animals; however, their physiological role in plants is unproven and may be different. (c) An equation depicting the complexity in accounting for respiration efficiency in plants. The four variables needed to solve the equation are connected by arrows to the corresponding metabolic components: a, ATP yield from glycolysis and the TCA cycle; b, proportional flux through complex I vs alternate dehydrogenases; c, proportional flux through the cytochrome vs the AOX pathway; d, mitochondrial inner membrane proton leakiness ratio, including UCP activity (proton leaked per proton pumped). H^+_{I}, and $H^+_{III\ IV}$ are the number of protons pumped across the membrane at complex I and III + IV, typically 4 and 6, respectively. H^+_{ATP} is the number of protons re-entering the matrix per ATP synthesized at ATP synthase, typically 3.

Plant respiratory redox balancing favours biosynthesis

Several mechanisms, such as alternative oxidase (AOX), alternative dehydrogenases and the uncoupling protein, allow plants to uncouple respiratory NAD(P)H oxidation from ATP synthesis (Fig. 3), thus allowing redox status to be regulated independently of ATP synthesis. These mechanisms constitute a key advantage in terms of metabolic flexibility under situations where supply of reductant is greater than demand for its use. For example, an oversupply of reductant is likely to occur during active photosynthesis/photorespiration when the plastid exports excess reducing equivalents. Occasional reductant oversupply is also possible during the rapid creation of C skeletons in support of biosynthesis or organic acid secretion; that is, overall biosynthesis can be a net producer of reductant. To maintain redox balance, cells have the option of slowing reductant production to match the rate of demand, or uncoupling reductant oxidation from its use in demand processes (e.g. ATP synthesis), thus releasing excess reducing power as heat. Plants would be greatly disadvantaged by fundamentally restraining photosynthetic metabolism or the respiratory production of C skeletons, essentially limiting growth. Rather, the occurrence of multiple uncoupling mechanisms suggests that plants have, in general, adopted a strategy of adjusting the efficiency of the mETC as needed so that biosynthesis can be maximized when growth conditions are optimal. Equally important is that high NAD(P)H/NAD(P)+ ratios increase the redox poise of electron carriers in the mETC and the plastid ETC (during photosynthesis), which, if left unchecked, increases the escape of single electrons to form reactive oxygen species (ROS), leading to cellular damage. Therefore, redox balancing is also required to maintain control over ROS concentrations.

Mitochondrial redox balancing influences daytime respiration rates

An example of the wider role of mitochondrial redox balancing in plants is that the mETC appears to support optimal amounts of photosynthesis by enabling high rates of both photorespiration and dissipation of excess photosynthetically derived reductant. To accomplish this role, leaf mitochondria express nonenergy-conserving components of the mETC in the light. Furthermore, inhibition of mitochondrial electron transport by chemical inhibitors or the genetic knockdown of the mitochondrial uncoupling protein limit photosynthesis under normal conditions; while loss of AOX activity impairs photosynthesis, particularly under drought conditions.

So what effect does mitochondrial redox balancing have on leaf respiration rates during the day? One key property of removing excess reductant produced by photosynthesis (not photorespiration) is that it theoretically involves a respiratory mode whereby mitochondrial O2 consumption becomes disconnected from respiratory CO2 production; essentially, mitochondria consume O2 but do not produce CO2. Although the true situation is complicated by inputs from photorespiration and other processes, redox balancing must nevertheless influence our interpretation of daytime O2 vs CO2-based respiration rates as they cannot be expected to be strictly related.

A second crucial influence on daytime leaf respiration is the ability of the plastid to export photosynthetically derived ATP to other cellular compartments. In contrast to mitochondria, direct export of ATP from the plastid is thought to be minimal. However, plastidial metabolite shuttles operate to indirectly export ATP; for example, the export of triose-phosphate and its conversion to 3-phosphogycerate and ATP by cytosolic glycolysis, followed by import of 3-phosphogycerate back into the plastid.

Although the situation remains poorly quantified, respiratory ATP generation is probably needed during photosynthesis to support ATP-consuming processes occurring outside the plastid (e.g. protein synthesis and sucrose synthesis/export). However, the observed activation of mETC uncoupling pathways in the light indicates that reductant supplied by photosynthesis and photorespiration to the mitochondria can, depending on conditions, replace the need for reductant formation via the 'classical' respiratory pathway involving the TCA cycle and CO2 release. Consistent with this idea, the measurement of the daytime metabolic fluxes using labelled C intermediates has revealed that the cyclic operation of the TCA cycle is largely absent in photosynthesizing plant cells.

Direct measurements of daytime respiratory fluxes remain technically difficult, but redox balancing and its associated ATP production provide a key part of the explanation for why respiratory CO_2 release, but not necessarily O_2 consumption, is often observed to be lower (and should be expected to be lower) under photosynthetic than under nonphotosynthetic conditions. Other factors related to decreased daytime respiratory CO_2 release include light-induced post-translational inhibition of key respiratory enzymes, including the pyruvate dehydrogenase complex and plastid glucose-6-phosphate dehydrogenase. Full elucidation of respiratory integration with photosynthesis remains an important research area, and excellent reviews specifically discussing many aspects of daytime respiration are available.

mETC uncoupling increases during many stresses

Redox balancing is also involved in plant responses to stress. Several stresses restrict the flux of electrons through the mETC, causing upstream mETC components to become increasingly reduced (e.g. ubiquinol : ubiquinone), consequently increasing ROS formation. Drought, flooding (hypoxia), cold, high salinity and metal toxicity can all lead to increased ROS production from the mETC (i.e. oxidative stress). Many studies have observed mETC uncoupling mechanisms, including AOX, alternative NAD(P)H dehydrogenases and uncoupling proteins, becoming activated during stresses, presumably to counteract ROS production, and details of the retrograde signalling cascades that facilitate these responses have been reported.

One might therefore expect higher respiratory rates during these stresses to compensate for the lowered efficiency of respiratory ATP production. However, any effects of lowered efficiency are generally masked by the large and unresolved changes in respiratory supply and demand conditions that accompany the onset of stress and also affect respiratory rate. Furthermore, the use of mutant lines to establish a link between respiratory efficiency and rate under stress has not been decisive.

Knockout of AOX should enhance respiratory efficiency, but Arabidopsis lines lacking the major AOX isoform do not show lower respiration rates, whereas knockdown of Complex I lowers respiratory efficiency but does not change respiratory rate. To better understand the response of respiration rate to stress, it will be necessary to distinguish the effects of changes in respiratory efficiency from changes to respiratory supply and demand conditions.

PRINCIPLE 2 – METABOLIC PATHWAY FLEXIBILITY UNDERLIES PLANT RESPIRATORY PERFORMANCE

One of the most characteristic features of respiration in plant cells, compared with other organisms, is the increased presence of alternative metabolic routes whereby C or electrons pass from point A to point B via different enzymatic pathways. Alternate routes exist among each of the three classical respiratory pathways: glycolysis, the TCA cycle and the mETC (Fig. **3**). For example, within eukaryotes, nonphosphorylating glyceraldehyde-3-phosphate dehydrogenase, phosphoenolpyruvate (PEP) carboxylase and AOX are each distinctive plant enzymes, the existence of which can create bypasses of classical respiratory enzymes. Furthermore, many respiratory reactions are duplicated in multiple plant subcellular compartments. For example, there are cytosolic and plastidial isoforms of most glycolytic enzymes, as well as mitochondrial and extra-mitochondrial isoforms of most TCA cycle enzymes. Previous reviews provide a detailed account of the existence and organization of alternative respiratory routes, and they will not be listed in full here. In brief, it is proposed that these instances of biochemical flexibility allow plant cells and tissues to optimize the energetic and biosynthetic performance of respiration (e.g. rate and stoichiometry) under a range of different and changing conditions, including variations in photosynthesis, and thus to better survive specific environments.

One important consequence of this biochemical flexibility is that the respiratory network is robust and able to tolerate genetic deletion or chemical inhibition of most but not all enzymes with seemingly minor phenotypic effects, presumably by circumventing the impairment. A useful analogy may be a river network with multiple channels, where damming one channel has little effect on overall water flow as other channels increase in flow rates. At a practical level, this robustness has made it difficult to understand the control of respiratory fluxes in order to construct quantitative models or to predict how genetic manipulations will alter respiratory behaviour.

Plant respiration has variable ATP production efficiency

So, how might metabolic flexibility contribute to variation in plant respiration rates seen in nature? The most obvious explanation is that the alternate metabolic routes of respiration often differ in the production of ATP and thus directly affect the energetic efficiency of respiration. In fact, every ATP-producing reaction in respiration can be bypassed (Fig. 3).

The result is that the ATP yield of respiration (ATP produced per C substrate consumed, or the ATP/O or ATP/CO_2 ratio) can and does vary widely, depending on the engagement of alternate routes and the permeability (leakiness) of the mitochondrial inner membrane to protons (Fig. 3c). In other words, the amount of respiration required to meet cellular ATP demands (e.g. maintenance respiration) is variable. This variation in ATP production, resulting from many independent mechanisms, greatly complicates efforts to construct general mechanistic models of plant respiration. Altogether, the prevalence of nonenergy-conserving alternate routes, many of which are absent in animal respiration, indicate that plant respiration is not organized to necessarily maximize ATP production efficiency.

The degree to which these alternate routes and enzymes throughout respiration are engaged in vivo remains poorly understood. In part, this is because engagement of ATP-bypassing reactions cannot be observed from intact tissues with standard gas exchange measurements. Competing electron fluxes through the mETC have been measured in select tissues using 18O/16O isotope discrimination methodology. The engagement of the nonenergy-conserving AOX pathway in typical nonstressed tissues ranges from 10% to 50% of total respiration, yet its function under nonstress conditions remains an open question. AOX engagement has been observed to vary greatly throughout development and with environmental conditions, including a number of stresses. The engagement of AOX while under stress is associated with controlling amounts of ROS production, but it is also hypothesized to support increased formation of C skeletons as compatible solutes for osmoregulation or as excreted organic acids for phosphate scavenging

Other measurements are possible which distinguish certain competing C fluxes through the respiratory pathways by using 13C-labelled substrates for metabolic flux analysis, but this has only been achieved in a few tissue systems. For example, the production of pyruvate from PEP can occur via multiple routes that differ in ATP production: two such routes are the classic glycolytic pyruvate kinase reaction, which produces ATP from ADP, and an alternative route via PEP carboxylase, then malate dehydrogenase, then malic enzyme which does not produce ATP (Fig. 3a).

The relative occurrence of the malic enzyme route of pyruvate production in mitochondria varies greatly between species and tissues under nonstress conditions (e.g. 9% in maize root tips, 40% in oilseed rape (Brassica napus) embryos), but no satisfactory explanation for this variation exists. By contrast, under certain stresses, alternative routes of C oxidation would appear to have a benefit. For example, during phosphate starvation in many tissues nonATP-producing (i.e. nonphosphate-requiring) glycolytic bypasses are upregulated, including the malic enzyme route of pyruvate production, presumably to circumvent crucial reactions that are phosphate-limited.

Altogether, measurements that resolve respiratory fluxes complement transcriptomic, proteomic, enzymatic and genetic data, which can indicate situations when nonenergy-conserving bypass enzymes are upregulated. However, despite some key advances, much more work needs to be done to understand when and why different alternative respiratory routes are engaged in different plant tissues so that efficiencies of ATP production can be better quantified, and to determine how to make such measures at higher biological scales, such as plants or managed plots of ecological sites.

Variation in C substrate use for plant respiration

Another aspect of respiratory flexibility is the use of alternative substrates besides carbohydrates, including organic acids, amino acids and fatty acids. Carbohydrates are the dominant respiratory C substrate in plants because: they are the main nutrient transported to heterotrophic tissues (usually sucrose); they are the primary product of photosynthesis (triose-phosphates); and they are the main storage form of C (starch, sucrose, hexoses, fructans).

However, along with carbohydrates, plant leaves also maintain robust diurnal cycles of organic acid accumulation during the day and draw down at night (e.g. malate, fumarate), which contributes to both the ATP and C skeleton production functions of respiration. By contrast, the substantial use of protein or fatty acids as respiratory substrates appears to be limited to conditions of C starvation (e.g. extended darkness), senescence and seed germination.

PRINCIPLE 3 – SUPPLY AND DEMAND INTERACT OVER TIME TO SET PLANT RESPIRATION RATE

A long-standing question in plant respiration research is whether respiration rate is limited by substrate supply or product demand. In the 'demand-centric' view, respiration rate is pulled by the activity of processes that consume respiratory products: ATP, NADPH, and C skeletons. The rate of ATP turnover to regenerate ADP has been a particular focus of demand limitation and is often referred to as adenylate restriction. Indeed, respiration rates often increase in experiments where ADP or mETC uncouplers are added to mimic increased ATP turnover. In the 'supply-centric' view, respiration rate is pushed by carbohydrate supply. Experimental evidence also supports supply limitation because respiration rates in various plant tissues correlate with carbohydrate concentration and can be stimulated by increasing the carbohydrate supply.

At the ecosystem and whole-plant scales, respiratory activity also appears to be correlated with both substrate supply and product demand. At the ecosystem scale, respiration increases linearly with both photosynthetic C gain and biomass production. At the whole-plant scale, the ratio of respiration to photosynthesis is observed to be remarkably constant in species from different habitats across moderate changes in temperature, and in a range of species under differing growth stages and temperatures. However, variation in the respiration-to-photosynthesis ratio has also been observed (e.g. between ecosystems) with higher values in boreal (0.7) and tropical (0.65) ecosystems than in temperate (c. 0.5) systems, and among several species of crops, grasses and trees, with values between 0.35 and 0.8. The patterns and causes of the variation in this ratio remain unclear.

The ratio of respiration to photosynthesis is also a factor influencing relative growth rates, as it decreases with increasing relative growth rate in some crops and wild plant species. Understanding the respiratory supply–demand relationship in whole plants therefore has important implications for agricultural as well as C budgets of natural ecosystems and future increases in atmospheric CO2.

Ultimately, the question of supply vs demand controlling respiration rate of intact tissues is problematic because supply and demand are not independent forces but rather are highly coordinated. There is ample evidence that multiple carbohydrate-sensing mechanisms exist which regulate energy- and C-consuming biosynthetic pathways, involving, for example, trehalose-6-phosphate, SnRK1 kinases and bZIP family transcription factors, as reviewed by researchers. There is also evidence of regulation in the opposite direction, that sink limitations in C use can lower photosynthetic substrate supply from the source.

Thus, as illustrated in Fig. 3, signals coordinating supply and demand operate alongside, and in addition to, established feedback and feedforward regulatory mechanisms within the respiratory enzyme network. In other words, increases in substrate levels need not affect respiratory enzymes/pathways directly but can nonetheless stimulate respiration indirectly. For example, in wheat, pea and spinach grown in a controlled environment, leaf night-time respiration correlates strongly with carbohydrate concentration amassed during the previous photosynthetic period, suggesting that respiratory substrate supply controls respiration.

However, the size of the carbohydrate reserve also regulates the rate of sucrose export during the course of the night, which itself is a major ATP demand, causing supply and demand to be linked. Each tissue will experience its own set of linked supply and demand conditions which will determine its respiration rate.

Importantly, the various mechanisms that coordinate supply and demand operate at different speeds and this can influence experimental observations. The most rapid regulatory responses, shown as metabolic regulation in Fig. 4, do not require transcription and translation of new proteins and involve regulation of enzyme activities by substrate–product ratios, allosteric effector concentrations, and post-translational modifications. At the level of metabolic regulation, the entire respiratory pathway of plants is mostly feedback-regulated, as opposed to feedforward-regulated. The accumulation of certain products of glycolysis (PEP, malate), the TCA cycle (NADH : NAD ratio) and the mETC (ATP : ADP ratio) all feedback-inhibit their respective pathways.

The result is that changes in respiratory fluxes over the very short term are largely driven by fluctuations in the use or turnover of these products, referred to here as demand. One important exception to the pattern of feedback inhibition is AOX, which is activated by upstream metabolites (e.g. pyruvate, NAD(P)H). However, in this case the feedforward regulation of the mETC serves to make less, not more, ATP product and is probably important to balance cellular redox status.

By relying on rapid feedback regulation and redox balancing, plants are able to effectively coordinate the separate respiratory demands for ATP, C skeletons and NAD(P)H, which all draw from common substrate pools. A well-described example of plant cell respiratory metabolic regulation in action comes from homogenous cultures of N-starved unicellular green algae grown under controlled conditions. Upon N resupply, the demand for energy and C skeletons to assimilate the N quickly depletes key feedback metabolites and thus, within seconds, stimulates the respiration of stored carbohydrate to fuel rapid N assimilation.

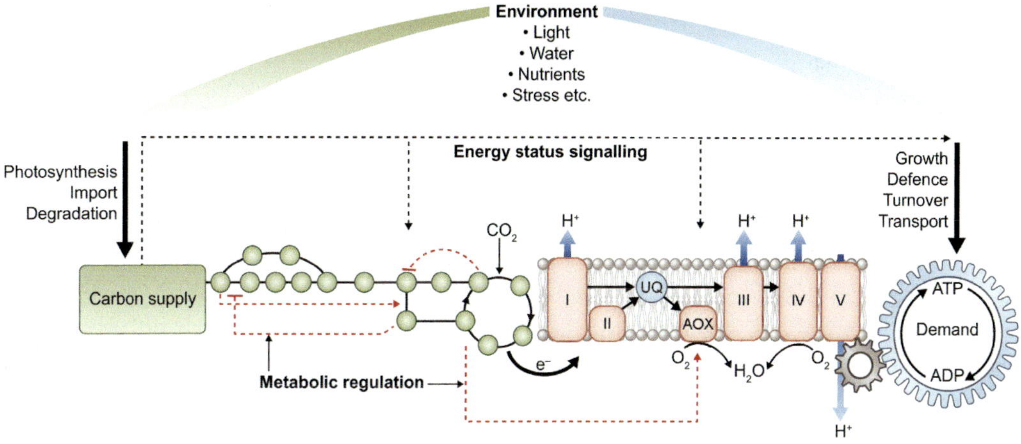

Fig. 4 The coordination of respiratory supply and demand is achieved at multiple levels. Metabolic regulation of respiration (red arrows) is rapid and involves post-translational mechanisms. Energy status signalling is a general term to describe signal transduction pathways originating from carbon supply which coordinate supply and demand through gene regulatory networks and therefore tend to be much slower than metabolic regulation. However, examples are emerging of respiratory enzymes whose regulatory properties are directly targeted by energy status signalling (e.g. phosphoenolpyruvate carboxylase

Over longer periods of time, plants will respond via changes in gene expression to a change in nutritional environment (shown as energy status signalling in Fig. 4). Increased resource availability will gradually induce increased use of those resources and resource limitation will signal an inhibition of energy-demanding pathways.

For example, following a change in carbohydrate supply via a step change in light intensities, it appears to take several diurnal cycles for a plant's gene expression adjustments to take full effect on metabolism. In another example, soybeans experiencing increased carbohydrate availability following growth under elevated CO2 displayed large changes in transcript abundance which were linked to an increase in respiration rate. In the opposite case, during the onset of C starvation, respiration rates in excised maize root tips gradually decrease but can be rapidly recovered to prestarvation levels by an mETC uncoupler.

This result demonstrated that, despite low carbohydrate availability in these roots, respiration was still demand-limited, and therefore the lowered respiration rate during starvation was attributable to downregulation of ATP turnover and other demands. The conclusion to be drawn here is that over longer periods of time carbohydrate supply will ultimately set the prevailing respiration rate because of the influence of carbohydrate concentrations on metabolic processes like growth which are supported by respiration.

Stresses and environmental conditions (e.g. less light) that lower photosynthesis lead to C depletion and downregulated respiratory demand, which persists long-term. For example, prolonged heat and drought events lowered ecosystem respiration in Europe in 2003 and this was attributable to decreased photosynthesis restraining growth and limiting C supply.

N content influences both supply and demand

In light of these observations, one can consider the empirical relationship between respiration and N that is often used to estimate respiration rates. N concentrations in plant tissues often display a positive linear correlation with dark respiration rate, at least when assessed on a tissue mass basis. The mechanism behind this relationship is probably multifaceted as a result of the scaling of N with enzymatic activities (i.e. metabolism) in all organs that influence both supply and demand. In leaves, a large proportion of leaf N is associated with the photosynthetic machinery and therefore C supply; for instance, the maximum carboxylation capacity of Rubisco (i.e. Vcmax) often correlates positively with leaf N content. Other prominent energy-demanding metabolic activities also scale with N because much of a cell's N is found in proteins/ enzymes.

For example, protein turnover was quantified and related to respiration rate and ATP yield in a seminal paper. Furthermore, N supply is related to respiratory demands for reductant, ATP and C skeletons needed for N assimilation into amino acids. In roots, there is a clear increase in respiratory activity upon commencement of root N assimilation because respiration must meet all three demands, and N2 fixers may have elevated respiration rates. The situation in leaves is more complicated because: the reductant and ATP needed for N assimilation are predominantly supplied by chloroplasts in the light; and the process of creating C skeletons for N assimilation is separated over a diurnal cycle and not necessarily simultaneous with the assimilation of N. It is thus unclear how daytime leaf N-assimilation rates contribute to leaf respiration rates when measurements are taken in the dark.

Plant size and composition influence respiration rate

Plant size is another factor determining respiratory activity that may help to explain the relationship among substrate supply, respiratory activity, and product demand at whole-plant to ecosystem scales. Respiration rates of individual plants scale with plant size, and the pattern of and explanation behind the scaling have attracted much attention and remain hotly debated on theoretical and empirical grounds. Observations show that whole-plant respiration rate scales with plant mass (R μ massa) with the value of the scaling exponent a mostly falling between 0.75 and 1.

The scaling exponent is near 1 in small plants and decreases to 0.75 in larger plants, but it varies with plant species, structure, and metabolically active cells, among other factors. Various explanations of the patterns have been proposed, including the decreasing proportion of metabolically active tissue or considerations of minimizing transport resistance through vasculature network while maximizing resource uptake.

The explanation based on vasculature network has been linked to plant trait correlation, relative growth rate and forest structure to indicate that, once plant size is accounted for, the ratio of respiration rate to photosynthesis at the ecosystem level is conserved across ecosystems with varying growing season temperature and differing species composition. This, together with observations that the ratio of respiration rate to net photosynthesis is conserved across moderate temperature changes at both leaf and whole-plant levels, suggests that the principle of coupled supply and demand sets respiration rate at whole-plant and ecosystem levels across plant communities.

Coordination of supply and demand may play a role across time as plants age. An increase in plant size also increases the proportion of nonphotosynthetic tissue within total plant mass. It was thought that respiratory costs required to maintain existing tissue would consume an ever larger fraction of photosynthetic C supply, leaving less for biomass production as trees grow older and larger.

However, the supply of respiratory substrates and the demand for products appear coordinated even here, as photosynthetic C supply, respiratory gas flux and biomass production all decline as trees grow larger. This correlation of respiratory supply, activity and product demand probably has complex mechanisms that include not only biochemical controls discussed earlier but also changes in transport costs of water and nutrients across ever greater distances.

PRINCIPLE 4 – PLANT RESPIRATORY ACCLIMATION INVOLVES ADJUSTMENTS IN ENZYME CAPACITIES

In addition to the supply and demand relationship, the total amount (capacity) of one or more respiratory enzymes is another factor that may limit respiration rate. Variations in the capacity of respiratory enzymes will influence the maximum attainable flux through the respiratory apparatus. Also, because enzyme activity is dependent on temperature, capacity limitations may occur only above or below certain temperatures. Therefore, we discuss the existence of capacity limitations and the extent to which respiratory acclimation to temperature is a result of changes in respiratory capacity.

During measurement at warm temperatures (*c.* 20–25°C), respiration rates from plant tissues are typically stimulated by chemically uncoupling the mETC to simulate increased ATP demand and/or by adding exogenous substrates, the latter more commonly in roots than leaves. Furthermore, respiration rates from isolated mitochondria under substrate saturation are greater than corresponding tissue respiration rates when both are expressed on a per-mitochondrial-protein basis. The general conclusion is that respiration *in vivo* operates below the maximum possible rate (i.e. below rates measured in the absence of adenylate and/or substrate limitations), suggesting that under warm conditions, enzymatic capacity of the glycolysis–TCA cycle–mETC system does not limit respiratory rates.

The strongest evidence for capacity limitations within respiration comes from measurements at low temperatures, as well as the commonly observed acclimation of tissue respiration rates to temperature. As mentioned previously, the response of respiration rate to environmental factors partly reflects the changes to the supply and demand components of respiration. However, respiratory acclimation may also involve adjustments in the capacity of respiratory enzymes, especially when faced with environmental conditions, such as cold, which affect enzymatic catalytic performance. When plants are shifted to lower growth temperatures, respiration rate is immediately reduced in a predictable fashion.

Acclimation of respiration rate to a new cold growth temperature is then often observed as a gradual increase in respiration rate. In some cases, and especially in new tissues that developed in the cold, thermal acclimation is associated with physiological changes to mitochondria and mitochondrial number, indicative of a capacity increase. Furthermore, alternative pathways, including AOX, alternative dehydrogenases and uncoupling proteins, have frequently been observed to be upregulated under cold temperatures. This suggests that as temperature is reduced, the cytochrome pathway or ATP synthase activity becomes insufficient to balance redox status, resulting in induction of alternative pathways, and perhaps also signals to increase capacity. However, what remains unclear is whether an adjustment in respiration rate following temperature acclimation is a result of an upregulated capacity in: (1) the respiratory pathway itself; (2) downstream energy-consuming pathways; (3) photosynthetic C gain; (4) or a combination of all three elements.

Regardless of the exact mechanism(s) involved, the process of respiratory acclimation to temperature has the effect of reducing respiratory rate variation through time in the face of changing environmental conditions. The short-term temperature response of respiration rate suggests near-exponential and universal increase in respiration rate with climate warming.

However, in the longer term, a trend towards homeostasis of respiration rate within moderate temperature ranges (e.g. 15–25°C) is often observed because metabolic processes acclimate to the extent that the rate at a new temperature becomes the same as it was at an initial temperature. One consequence of this is that, globally, leaf respiration rates measured at local growth temperatures increase only two-fold as local growing season temperatures increase from 8 to 28°C.

This increase is only approximately half of what would be predicted by the known scaling of leaf respiration rate with instantaneous changes in temperature, where rate approximately doubles for each 10°C increase. Furthermore, a field study of boreal and temperate tree species experiencing an artificial change in growth temperature revealed that acclimation eliminated 80% of the expected increase in leaf respiration rates as a result of a 3.4°C increase in temperature.

The existence of seasonal thermal acclimation of leaf dark respiration is also supported by measurements of field-grown deciduous and evergreen species in a variety of ecosystems. Therefore, it may be that one aspect of acclimation is to reduce capacity limitations when they occur (e.g. during cold).

PRINCIPLE 5 – RESPIRATION IS A COMPLEX TRAIT THAT HELPS TO DEFINE, AND IS IMPACTED BY, PLANT LIFESTYLE STRATEGIES

Dark respiration rate is a commonly measured trait in plant organs and provides an indication of the current rate of energy expenditure. As such, respiration measurements have made an important contribution to the development of the economic spectrum model for leaves, roots and stems, where respiration strongly covaries with other traits along a spectrum from quick return on investment (high metabolic rates in relatively cheaply constructed and short-lived tissues) to a prolonged return on investment (low metabolic rates in relatively costly and long-lived tissues).

The plant economic spectrum illustrates an important attribute of respiration as a trait: it is connected to, and highly correlated with, other traits. For example, the ratio of leaf photosynthesis to respiration in many species is generally stable across different growth temperatures; that is, both processes acclimate concertedly. Nevertheless, the correlation between leaf respiration rates and photosynthetic rates or leaf structural traits does vary among species, across seasons, and with soil age, reflecting strong environmental as well as genetic influences on respiratory relationships.

Because of its highly feedback-regulated nature, in many scenarios, respiration probably acts to accommodate the energy and C-skeleton demands of other processes like growth and transport, rather than acting to set the rate of other processes. As a result, many genetic determinants that influence processes of energy capture and use likely have an indirect influence on respiration rate. For example, shade tolerance among coinhabiting tree species is linked to lower respiration rates at a standardized light intensity.

In this case, the genetically based differences in respiration rate may relate to evolved differences in the regulation of respiration itself, such as in respiratory enzyme expression levels or regulatory properties; in light capture efficiency (i.e. photosynthetic adaptation affecting respiratory supply); or in energy-use processes (e.g. growth rate). Within each category, multiple individual genes may each contribute to the observed respiratory difference, thus illustrating the potential complexity of respiration as a genetic trait. It is also important to note that, although the demand for respiratory products probably sets the respiratory flux in many scenarios, changes in respiratory flux/efficiency that occur in response to environmental change, in turn, have the potential to influence other traits relevant to the leaf economic spectrum.

FUTURE DIRECTIONS

Looking forward, we see exciting opportunities for ways in which the five principles outlined in this review could be used to rationalize future experiments. In the following sections, we outline three examples of how the principles can guide researchers with interests in understanding and modelling biochemical pathways; in crop breeding, where there is a need to know the limits of respiratory behaviour and relationships between respiration and other traits; and in designing experiments to help improve parameterization of respiration in ecosystem and climate models.

BIOCHEMICAL MODELLING

The need to understand regulatory mechanisms within respiration is central to the development of metabolic models that predict C fluxes throughout the respiratory network under a given physiological and environmental condition. Such models open the door to improved metabolic engineering of respiratory traits, such as increased ATP yield and the diversion of C towards desired end products. However, it is difficult to understand which of the many potential regulatory mechanisms are most relevant to changing the behaviour of the respiratory network.

One solution is to identify differences in the regulatory properties of respiration that have coevolved in nature along with other traits and to attempt to categorize this variation or place it along a spectrum. This approach has been taken with regard to different modes of photosynthesis (e.g. C_3 vs C_4). More specifically, variation in the catalytic properties of Rubisco and PEP carboxylase has been identified among photosynthesis subtypes and causally related to differing photosynthetic performance. To apply this approach to respiration, one could look for natural genetic variation in catalytic and regulatory properties of individual respiratory enzymes between species and ask whether the occurrence of enzyme variants relates to differences in respiratory performance. Furthermore, different respiration subtypes probably occur within the same plant under different situations, and could be related to expression of specific isozymes with relevant properties. Key respiratory enzymes encoded by small gene families with catalytic and regulatory properties known to vary within and among species include AOX and PEP carboxylase. Once it is understood which regulatory mechanisms are needed to produce different types of respiratory behaviour seen in nature, such as for thermogenesis in aroid flowers, mass organic acid exudation in proteoid roots or rapid biosynthesis of oil and protein from sugars in oilseeds, these properties can be reproduced or avoided elsewhere to the advantage of metabolic engineering.

Crop improvement

Is respiration a wasteful process that should be selected against to improve C-use efficiency in crops? The answer to this question depends on how exactly respiration is functioning, which may differ between tissues. As outlined in Principle 1, increased respiration can be linked to increased rates of biosynthesis, which would generally be beneficial for growth. Alternatively, increased respiration may accompany stress and redox imbalance in the metabolic network not linked to biosynthesis and leading to lower C-use efficiency. How can we sort the good from the bad?

To observe variation in respiratory functionality, it will be necessary to further develop experimental approaches that integrate multiple measurements. Examining differences in single gas exchange rates (e.g. CO2 production) provides limited information about the biochemical nature of respiratory variation. However, combinations of gas exchange with other measurements can be interpreted with regard to, for example, the operation of alternative pathways or the correlation between tissue respiration and yield. Existing combined measurements include O2 isotope fractionation to determine cytochrome vs alternative pathway flux, and calorespirometry to determine CO2 and heat production rates; however, practical improvements in both measurement techniques should be pursued to improve ease of use.

Other potential measurements can be drawn from techniques used to resolve respiratory variation in the animal field. One example is the simultaneous analysis of glycolytic flux and mitochondrial oxygen consumption, which in animal cells can distinguish the relative contribution of each pathway to total ATP production, thereby revealing different energetic modes of respiration, some associated with stress. Additional techniques include the use of fluorescent protein sensors to provide subcellular measurements of the ATP : ADP ratio and cellular redox status, thereby providing a novel opportunity to assess whether respiration is operating efficiently. The challenge will be to adapt the use of these multiparameter metabolic measurements, both theoretically and practically, to plant tissues so that they can provide precise phenotypic data in a relatively high-throughput manner. Fortunately, owing to the recent advent of high-throughput methodology, tissue respiration rates can now be included in large multivariable studies, which examine correlations between plant metabolic/physiological properties and yield or other assessments of plant performance.

Ecosystem/climate modelling

Most plant, ecosystem and terrestrial biosphere models represent plant respiration using the growth and maintenance paradigm, relating respiration rate (CO_2 efflux rate) to temperature and some combination of mass, N, photosynthetic capacity, photosynthetic C gain and growth rate. The models are used to estimate and predict respiration rate of an individual or a group of individuals at a site under past, current and future environmental conditions. The parameters currently used in these model frameworks are taken from empirical relationships that describe general trends often observed among species and between variables measured at differing points in time.

Looking forward, one option is simply to expand the availability of data to further improve the generality of empirical relationships used to predict patterns of plant respiration. Although useful for assessing current patterns, such an approach may not be sufficiently predictive for future, novel climates or apply to all species. An alternative is to use our understanding of the five principles to design experiments that will enable a more principles-based prediction of variability in respiratory rates.

Principle 1 (three functions) and Principle 4 (enzymatic control of acclimation) suggest that the predictive ability of empirical relationships may be greater if the large variation around the relationships is explained in terms of the knowledge of the three functions and enzyme capacity of respiratory metabolism. Respiration models based on the growth and maintenance paradigm largely assume that the relative contributions of the functions are at equilibrium and that enzyme capacities do not change with time; hence, the quantity of ATP generated and CO_2 released should be constant for a given amount and composition of existing and newly generated biomass. This assumption remains poorly examined across space and through time, but studies probing these relations are becoming increasingly possible with the introduction of high-throughput methods of O_2 uptake and developments in our capacity to quantify engagement of phosphorylating and nonphosphorylating electron transport pathways.

Principles 1 and 4 also suggest that empirical relationships may be considered general if the relative contributions of each function and enzymatic capacities to respiration rate are at equilibrium, or reach an equilibrium relatively quickly. We suggest that future work should focus, in part, on assessing over what timescales equilibrium of the three distinct functions is, and is not, achieved when plants experience sustained changes in temperature and other environmental factors (particularly daily photon input and soil moisture).

The interdependence of respiratory supply and demand illustrated in Principle 3 suggests that respiratory models merit greater confidence if they adequately represent feedback loops (e.g. coupling between respiration and photosynthesis, and with growth and demand). Principle 3 indicates that the flexibility of respiratory metabolism is constrained through those feedback loops. Importantly, when incorporating Principle 3 into future models, there will need to be an allowance for metabolic flexibility (Principle 2); that is, the coefficients that are used to scale respiration to supply and demand need to be flexible to allow for variability through time, space and among contrasting genotypes. Such models may describe all of the necessary mechanisms of feedback loops (but noting that complexity can make such models more difficult to deploy) or use optimization principles such as those developed for stomatal conductance.

Conclusion

This review has attempted to organize the major concepts in respiratory variation in plants so that conceptual links between different fields of study can be more easily achieved. Observations of variation in plant respiration at different biological scales of size and time should be describable in terms of one or more of the principles outlined here. The principles, in turn, arise from the interactions of mechanisms that determine respiratory rate and function under the constant influence of the environment. In the future, it may be necessary to modify or extend the principles and mechanisms introduced here and we welcome such discussion.

CHAPTER

SYNTHESIS AND METABOLISM OF STORAGE AND STRUCTURAL CARBOHYDRATES

13

INTRODUCTION

Carbohydrates are the major components of plant tissue, making up to 60% to 90% of the dry matter (DM). Carbohydrates contain carbon, hydrogen, and oxygen in the proportion found in water (CH_2O) and are hence hydrates of carbon. Carbohydrates are the basic energy source in animal cells. Dietary carbohydrates obtained from plant-based products serve as a major source of energy for the animal. The chlorophyll in plant cells traps solar energy and produces carbohydrates using carbon dioxide and water and gives off oxygen, as shown in the following equation:

$$\text{solar energy} + 6\ CO_2 + 6\ H_2O \rightarrow C_6H_2O + 6O_2.$$

In the plant cell, carbohydrates could be present in the cell content as sugar or starch, or they could be associated with the cell wall structure (e.g., cellulose). When animals eat plant materials (e.g., cereal grains, grass, fodder), energy in the feed's carbohydrates is made available through metabolic processes in the animal cell. Overall, animal metabolism produces energy in a reverse process to that of photosynthesis.

STRUCTURE AND CLASSIFICATION

One method of classifying carbohydrates is based on the number of carbon atoms per each molecule of a carbohydrate and on the number of molecules of sugar in the compound. Based on the number of carbon atoms, a carbohydrate can be classified as triose (3 C), tetrose (4 C), pentose (5 C), and hexose (6 C). The suffix "*ose*" at the end of a biochemical name flags the molecule as a "sugar." Among these, pentoses (e.g., ribose in ribonucleic acid (RNA)) and hexoses (e.g., glucose, or blood sugar) are the most common sugars in animal tissues. Based on the number of molecules of sugar in the compound, carbohydrates can be classified as (1) monosaccharide, one unit of sugar; (2) disaccharide, two monosaccharides; (3) oligosaccharide, three to fifteen monosaccharides; and (4) polysaccharides, large polymers of simple sugars.

A. Monosaccharides are often referred to as simple sugars (e.g., glucose) and cannot be hydrolyzed into simpler compounds.

Monosaccharides can be subdivided based on the number of carbon (C) atoms. The following list shows the prefixes for numbers of carbons in a sugar.

- Triose (3 C)
- Tetrose (4 C)
- Pentose (5 C; e.g., Xylose and Ribose)
- Hexose (6 C; e.g., glucose, fructose, galactose, and mannose)
- Most monosaccharides in animal tissues are of 5 C and 6 C sugars. Simple sugars are also subdivided into aldose, a sugar that contains an aldehyde structure, or ketose, a sugar that contains a ketone group. Both glucose and fructose have the same molecular formula C6H12O6 and are hexoses (6 C). But glucose is an aldose (also called aldohexose) and fructose is a ketose, or a ketohexose.
- The three hexoses that are nutritionally and metabolically important are glucose, fructose, and galactose (see Fig. 1).

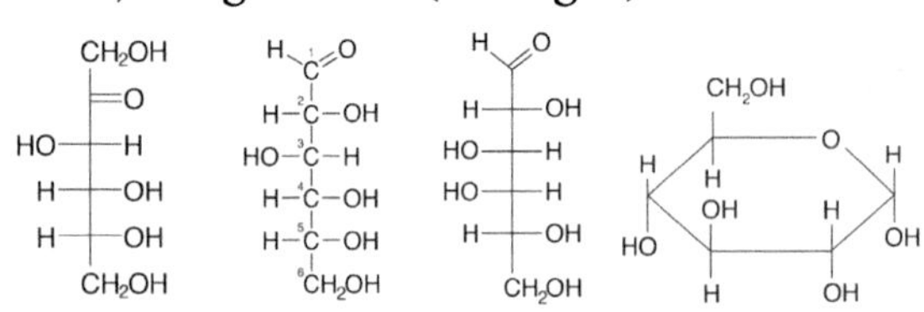

Fig. 1. Structure of simple sugars (Source: Wikipedia)

The chemical structure of glucose can be represented as a straight chain form (Fig. 1) and in cyclic form (also shown in Fig. 1). In a biological system, glucose exists primarily as a cyclic form and very rarely in a straight form (in aqueous solution). Glucose is the form of carbohydrates found in circulating blood (blood sugar) and is the primary carbohydrate used by the body for energy production. Fructose, or "fruit sugar," is found in ripened fruits and honey and is also formed by digestion of disaccharide sucrose. Galactose is found along with disaccharide lactose in mammalian milk and is released during digestion.

Glucose can exist as α and β isomers and has immense animal nutritional implications. These two isomers differ in their orientation of OH on C #1 (shown in red in Fig. 1).

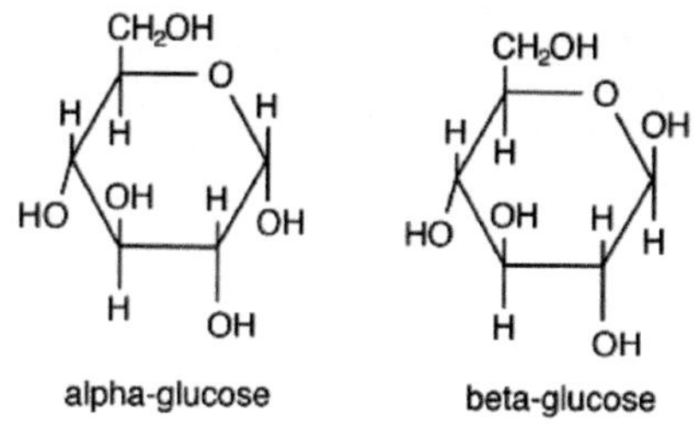

Fig. 2. Alpha- and beta-glucose structure

For example, starch contains α-D-Glucose, while cellulose has rigid polymers with β-D-Glucose. Nutritionally important sugars are of the D-form (not the L-form). D and L refer to stereo-orientation at asymmetric carbon position 5 in a hexose or carbon position 4 in a pentose.

B. Disaccharides are made up of two monosaccharides bonded together by a glycosidic (covalent) bond. The following are some of the common disaccharides:

- Sucrose-glucose + fructose (e.g., table sugar)
- Lactose-glucose + galactose (milk sugar)
- Maltose-α-D-Glucose + β-D-Glucose (malt sugar)
- Cellobiose-β-D-Glucose + β-D-Glucose (cellulose)

Among the different disaccharides, lactose (milk sugar) is the only carbohydrate of animal origin. However, cellobiose as a component of cellulose is important in animal nutrition.

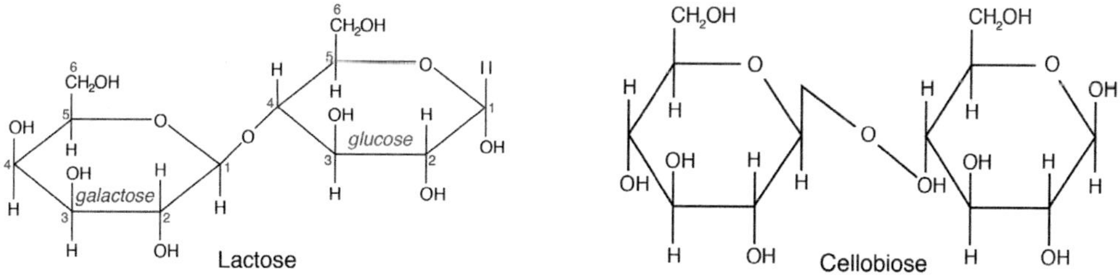

Fig. 3. Important disaccharides in animal nutrition and feeding, lactose and cellobiose

Monogastric animals cannot digest cellulose because they do not produce the cellulase enzyme that can split β-D-Glucose.

C. Oligosaccharide are made by bonding together three or more (3 to 15) monosaccharides bonded together.

- Raffinose (glucose + fructose + galactose; 3 sugars)
- Stachyose (glucose + fructose + 2 galactose; 4 sugars)

In animal diets, oligosaccharides are commonly found in beans and legumes. Some oligosaccharides are used as substances to enhance the growth of good microbes (prebiotics). Recently, there has been an increased interest in the use of different oligosaccharides as feed additives to enhance hindgut health (e.g., fructooligosaccharides, mannan oligosaccharides).

D. Polysaccharides, as their name implies, are made by joining together large polymers of simple sugars.

Polysaccharides are the most important carbohydrate in animal feed. Polysaccharides are composed of many single monosaccharide units linked together in long, complex chains. The functions of polysaccharides include energy storage in plant cells (e.g., seed starch in cereal grains) and animal cells (e.g., glycogen) or structural support (plant fiber). Components of cell wall structure are also called nonstarch polysaccharides, or resistant starch, in animal nutrition, as they cannot be digested by animal enzymes but are fermented by hindgut and rumen microbes.

Polysaccharides can be homopolysaccharides or heteropolysaccharides.

- Homopolysaccharide
- Heteropolysaccharide

Homopolysaccharide: Contains only one type of saccharide unit.

Examples of homopolysaccharides that are important in animal nutrition include starch (nonstructural form), glycogen (animal form), and cellulose (plant structural form).

- Starch: Principal sugar form of carbohydrate in cereal grains (seed energy storage). The basic unit is a-D-Glucose. Forms of starch in cereal grains include
  - Amylose-α 1,4 linkage-straight chain, nonbranching, helical structure
  - Amylopectin-α 1,4 linkage with alpha 1,6 linkage at branch points

Amylose is the simplest of the polysaccharides, being comprised solely of glucose units joined in an alpha 1,4 linkage (Fig. 4). Amylose is water soluble and constitutes 15% to 30% of total starch in most plants.

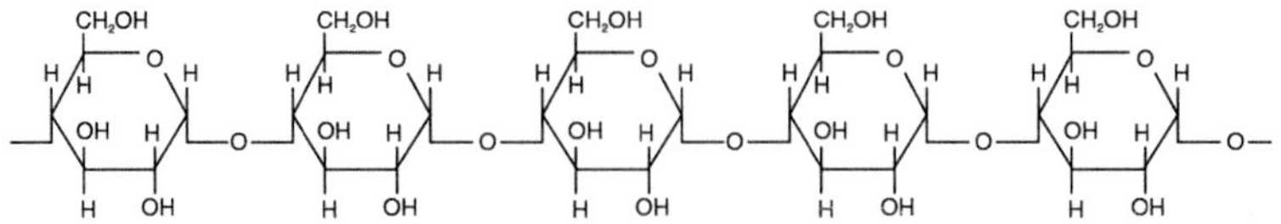

Fig. 4. Amylose structure showing straight a 1,4 linkage

Amylopectin differs in how the glucose units are joined together. Alpha 1,4 linkages predominate, but a "branch" arises from an alpha 1,6 linkage. Such branches make the structure of amylopectin more complex than that of amylose. Amylopectin is not water soluble and constitutes 70% to 85% of total starch in plant cells.

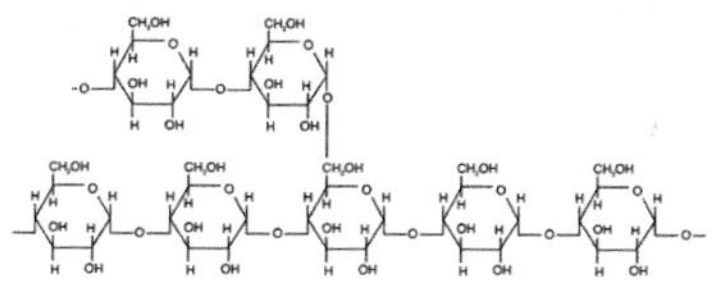

Fig. 5. Amylopectin structure showing straight a 1,6 linkage

Glycogen is a form of starch found in animal tissue and is hence called animal starch. Glycogen is a polysaccharide that is physically related to amylopectin with basic alpha-D-Glucose but has a mix of a 1,4 and a 1,6 bonds. Glycogen exists in a small amount (< 1%) in liver and muscle tissue.

Cellulose is the most abundant carbohydrate in nature. It provides structural integrity to plant cell walls. The basic unit is b 1,4 linkage, straight chain, nonbranching (Fig. 3). Cellulose is highly stable. No animal enzyme can break it; only microbial cellulase can degrade it.

Ruminant animals such as cattle, however, have bacteria in their rumen that contain the enzyme cellulase. It breaks the beta 1,4 links of the glucoses in cellulose to release the sugar for energy.

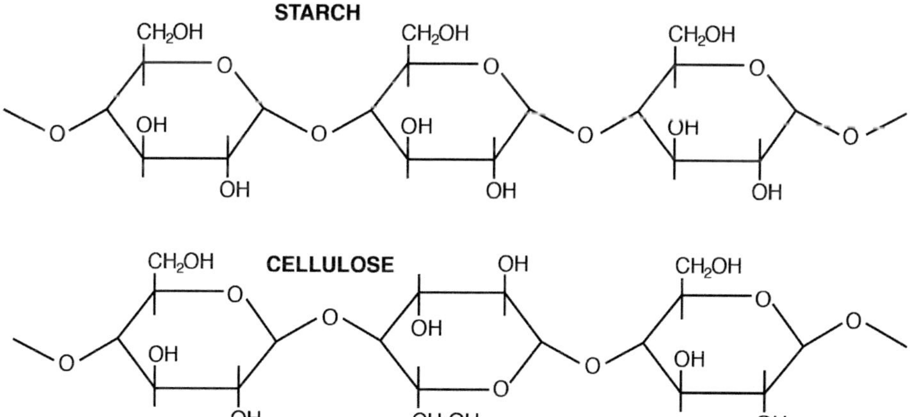

Fig. 6

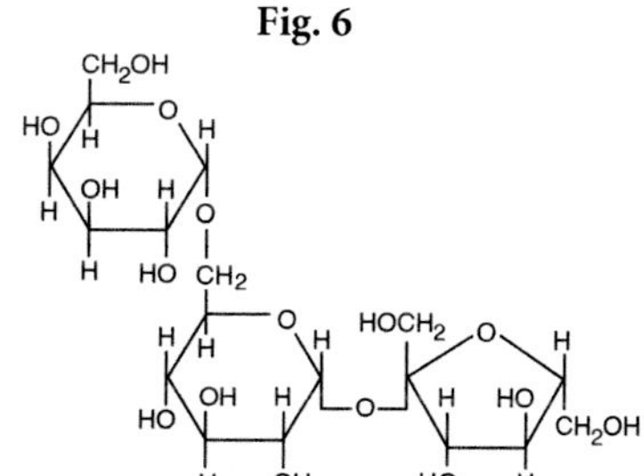

Fig. 7. Heteropolysaccharide structure showing a mix of 6 and 5 C sugars

b: Heteropolysaccharide: A component of plant cell walls with a mix of 5 C and 6 C sugars (e.g., hemicellulose and pectin, a mixture of pentose and hexose units).

CARBOHYDRATE METABOLISM IN PLANT CELLS

Plants obtain their energy and resources via an autotrophic means. All their organic molecules are synthesized from inorganic elements such as CO_2 and primarily from diverse compounds including nitrogen, sulfur, and phosphorus, among others. In addition to their rich secondary metabolism, plants produce a multitude of primary metabolites including carbohydrates, lipids, and amino acids.

The group of carbohydrates of plant origin comprises a wide range of simple sugars such as mono- and disaccharides, sugar alcohols, and polymers such as starch and cellulose. Pathways of plant carbohydrate metabolism are well known. From the generated-Calvin cycle triose-phosphate pool, the biosynthetic routes of fructose, glucose, and other simple carbohydrates are well delineated.

In plants, sucrose metabolism is crucial to feed phloem transport from source parts to sink parts. Gluconeogenesis and pentose phosphate shunt are the major pathways where simple carbohydrate biosynthesis is involved. Other important compounds such as polymers (starch, cellulose, and derived) and sugar alcohols/ polyalcohols (glycerol, myo-inositol, sorbitol, manitol, etc.) are synthesized from the formers. Carbohydrates, polyalcohols, and some amino acids (especially proline), in addition to their nutritional function in cells, have an interesting role as osmoregulatory compounds, especially in stressful situations.

BIOSYNTHESIS OF MELATONIN IN PLANTS

Melatonin (N-acetyl-5-methoxytryptamine) is a tryptophan-derived compound discovered in plants in 1995. Melatonin is a highly studied biomolecule due to its known role in mammals as a regulating hormone of sleep-wake cycles, and other functions in endogenous rhythms, mood, metabolism, and immunological responses. In addition, it has been investigated as to its therapeutic efficacy in Alzheimer's disease, Parkinsonism, cancer, diabetes, and SARS-CoV-2.

Melatonin biosynthesis in plants originates with the amino acid tryptophan, which is endogenously synthesized in plant cells in the chorismate pathway. Five enzymes are involved in the conversion of tryptophan to melatonin; these are tryptophan decarboxylase (TDC), tryptamine 5-hydroxylase (T5H), serotonin N-acetyltransferase (SNAT), acetylserotonin methyltransferase (ASMT), and caffeic-O-methyltransferase (COMT).

These enzymes catalyze the conversion of the indolic compounds tryptophan, tryptamine, serotonin, 5-methoxytryptamine, and N-acetylserotonin to melatonin, as illustrated in the biosynthetic pathway shown in Fig. 8.

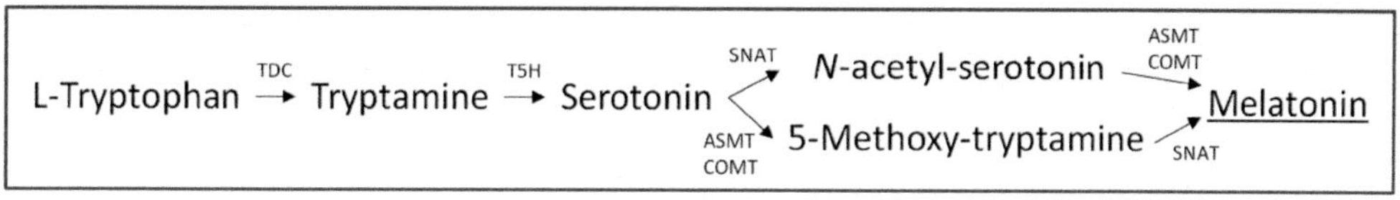

Fig. 8. Biosynthesis of melatonin in plants.

However, this primary melatonin biosynthetic pathway may present alternatives such as serotonin biosynthesis through 5-hydroxytryptophan, although this possibility seems specific to animals since the responsible enzyme (tryptophan hydroxylase) has not been detected in plants. In addition, a conversion of N-acetylserotonin to serotonin by the enzyme N-acetylserotonin deacetylase has been described. With respect to the subcellular localization, several studies in arabidopsis and rice plants indicated that the involved enzymes act in the cytoplasm (TDC, ASMT and COMT), endoplasmic reticulum (T5H), and chloroplasts (SNAT). In addition, the participation of mitochondria has been described, through arylalkylamine N-acetyltransferases (AANAT) and hydroxyindole-O-methyltransferases (HIOMT), observing that, when the melatonin pathway is artificially blocked in chloroplasts, melatonin biosynthesis shifts to the mitochondria to maintain melatonin generation. Generally, stressors induce melatonin biosynthesis in plants through the upregulation of diverse biosynthesis isozyme transcripts, increasing endogenous melatonin production.

ROLES OF MELATONIN IN PLANTS

Melatonin is a pleiotropic molecule in plants. Melatonin has many beneficial actions, generally improving physiological responses such as seed germination and growth, photosynthesis (pigment content, photorespiration, stomatal conductance and water economy), seed and fruit yield, osmoregulation, and the regulation of the different metabolic pathways (carbohydrates, lipids, nitrogen compounds, sulphur, and phosphorus cycles). With respect to secondary metabolism, melatonin induces the biosynthesis of simple phenols, flavonoids, anthocyanins, carotenoids, and several terpenoids. Melatonin promotes rooting processes and also delays leaf senescence. In postharvest fruit, it regulates ethylene and lycopene content, as well as general ripening metabolism and induces parthenocarpy during fruiting. It also preserves cut flowers. In pathogen infections, melatonin slows damage, stimulating systemic acquired resistance (SAR) and contributes to crop health. Due to this high number of actions, melatonin has been referred to as a plant master regulator, mainly due to its role as a plant hormone regulator, with a substantial influence on auxin, gibberellins, cytokinins, abscisic acid, ethylene, jasmonic acid, salicylic acid, and brassinosteroids.

Melatonin displays a relevant role in the stress responses. Similar to what occurs in animal cells, melatonin acts as an excellent scavenger of reactive oxygen species (ROS) and reactive nitrogen species (RNS) in plants. This antioxidant capacity has been extensively studied. The data show that melatonin acts as a direct antioxidant, neutralizing several ROS/RNS and other radical species harmful to the cell, and also acts as an activator of the antioxidant response, upregulating various transcription factors that trigger the activity of antioxidant enzymes such as superoxide dismutases, catalases, peroxidases, and those involved in the ascorbate-glutathione cycle, among others. Via these means, melatonin acts as a master regulator of the responses of the redox, hormonal, and osmoregulatory systems. In summary, as can be seen in Fig. 9, through the redox and hormonal network, melatonin regulates photosynthesis, primary and secondary metabolism, and pathogenic response to increase abiotic/ biotic tolerance and, as a result, crop yield.

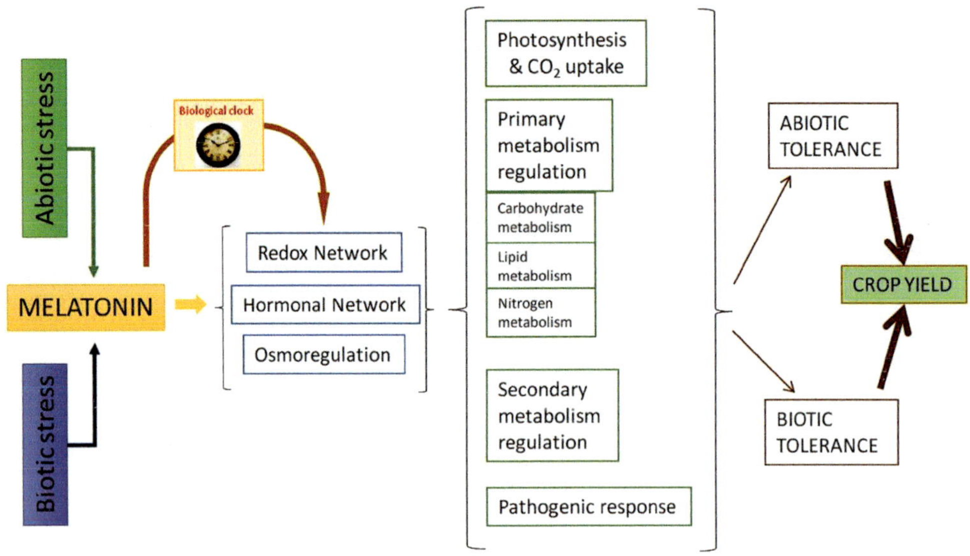

Fig. 9. Melatonin actions as a response to abiotic and biotic stressors.

One of the most interesting aspects is the ability of melatonin to regulate the carbohydrate metabolism and its relationship with the osmoregulatory response, which is a key in stressful situations of plants.

EFFECT OF MELATONIN IN SIMPLE CARBOHYDRATES, STARCH, AND POLYALCOHOLS

The term phytomelatonin refers to melatonin of plant origin as opposed to the animal hormone, but they have the identical chemical structure. The first studies on the role of phytomelatonin in plants appeared at the end of the last century and the beginning of the present one. In this study, exogenous melatonin added to the culture media induced plant growth and rooting in shoot tip explants; in addition, an elevation in endogenous levels of total soluble sugars in 9-week-old plants, both in leaves and roots, and in chlorophylls, carotenoids, and proline level were also observed. These findings indicate an improvement in plant primary carbon metabolism, with a melatonin-concentration dependent response.

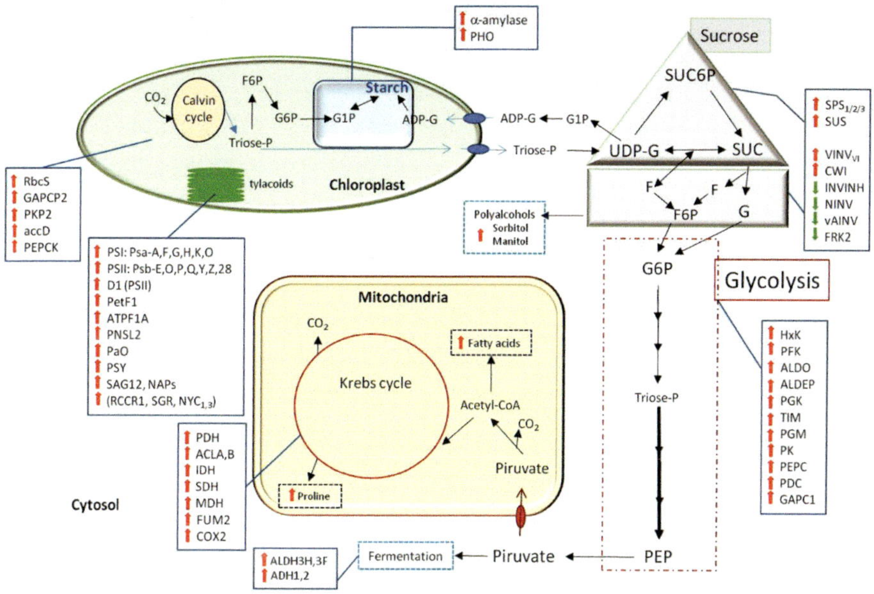

Fig 10. Schematic model representing melatonin's action in different metabolic pathways.

Also in apple trees, melatonin treatment of leaves produced an increase in the levels of monosaccharides, sucrose, starch, and sorbitol as well as an improvement in the photosynthetic rate and a reduction in foliar senescence and autophagy. Other studies were focused on improving the plants' tolerance to certain stresses. Thus, melatonin treatments enhanced saline tolerance in soybean, tomato, and bermudagrass plants, accompanied by an activation of carbohydrate metabolism and, in some cases, lipid and ASC-GSH metabolism as well.

There are many studies on the promotional effect of fruit development after the application of melatonin in leaves and/or roots. One of the first was carried out in tomato plants, where melatonin applications induced photosynthetic processes with a higher yield in biomass and a greater number of fruits which were of greater caliber and exhibited optimal ripening. In pear trees, 100 μM melatonin treatments induced higher total sugars and starch levels and better fruit sizes which were of high quality.

In addition, postharvest melatonin treatments in various fruits gave rise to higher quality fruits with an increased content of sugars, starch, organic acids, and pigments, as had been demonstrated in tomato and banana, and other fruits such as peach, strawberry, pear, plum, and litchi. In one comprehensive study, melatonin treatments induce innate immunity in *Arabidopsis* with the accumulation of various sugars and glycerol, as well as increasing disease resistance against *Pseudomonas syringe*. In general, plants treated with melatonin exhibit increases in the levels of simple sugars, sucrose, starch, and some polyalcohols.

REGULATORY ACTION OF MELATONIN ON CARBOHYDRATE METABOLISM

In general, melatonin improves photosynthetic and related parameters, such as photosynthetic rate, transpiration rate, stomatal conductance, leaf area, relative water content, and levels of chlorophylls and carotenoids, and also delays leaf senescence. Melatonin has a protective role against oxidative stress, reducing the levels of superoxide anion, hydrogen peroxide, and malondialdehyde, and improving membrane stability indexes. It also induces the expression of genes for antioxidant enzymes such as superoxide dismutases, catalases, guiacol-, and ascorbate peroxidases, which in turn raises ascorbate and glutathione levels.

Also relevant is the melatonin-mediated improvement in the uptake of mineral nutrients, which induces the expression of mineral transporters. Collectively, the up- and downregulated genes following melatonin treatment functions to aid plants in physiologically overcoming negative stress situations and to increase tolerance to multiple abiotic stressors such as drought, waterlogging, salinity, heavy metals, extreme temperatures, radiation, etc., including osmoregulatory responses.

In 2014, Guo and colleagues performed transcriptional studies and were the first to detect changes in the expression of genes involved in carbohydrate metabolism due to melatonin treatment. In addition, in an excellent study on the effect of melatonin in salinity tolerance of soybean plants, a detailed transcriptomic analysis on primary metabolism was presented. Melatonin clearly over-expressed the transcripts of many enzymes related to photosynthesis, starch, sucrose, glycolysis, fermentation, the Krebs cycle, and other metabolic pathways. Figure 10 diagrammatically summarizes some of the genes up or downregulated by melatonin that are related to carbohydrate metabolism.

In bermudagrass, melatonin-pre-treated plants exhibited significantly higher levels of several metabolites than non-treated plants under abiotic stress conditions (salinity, drought, and cold). These primary metabolites included 10 amino acids, 18 carbohydrates (allose, arabinose, cellobiose, fructose, galactose, gentiobiose, glucose, lactose, lactulose, maltose, manobiose, manose, melibose, sucrose, tagatofuranose, talose, trehalose, and turanose) and five polyalcohols (dulcitol, galactinol, glycerol, myo-inositol, and sorbitol), involved in glycolysis, oxidative pentose phosphate pathway, and the Krebs cycle.

With respect to gene regulation, 2361 upregulated and 1572 downregulated transcripts were differentially expressed in melatonin-treated plants versus control.

Some examples of carbohydrate-related transcripts regulated by melatonin are: glycosyl –transferases, -hydrolases, -phosphatases, -epimerases, -invertases and -mutases, hexokinases, mannosidases, a- and b-amylases, a- and b- glucan related-enzymes, and several dehydrogenases (3-phosphoglycerate-, UDP-glucose-, alcohol- and aldehyde-), among others.

The regulation of carbohydrate metabolism by melatonin is accompanied by an activation of chloroplast metabolism and an improvement in the function of these organelles. As seen in Fig. 10, there are many factors upregulated by melatonin in the photosynthetic light apparatus and in the Calvin cycle. Melatonin also stimulates the biosynthesis and mobilization of starch and of sucrose destined for the phloem.

The role of melatonin in sucrose metabolism has received significant attention. In maize plants, low doses of melatonin (1–10 μM) induce sugar metabolism, photosynthesis, and sucrose phloem loading. The authors demonstrated that high doses of melatonin inhibit seedling growth by inducing the excessive accumulation of sucrose, hexose and starch, suppressing photosynthesis and sucrose phloem loading. The role of melatonin in improving sucrose biosynthesis was also confirmed in pear, grape, and rice plants.

Intimately related to carbohydrates is the osmoregulatory response in plants subjected to stress. Melatonin clearly mediates these responses by increasing the levels of carbohydrates and polyalcohols, as already described. One of the key metabolites in the osmoregulatory response is proline, an amino acid that accumulates in the presence of various stressors, especially drought and salinity. Melatonin raises proline levels in stressed plants (Fig. 10), which has been demonstrated in various species and situations. Several reviews in this regard can be consulted since this aspect is outside the present review on carbohydrates.

SUCROSE

Sucrose is the end product of photosynthesis and the primary sugar transported in the phloem of most plants. Sucrose synthase (SuSy) is a glycosyl transferase enzyme that plays a key role in sugar metabolism, primarily in sink tissues. SuSy catalyzes the reversible cleavage of sucrose into fructose and either uridine diphosphate glucose (UDP-G) or adenosine diphosphate glucose (ADP-G). The products of sucrose cleavage by SuSy are available for many metabolic pathways, such as energy production, primary-metabolite production, and the synthesis of complex carbohydrates. SuSy proteins are usually homotetramers with an average monomeric molecular weight of about 90 kD (about 800 amino acids long).

Plant SuSy isozymes are mainly located in the cytosol or adjacent to plasma membrane, but some SuSy proteins are found in the cell wall, vacuoles, and mitochondria. Plant *SUS* gene families are usually small, containing between four to seven genes, with distinct exon-intron structures. Plant *SUS* genes are divided into three separate clades, which are present in both monocots and dicots. A comprehensive phylogenetic analysis indicates that a first *SUS* duplication event may have occurred before the divergence of the gymnosperms and angiosperms and a second duplication event probably occurred in a common angiosperm ancestor, leading to the existence of all three clades in both monocots and dicots.

Plants with reduced SuSy activity have been shown to have reduced growth, reduced starch, cellulose or callose synthesis, reduced tolerance to anaerobic-stress conditions and altered shoot apical meristem function and leaf morphology. Plants overexpressing *SUS* have shown increased growth, increased xylem area and xylem cell-wall width, and increased cellulose and starch contents, making *SUS* high-potential candidate genes for the improvement of agricultural traits in crop plants. This review summarizes the current knowledge regarding plant SuSy, including newly discovered possible developmental roles for SuSy in meristem functioning that involve sugar and hormonal signaling.

Sucrose Definition

Sucrose, commonly known as "table sugar" or "cane sugar", is a carbohydrate formed from the combination of *glucose* and *fructose*. Glucose is the simple carbohydrate formed as a result of *photosynthesis*. Fructose is nearly identical, except for the location of a double-bonded oxygen. They are both six-carbon molecules, but fructose has a slightly different configuration. When the two combine, they become sucrose.

Plants use sucrose as a storage molecule. For quick energy, cells may store the sugar for later use. If far too much is accumulated, plants may begin to combine the complex sugars like sucrose into even large and denser molecules, like *starches*. These molecules, and oily lipids, are the main storage chemicals used by plants. In turn, animals eat these sugars and starches, break them back down into glucose, and use the energy within the bonds of glucose to power our cells.

Sucrose has been an important sugar for humans because it is easy extracted from plants such as sugar cane and sugar beets. These plants tend to store an excess of sugar, and from this we produce the majority of the sugar that we use. Even most "natural" sweeteners, which claim to be healthier than sucrose, are simply a different version of glucose combined in a different manner by plants.

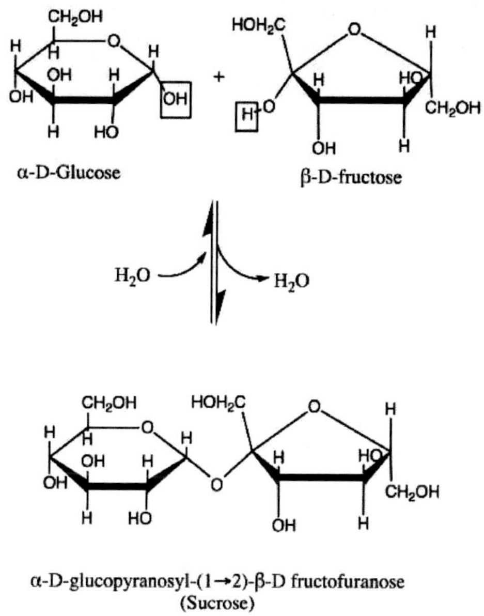

Fig. 11. Sucrose condensation

Sucrose Structure

As mentioned above, sucrose is *disaccharide*, or a molecule made of two *monosaccharides*. Glucose and fructose are both monosaccharides, but together they make the disaccharide sucrose. This is an important process for the storage and compression of energy. Plants do this to make it easier to transport large amounts of energy, via sucrose. This process can be seen in the following image.

Glucose is seen on the left. Glucose is known as an *aldose*, meaning the *carbonyl group* (carbon double bonded to an oxygen) is found at the end of the chain of carbons. When the molecule creates a ring back on itself, it forms a 6-sided ring. Fructose, on the other hand, is a *ketose*. This means that the carbonyl group is found in the middle of the middle of the molecule. In this case, it forces fructose into a five-sided ring.

In a plant creating sucrose, an enzyme comes along to smash these two rings together, and extract a molecule of water. This process is called a *condensation* reaction, and forms a *glycosidic bond* between the two molecules. As you can see in the image, the reaction can also go the other way. To dissolve sucrose into fructose and glucose, a molecule of water can be added back in. This is what happens to sucrose as you digest it.

Sucrose Uses

Sucrose is the most common form of carbohydrate used to transport carbon within a plant. Sucrose is able to be dissolved into water, while maintaining a stable structure. Sucrose can then be exported by plant cells into the phloem, the special vascular tissue designed to transport sugars. From the cells in which it was produces, the sucrose travels through the intercellular spaces within the leaf. It arrives at the vascular bundle, where specialized cells pump it into the phloem.

The xylem, or vascular tube which carries water, adds small amounts of water to the phloem to keep the sugar mixture from solidifying. The sucrose mixture then makes its way down the phloem, arriving at cells in the stem and roots which have no chloroplasts and rely on the leaves for energy.

The sucrose is absorbed into these cells, and enzymes begin breaking the sucrose back into its constituent parts. The six-carbon glucose and fructose can be broken down into 3-carbon molecules, which are imported into the mitochondria, where they go through the citric acid cycle (AKA the Krebs Cycle).

This process reduces coenzymes, which are then used in oxidative phosphorylation to create ATP. The energy within the bonds of ATP can power many of the reactions these cells need to complete in order to maintain the stem and roots.

Likewise, all other life on Earth is dependent upon sucrose and other carbs produced by plants. Sucrose was one of the first substances to be extracted from plants on a mass-scale, creating the white table sugar we know today.

These sugars are extracted and purified from large crops, including sugar cane and sugar beets. To extract the sugar, the plants are usually boiled or heated, releasing the sugar. "Sugar in the Raw" is sugar which has not been treated further, while white table sugar undergoes more purification.

ROLE OF SUCROSE SYNTHASES IN PLANTS

Role of SuSy in Sink Strength

Sink strength can be defined as the ability of an organ to import photoassimilates. Sink strength mainly relies on two parameters, sink organ size as a physical constraint and activity as a physiological constraint. Although imported photoassimilate can be used for respiration, sink-strength estimations are mainly based on net weight gain. Phloem loading is thought to be highly important for defining sink strength and the breakdown of Suc in sink organs may also contribute to sink strength. Work with promoter-GUS fusions has revealed *SUS* promoter activity in the phloem of many plant species, including potato, Arabidopsis, maize, rice, tomato and *Craterostigma plantagineum.*

Work with isoform-specific antibodies has revealed that specific SuSy isoforms are more abundant in the phloem. In Arabidopsis, AtSUS5 and AtSUS6 are reported to be phloem specific. SuSy proteins have also been detected in citrus (Citrus paradisi) and maize phloem companion cells using immunohistological analysis. However, SuSy appear to localize to the phloem not only in the Suc unloading zones, but also in loading zones in mature leaves of citrus, maize, rice, Arabidopsis, and poplar, which suggests that SuSy probably play a general role in the phloem that is not limited to phloem unloading.

In the phloem, SuSy may play a role in the maintenance of equilibrium between Suc and its breakdown products, supplying hexoses for energy production in companion cells and substrates for complex carbohydrates, like callose. There is plenty of evidence that SuSy, and not INV, is the primary active enzyme in actively growing sink organs of different species, such as potato tubers, cassava (Manihot esculenta) roots, lima bean (Phaseolus lunatus) seeds, and tomato fruits. Numerous studies involving plants that have altered SuSy activity and exhibit altered growth rates and weight gain in their sink organs support the putative role of SuSy in sink strength.

In maize, the shrunken (sh) mutant is characterized by a 90% reduction in endosperm SuSy activity, low seed weight and a shrunken-seed phenotype. Pea (Pisum sativum) SuSy mutants (rug4) also exhibited significantly reduced SuSy activity in their embryos, reduced seed weight and a wrinkled-seed phenotype. Transgenic potato plants with reduced SuSy activity only in tubers exhibited reduced tuber dry weight, further supporting the correlation between SuSy activity and sink strength. SuSy activity was also suggested as an indicator for high rice grain yield in rice breeding programs. A recent study found that the Suc-cleavage activity of a castor bean SuSy, RcSUS1, is inhibited by trehalose 6-phosphate, suggesting another mechanism by which Suc flux can be controlled in heterotrophic tissues.

Other studies have produced somewhat contradictory evidence for the role of SuSy in sink strength. In transgenic tomato plant with antisense suppression of SUS, reduced Suc import was observed only in very young fruits (7 days after anthesis) and not during the later starch-accumulation phase (23 DAA). In another transgenic tomato with reduced SUS expression only in fruits, there were no reported effects on fruit development or the accumulation of starch and sugar in young green fruit, challenging the suggested importance of SuSy for sink strength.

In Arabidopsis, AtSUS2 and AtSUS3 mutants had altered metabolism and accumulated less transient starch during seed development, with no effect on agronomic traits like seed weight and oil content (Angeles-Nunez and Tiessen, 2010). Interestingly, even a double mutant (sus2 and sus3) and a quadruple mutant (sus1, sus2, sus3, and sus4) did not show any seed-related phenotypes, suggesting that Arabidopsis SuSy are not important for seed sink strength. Overall, the data suggest that SuSy might be important for sink strength, especially in starch-accumulating organs, although that role is probably not conserved in all plants.

Role of SuSy in Starch Synthesis

For many decades, researchers have thought that SuSy may play important roles in the conversion of Suc into starch. The first genetic evidence for this came from the characterization of the maize *sh* mutant. The *sh* mutant exhibits a 90% reduction in SuSy activity in the endosperm, with no effect on SuSy activity in the embryo, and is also characterized by a 40% reduction in starch accumulation in the endosperm during development and at maturity. Similarly, a pea SuSy mutant (*rug4*) also showed reduced seed starch content.

Other studies involving transgenic plants with suppressed SUS expression have demonstrated reduced starch accumulation in potato tubers and carrot (Daucus carota) taproots, further supporting the hypothesized connection between SuSy and starch synthesis. Suc can be converted into starch via different pathways, which also differ between chloroplasts and heterotrophic tissues.

Adenosine diphosphate glucose is the main molecule converted into starch by the starch synthases in plastids. In chloroplasts, the main starch-synthesis pathway starts with two molecules of triose-P produced by photosynthesis, which yield F1,6BP. F1,6BP is transformed to F6P by fructose bisphosphate phosphatase to F6P, which is then converted into G6P by phosphoglucoisomerase (PGI). G6P is converted into glucose 1-phosphate (G1P) by the plastidic phosphoglucomutase (PGM). G1P is then converted into ADP-G, in a reaction catalyzed by ADP-G pyrophosphorylase (AGPase).

This chloroplast starch synthesis pathway does not require Suc cleavage and, therefore, cytosolic SuSy is not expected to play an important role in leaf starch synthesis. The main genetic evidence supporting this claim is that the plastidic PGI, PGM and AGPase mutants, and plants with reduced expression of these genes are either starchless or contain very low levels of starch in their leaves. However, there is also growing line of evidence suggesting that SuSy might play some role in leaf starch synthesis.

Although UDP is considered to be the preferred nucleotide phosphate used by SuSy for Suc cleavage, ADP can also be effectively used by SuSy to form Fru and ADP-G. Potato and Arabidopsis plants expressing the ADP-G hydrolase in their cytosol accumulate less ADP-G in their leaves, indicating that there is a cytosolic source of ADP-G, probably from cleavage of Suc by SuSy. ADP-G levels in leaves of AGPase- and pPGM-mutant plants are comparable with those seen in the WT, indicating that AGPase is not the only source of ADP-G.

Indeed, overexpression of Arabidopsis *SUS* in tobacco results in increased leaf starch. Based on this and many other studies, researchers suggested a chloroplast starch synthesis model in which (1) Suc cleavage by SuSy produces cytosolic ADP-G which is transported to the chloroplast for starch synthesis and (2) plastidic PGM and AGPase are recycling Glc units derived from starch breakdown back to starch. According to this model, the rate of starch accumulation is determined by the rate of cytosolic SuSy activity that yields ADP-G, cytosolic ADP-G transport to the chloroplast, starch synthesis, starch breakdown and the efficiency of the recycling of the products of the breakdown of starch.

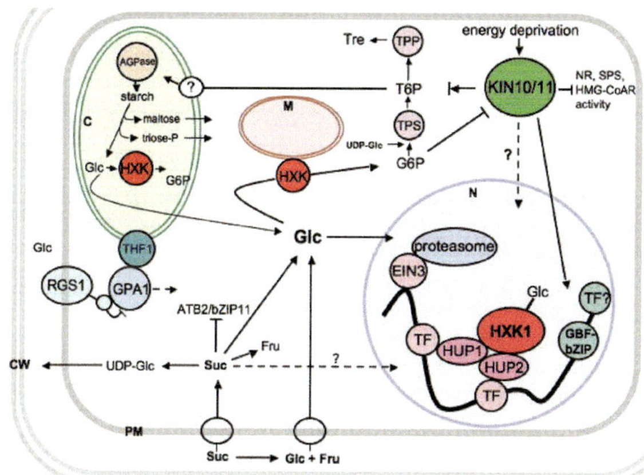

Fig. 12. Summary of sugar and energy sensing and signaling models in Arabidopsis.

There is much more evidence linking SuSy to starch synthesis in non-photosynthetic tissues or sink tissues. For example, a reduction in SuSy activity reduced starch content in potato tubers, carrot taproots and maize endosperm. In addition, five maize starch-deficient endosperm mutants were screened for metabolic enzyme activity and all showed reduced SuSy activity.

However, because 90% suppression of SuSy activity in maize endosperm resulted in only a 40% reduction in starch, doubts were raised as to whether or not SuSy is directly involved in starch synthesis in sink tissues. SuSy from potato tubers and barley endosperm were shown to have similar Km values for the nucleotides UDP and ADP at saturated Suc levels. This fact, the greater activity of SuSy, as compared with AGPase, in barley seeds, and the reduced levels of ADP-G observed in potato tubers with repressed *SUS* together suggest that Suc cleavage by SuSy directly supplies ADP-G for starch synthesis.

Overexpression of *SUS* in potato tubers increased UDP-G and ADP-G levels and increased starch accumulation and yield. Similarly, overexpression of *SUS* resulted in higher ADP-G levels and starch accumulation in maize endosperm, and Arabidopsis T-DNA mutants for *AtSUS2* and *AtSUS3* exhibited reduced transient starch accumulation in seeds during early to mid-development. All these observations support the role of SuSy in starch accumulation.

Role of SuSy in Vascular Tissues

There is a lot of evidence that *SUS* are highly expressed in vascular tissues. *SUS* promoters are expressed in the vasculature of many plant species, mostly in the phloem. *In situ* hybridization has revealed the accumulation of *AtSUS5* and *AtSUS6* specifically in the phloem of Arabidopsis roots and hypocotyls. SuSy protein has also been immunolocalized to the phloem companion cells in citrus and maize and in leaves of 9-day-old barley seedlings. SuSy is also the main enzyme metabolizing Suc in the phloem of *Ricinus communis*. Sucrose synthase may play a number of different roles in the phloem involving the regulation of sink strength and phloem unloading, supplying hexoses for companion cell respiration and supplying precursors for complex carbohydrates, such as callose and cellulose.

Only a few studies have used mutant and transgenic plants to elucidate the roles of SuSy in phloem. In cucumber (*Cucumis sativa*), transgenic plants with suppressed *CsSUS3*, which is mainly expressed in root phloem companion cells, were found to be more sensitive to hypoxic stress caused by flooding. In contrast, the Arabidopsis double mutant of the two phloem-specific *SUS* (*sus5 sus6*) exhibited no specific phenotype, even under hypoxic stress.

However, the double mutant had less callose in its sieve plates and in response to wounding, as compared with WT or quadruple-mutant (*sus1*, *sus2*, *sus3*, and *sus4*) plants, suggesting that *AtSUS5* and *AtSUS6* are essential for callose synthesis and indicating a possible role for phloem SuSy in callose synthesis.

There is also sufficient evidence for the localization of SuSy to xylem tissues. Quantitative PCR analysis found the highest expression levels of *SUS2* in a poplar hybrid (*Populus simonii* × *Populus nigra*) in the xylem and phloem; those levels were 4-fold and 3-fold higher than the expression levels in the cambium, respectively. In transgenic tomato plants expressing GUS under the *SlSUS1* promoter, expression in the stem was observed mainly in the xylem. In the parasitic plant *Phelipanche ramosa*, *PrSUS1* transcript was found in the xylem of developing roots.

Sucrose synthase activity was found in immature metaxylem and the central vessel in the elongation zone of wheat seedlings following hypoxia. Relatively high mRNA levels and activity were also reported in carrot tap root xylem. High SuSy activity and protein levels were reported in differentiating xylem of *Robinia pseudoacacia* during the spring.

Sucrose synthase activity in developing xylem vessels or fibers may be particularly important for the cellulose synthesis needed for the construction of thick secondary cell walls. Work with a cell culture of *Zinnia elegans* revealed that SuSy is highly enriched in differentiating tracheary elements near the plasma membrane, where secondary cell-wall thickening occurs. Overexpression of *SUS* in several plant species increased the thickness of xylem cell walls, further supporting the proposed roles of SuSy in xylem development.

Roles of SuSy in Cellulose and Callose Metabolism

Because Suc cleavage by SuSy yields UDP-G, which is a direct substrate for both cellulose (β1-4) and callose (β1-3) glucans, it is widely assumed that SuSy plays a role in the synthesis of both of these polysaccharides. There is sufficient supporting evidence for these proposed roles from the SuSy subcellular localization to cell walls and adjacent to plasma membranes. Immunolocalization of the cotton fiber SuSy revealed an arrangement pattern similar to that of cellulose microfibril deposition. It was later found that a cellulose synthase rosette-like structure, isolated from azuki beans, lacks cellulose-synthesis activity in the absence of another particle referred to as the catalytic unit. This catalytic unit was enriched with a 78 kDa active SuSy, further supporting the suggested role of SuSy in supplying UDP-G for cellulose synthesis. Another immunolocalization study also demonstrated that cotton fiber SuSy is co-localized with callose, suggesting a dual role for SuSy in cellulose and callose synthesis.

Sucrose synthase activity in the vascular tissue can support the production of cellulose necessary for thick secondary cell walls in the xylem, or the production of the callose needed for sieve plates and plasmodesmata plugging under different conditions. Evidence for a role of SuSy in callose deposition was found in an Arabidopsis double mutant of phloem-specific SUS (*sus5 sus6*). That double mutant had less callose in its phloem plasmodesmata and in response to leaf wounding, as compared with WT or quadruple mutant (*sus1*, *sus2*, *sus3*, and *sus4*) plants.

The role of SuSy in the synthesis of cellulose and callose has been thoroughly investigated in cotton, with cotton fibers serving as a model for these processes. The development of cotton fibers starts with the initiation and elongation of the epidermal cells, followed by secondary growth and maturation marked by massive cellulose production. A fiberless cotton mutant lacking SuSy protein and activity at anthesis was identified, indicating that SuSy might be crucial for fiber initiation. It was later shown that transgenic cotton plants with *SUS* suppression exhibit reduced fiber initiation and elongation and that SuSy may play a role in both the synthesis of cellulose for cell walls and the deposition of callose in the plasmodesmata, which is required for the turgor build-up necessary for cell elongation. In the secondary growth phase of cotton fibers, cellulose synthesis can increase 100-fold relative to the elongation phase and this process probably involves SusC and SusA.

It would be very interesting to see whether the proposed roles of the cell wall SuSy in cellulose and callose synthesis could be observed in transgenic cotton plants with SusC suppression or overexpression. Different reports also support the roles of SuSy in cellulose synthesis in other plant species. Although overexpression of cotton *SUS* in tobacco plants did not affect cellulose content, its overexpression in poplar trees did increase their cellulose content, as well as cell-wall thickness and wood density. Similarly, overexpression of poplar xylem *SUS* in tobacco plants also resulted in increased cellulose content and xylem cell-wall thickness.

In contrast, *SUS* suppression in the developing wood of hybrid aspen (*Populus tremula* L. × *tremuloides* Michx.) did not affect cellulose synthesis specifically, but instead reduced the total amount of carbon incorporated into wood cell walls.

Role of SuSy in Low-Oxygen Environments

Oxygen deficiency (hypoxia) and a complete absence of oxygen (anoxia) are forms of serious abiotic stress that often cause reduced plant growth and productivity. Low-oxygen stress in plants is often caused by flooding, but may also occur naturally in dense, bulky and inner organs and tissues or in very rapidly growing tissues in which oxygen consumption is high. Oxygen is the final acceptor in the mitochondrial electron transport chain and the absence of oxygen blocks electron transfer and cellular ATP production. Some of the plants' responses to oxygen deficiency can occur very rapidly and may involve changes in the transcription and activity of metabolic enzymes.

One of the enzymes thought to be involved in plant responses to hypoxia is SuSy. Transcript levels of some *SUS* genes have been found to increase in response to low levels of oxygen in potato, maize, rice, carrot, Arabidopsis, wheat, beet, pigeon pea and pondweed (*Potamogeton distinctus*).

Oxygen deficiency has also been shown to increase SuSy protein levels in Arabidopsis roots and leaves and in maize roots. In beet, *SUS1* mRNA levels increased under anaerobic conditions, but there was no increase in SuSy1 protein. Increased SuSy activity under low-oxygen conditions has been noted in many plants and is often seen in combination with reduced INV activity in rice seedlings, maize seedlings, Arabidopsis roots, wheat roots and potato tubers.

The possible role of SuSy in metabolism under reduced-oxygen conditions is further supported by the findings of studies with *SUS* mutants and transgenic plants. In maize, the root tip of a double *SUS* mutant (*Sh*, *SUS1*) was shown to be more sensitive to anoxia after a hypoxia pretreatment. Potato tubers of a *SUS* antisense transgenic line were more sensitive to hypoxia than control plants. A study of potato tubers of transgenic plants overexpressing INV or Suc pyrophosphorylase, which allows a way to bypass the degradation of Suc by SuSy, revealed a steeper reduction in oxygen levels inside the tubers, reduced starch synthesis and a lower ATP to ADP ratio, underscoring the importance of SuSy under low-oxygen conditions.

In Arabidopsis, a double-knockout mutant (*sus1* and *sus4*) was found to be more sensitive to flooding than the control, but did not differ from the WT in its responses to hypoxia or anoxia. In cucumber, antisense suppression of *CsSUS3* led to increased sensitivity to hypoxic stress. SuSy activity has also been found to be correlated with rice coleoptile length under submerged conditions, further indicating the advantage of Suc metabolism that involves SuSy under anoxic conditions.

Overall, the data show that some SuSy isoforms may indeed play a vital role in metabolism under low-oxygen conditions. Suc breakdown by SuSy may be more energy efficient than Suc breakdown via INV, as it may save up to two ATP molecules for each Suc molecule converted into hexose monophosphates.

Role of SuSy in the Shoot Apical Meristem

Sucrose synthase may play another, less studied role in the development of shoot apical meristem (SAM). The SAM receives Suc from the phloem and there is evidence that SUS are expressed in the SAM. The SlSUS4 promoter GUS fusion showed activity in young meristematic areas, including the SAM. RNAseq data obtained by researchers indicate that out of the six tomato SUS genes, SlSUS1, SlSUS3, and SlSUS4 transcripts are found in both meristems and in primordia at all development stages in different ratios.

The *SlSUS4* transcript was shown to be present asymmetrically and localized to the primordia from very early stages of development using *in situ* hybridization with an *SlSUS4* antisense probe. *SlSUS4* transcript levels around the SAM also seemed to increase in response to Suc, Glc and brassinosteroid treatments and to decrease when Fru was applied. *In situ* hybridization also revealed the presence of *SUS* transcript in young maize leaf primordia, suggesting a role for SuSy in early leaf development.

Functional studies also suggest that *SUS* genes play significant roles in the SAM and in early leaf development. Transgenic tomato lines with suppressed *SlSUS1*, *SlSUS3*, and *SlSUS4* exhibited abnormal cotyledons and leaf morphology, altered expression of auxin-related genes in the SAM and altered auxin transport, indicating the importance of SuSy for meristem and primordia function in tomato.

Other work involving transgenic plants that overexpress *SUS* genes has revealed altered growth rates that may suggest some possible effects of these genes on SAM function. Overexpression of potato *SUS* in cotton plants led to increased vegetative growth. Similarly, overexpression of aspen (*Popolus tremuloides*) *SUS* in Arabidopsis resulted in an increased growth rate and increased plant biomass, and also induced early flowering. Overexpression of cotton *SUS* in tobacco also led to an increased growth rate and taller plants. Although the mechanism by which *SUS* overexpression speeds up the growth rate is not clear, it is tempting to speculate that increased SuSy activity in the meristem may facilitate increased cell proliferation.

In another study, researchers examined transgenic tobacco plants overexpressing each of the six Arabidopsis *SUS* genes (*AtSUS1-AtSUS6*) and found that all of those lines grew more quickly and were taller and thicker than the WT plants. The transgenic plants overexpressing *AtSUS1* showed increased chlorophyll levels, as well as increased photosynthesis, TSS (total soluble sugars), starch, Suc and Fru, as well as increased enzymatic activity of SPS and SPP in leaves, indicating increased sugar production in the transgenic plants.

In addition, all six lines had increased Suc and Fru levels in their shoot tips and had a high percentage of dichotomous branches (30–40%, as compared to none in the WT), indicating a developmental event in the SAM that divided it into two SAMs and caused the formation of two branches. The results also revealed increased expression of both *WUSCHEL* (*WUS*) and *CycD3*, leading to the increased proliferation of meristematic cells. Since exogenous Suc has been shown to promote *WUS* expression, the increased Suc levels in the SAM of the transgenic plants may have affected *WUS* and *CycD3* expression.

In addition to serving as energy resources and structural components, sugars such as Suc, Glc, and Fru may also act as signaling molecules to regulate developmental processes and responses to environmental changes. These sugars have also been shown to rapidly affect the expression of many genes, even at concentrations as low as 1 mM. *AtSUS1* expression was found to be regulated by Glc via a HXK-dependent pathway, meaning that not only does *SUS* regulate Suc metabolism, but that its own expression is also regulated by sugars. The role of sugars as signaling molecules in the SAM is a subject of lively debate and it is not always easy to differentiate between their signaling function and their metabolic role. In work with Arabidopsis seedlings conducted by researchers, it was shown that both sucrose and light affect *WUS* expression in the SAM, although Suc is unable to release SAM dormancy in the dark. Those authors also found that a non-metabolizable Suc analog, palatinose, has no effect on *WUS* expression in the dark, possibly indicating that Suc *per se* does not act as a signaling molecule in the SAM during seedling establishment.

Other studies have found correlations between Suc treatments or levels and flowering, suggesting that Suc may play a signaling role in the development of SAM into flowers. The Suc signal for flowering may be mediated by trehalose 6-phosphate (T6P). T6P is generated from UDP-G and G6P by trehalose phosphate synthase (TPS) and is thought to be a signaling molecule rather than a metabolic substrate, because it exists at very low levels. T6P levels is an indicator of high Suc levels and expression of *TPS1* in the SAM of Arabidopsis plants induces early flowering, further supporting the proposed roles of Suc and T6P in SAM in flowering.

Another reason to believe that Suc and SuSy may play some regulatory function rather than just a metabolic one comes from tomato plants in which the expression of three *SUS* genes was suppressed. These plants exhibited abnormal leaf development and irregular auxin patterning, suggesting that altered sugar signaling in the SAM or primordia, rather than lower sugar metabolism, is likely to be the cause of these developmental changes.

Other Roles of SuSy

Sucrose synthase may also play other important roles, in addition to its role in Suc cleavage. The localization of SuSy to mitochondria and its possible interaction with a high voltage-dependent anion channel suggest that these SuSy may play a role in regulating solute fluxes between the mitochondria and the cytosol.

Plant SuSy have also been found to play a role in mutualism with symbiotic organisms like Rhizobium bacteria. In pea, a SUS mutant (rug4) exhibited a 90% reduction in SuSy activity in its root nodules. Those plants were incapable of effective nitrogen fixation, even though the nodules appeared normal. Although nitrogenase protein levels were normal, there was no nitrogenase activity. It was suggested that Susy activity might be essential for nitrogen fixation in root nodules, due to the low-oxygen environment in the nodules. Interestingly, the mutation did not appear to affect the plant's symbiotic relationship with arbuscular mycorrhizae.

Sucrose synthase may also play a role in metabolism under heat stress. A recent study found that a SUS3 allele that is highly expressed during seed ripening may confer resistance to the chalky grain phenotype of brown rice caused by heat stress. The expression of the SUS3 gene was found to be higher in the resistant line under heat stress. Interestingly, transgenic plants of a commercial rice cultivar expressing SUS3 showed a decreased percentage of chalky grains under heat stress only when both the promoter and the cDNA of the heat-tolerant allele were introduced, indicating not only the importance of the SUS3 protein, but also the response rate to heat stress in terms of gene expression. Another potential heat-tolerant SuSy was purified from a heat-tolerant line of wheat (WH-1021). That SuSy exhibited optimum activity at 37°C and was stable at temperatures up to 50°C, unlike other SuSy proteins, whose stability decreases at temperatures above 30°C.

In strawberry (*Fragaria ananassa*), SuSy may play an important role in fruit ripening. Strawberry fruits with RNAi suppression of *FaSUS1* by virus-induced gene-silencing exhibited delayed fruit ripening, maintained their firmness and exhibited delayed anthocyanin accumulation.

SUGAR SENSING AND SIGNALING IN PLANTS

Introduction

Since their appearance, plants have played a crucial role in the evolution of life on earth through the production of energy-rich sugar molecules and oxygen by photosynthetic carbon fixation. Sugars are the prime carbon and energy source to build and fuel cells, and early in evolution also acquired important regulatory functions in controlling metabolism, stress resistance, growth and development. For free-living microorganisms, nutrient availability is a major factor controlling growth in a constantly changing environment, while multi-cellular organisms need to maintain nutrient and energy homeostasis within cells and tissues. Sugars can complement and interact with a variety of hormone and growth factor signaling mechanisms to modulate metabolism and growth in complex systems.

In sugar-producing and sessile plants, monitoring the sugar and energy status is particularly critical, enabling the integration of metabolic, environmental and developmental cues required for their physiological and developmental plasticity. Unlike hormones and growth factors, however, sugars are present in the millimolar range and actively participate in metabolism, making the investigation of their dual functions in metabolism and signaling very challenging. Still, in recent years, the pivotal role of sugars in plant growth and development and key players in the sugar signaling network have been uncovered using Arabidopsis as the prime model system. Several comprehensive reviews have been published on the subject. This chapter presents a primer of this rapidly evolving field, with an integrated view on the plant sugar signaling network, and an update on the most recent literature.

Generating Sugar Signals

As photoautotrophic organisms, plants generate their own sugars through the process of photosynthesis. During the day, photosynthetic source tissue converts CO_2 and water to carbohydrates and oxygen, using sunlight as an energy source. Carbohydrates, generated in the chloroplasts, are then exported to the cytosol, mainly as triose-phosphates, where they can be converted to hexose-phosphates or sucrose for local use or storage in the vacuole (Fig. 12).

In addition, sucrose is transported to non-photosynthetic sink tissues. There, sucrose is taken up and converted to different hexoses by invertases and sucrose synthases or stored in vacuoles and in amyloplasts as starch for longer term storage. To ensure a continuous supply of sugars, chloroplasts also store excess photosynthate as transitory starch during the day to be remobilized and exported as maltose and glucose during the subsequent night (Fig. 12). Thus, a variety of sugars are produced by plant metabolism at different times and locations. In order to coordinate all these processes and respond appropriately to changes in the environment (affecting energy metabolism) and altering metabolic and energy demands during development, plants have developed an array of mechanisms to specifically sense diverse 'sugar signals'. An efficient communication between activities in different organs, tissues and subcellular compartments is crucial for an optimal use of the available resources for growth and development.

Carbon Regulation of Source-Sink Status

Interactions between source tissues (producing and exporting sugars, e.g. rosette leaves) and sink tissues (importing sugars, e.g. roots) are an important regulatory mechanism in plants. Feedback inhibition of photosynthetic carbon assimilation as a result of decreased sink demand or nitrogen limitation has been investigated extensively and different experimental approaches have established the pivotal role of sugars in the repression of photosynthetic gene expression. Conversely, sugar depletion results in the activation of photosynthetic gene expression and increases photosynthetic capacity.

Invertases, converting sucrose to glucose and fructose, are key players in the regulation of carbon allocation and different isoforms in distinct subcellular compartments are differentially controlled by sugar signals. Stress and hormones play important roles in fine-tuning sink-source interactions. For example, invertase expression can be induced by microbial elicitors and the growth stimulating cytokinin and brassinosteroid hormones. In contrast, ethylene represses the expression of extracellular invertase.

When assimilates are generated in excess of sink demand by high rates of photosynthesis, sugar accumulation in leaves has direct effects on ADP-glucose pyrophosphorylase (AGP) activity, the first committed step in starch synthesis, and other carbohydrate-responsive enzymes of sucrose and starch metabolism. For instance, sucrose can induce AGP large subunit (APL3) gene expression and redox-dependent activaton of AGP in Arabidopsis rosette leaves. During the night, reduced leaf sucrose content can trigger starch degradation. In addition, sucrose feeding in sugar beet leaves represses the expression of a sucrose symporter gene in the phloem companion cells, which may provide a regulatory mechanism in the coordination between source photosynthetic activity and sink utilization of photoassimilate. These examples illustrate the enormous complexity of source-sink regulation, the generation and sequestration or storage of sugars and sugar signals, and the delicate balance required for optimal growth.

SUGAR SENSING

HXK1 as a Glucose Sensor

Glucose is a universal carbon and energy source, by far preferred by organisms ranging from unicellular microbes to plants and animals. While sucrose is the most important transport sugar in plants, most regulatory effects can be ascribed to glucose. In many organisms, HXKs and other sugar kinases are the most ancient, conserved sugar sensors. A regulatory role for plant HXKs in plant sugar sensing was first suggested by the marked effects of various sugars on photosynthetic and glyoxylate cycle gene expression in *Chenopodium* and cucumber cell cultures and maize mesophyll protoplasts. The reduction in photosynthesis related gene expression by low amounts of glucose (1–10 mM) but not glucose-6-phosphate (G6P) in a maize protoplast transient expression system, supported a signaling role for HXK.

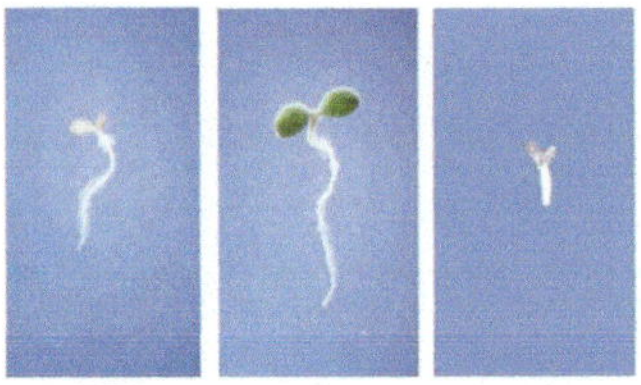

Fig. 13. Arabidopsis WT, *glucose insensitive* (*gin*) and *glucose oversensitive* (*glo*) seedling growth phenotypes on MS medium containing high (6%) exogenous glucose levels. Wild type seedlings are affected in cotyledon greening and expansion and show reduced hypocotyl and root elongation.

Other substrates for HXK phosphorylation, like fructose, mannose and 2-deoxyglucose (2-dGlc) also trigger repression. Moreover, the observed repression can be blocked by the HXK-specific competitive inhibitor mannoheptulose. Conversely, sugars that cannot be transported into the cell (L-glucose), or that are taken up but cannot be phosphorylated by HXK (6-deoxyglucose and 3-O-methylglucose), as well as intermediates of glycolysis and sugar phosphates (delivered into the protoplasts by electroporation), do not cause the same repression. The possibility that depletion of Pi and ATP caused the reduced gene expression is excluded by the inability of P_i and ATP to relieve the glucose repression.

The hypothesis that HXK acts as a glucose sensor was further substantiated by the characterization of transgenic Arabidopsis *HXK* sense and antisense lines. Wild type Arabidopsis seedlings grown under continuous light on MS medium containing 6% (approximately 333 mM) glucose exhibit developmental arrest with repression of cotyledon greening and expansion, leaf development, and hypocotyl and root elongation. Plants carrying the antisense construct of *HXK1* are hyposensitive to glucose repression, while plants overexpressing *HXK1* are hypersensitive (Fig. 13).

No apparent differences are observed with the same concentration of mannitol or 3-O-methylglucose as osmotic controls. Further evidence for HXK1 as a sugar sensor is provided by the isolation of two *gin2* (*glucose insensitive 2*) mutants mapped to the *HXK1* gene. Importantly, two catalytically inactive HXK1 alleles (G104D and S177A) can mediate developmental arrest on 6% glucose and glucose repression of chlorophyll accumulation and photosynthetic gene expression in the *gin2* mutant background. These results unambiguously show that HXK1 is a true glucose sensor in plants and that glucose signaling can be uncoupled from glucose metabolism.

The *gin2* mutants provide a valuable tool to investigate the physiological importance of glucose sensing and signaling during normal plant growth, without the manipulation of exogenous sugars. In wild type plants, high light conditions (200–300 μE) can boost photosynthesis and sugar production, leading to accelerated growth with earlier leaf senescence.

However, *gin2* plants remain small and dark green and show little cell expansion, supporting the role of HXK1 in promoting cell expansion and growth in roots, leaves and inflorescences under enhanced light conditions. In transgenic tomato plants, displaying higher sugar sensitivity than Arabidopsis plants, *HXK1* overexpression inhibits growth and chlorophyll accumulation and promotes senescence. Thus, HXK1-mediated signaling can promote or repress growth, depending on the intrinsic glucose levels and glucose sensitivity of the plants.

In the Arabidopsis genome, six HXK and HXK-like (HKL) genes can be found, serving a variety of physiological functions and likely controlled by tissue-specific expression patterns, sub-cellular localization and protein complex formation. Proteomic and GFP fusion analyses indicate that several plant HXKs are located on the outer mitochondrial membrane, where a completely functional glycolytic metabolon can be found.

In addition, plant HXKs are found in plastids. Unique in moss, the chloroplast stromal HXK accounts for 80% of the total hexose kinase activity and is responsible for glucose-mediated growth and formation of caulonemal filaments. Several recent papers also report diverse functions for mitochondrial HXKs in processes ranging from programmed cell death to actin filament reorganization and glucose-mediated gene expression. It will be interesting to determine how plant HXKs integrate glycolysis and cell death without the proapoptotic protein, associated with the mitochondrial glucokinase in mammalian cells. The functional significance of HXK interactions with porin, actin and voltage-dependent anion channels deserves further investigation.

HXK1 is also found in high molecular weight complexes in the nucleus. Apparently, HXK1 regulates transcription by direct binding to the promoter of glucose-repressed genes. Chromatin immunoprecipitation experiments with HXK1 show that this complex specifically binds to *cis*-regulatory elements upstream of the *CAB2* (chlorophyll a/b binding protein 2) and *CAB3* coding regions. This nuclear activity requires two HXK1 unconventional partners, HUP1 and HUP2, components shown to be part of the plant vacuolar H^+-ATPase (VHA-B1) and the 19S regulatory particle (RPT5B) of the proteasome complexes, respectively. Interestingly, both interaction partners have closely related homologues that do not interact with HXK1 in Arabidopsis. In addition, *vha-b1* and *rpt5b* mutants display similar seedling and adult plant phenotypes as observed for *gin2* plants, confirming the vital role of their interaction with HXK1 in glucose signaling.

The HXK-Independent Pathway

Analysis of sugar-regulated gene expression and growth of wild type, *35S-HXK1* and *35S-antiHXK1*, and *35S-ScHXK2* (yeast *Saccharomyces cerevisiae HXK2*) Arabidopsis plants confirms the existence of both HXK1-dependent and -independent signaling pathways. Interestingly, the effect of glucose on the expression of genes encoding AGP, CHS (chalcone synthase), PAL (phenylalanine ammonialyase) and ASN (asparagine synthetase) is similar in wild type plants and transgenics, indicating that this effect is independent of HXK1.

Furthermore, seed germination can be inhibited by glucose in concentrations below those known to inhibit early seedling development. This inhibition operates on imbibed seeds and is independent of HXK function (Dekkers et al., 2004). Interestingly, the HXK-dependent and HXK1-independent pathways can cooperate in glucose regulation, acting at different levels. For example, glucose repression of the grape monosaccharide transporter *VvHT1* expression is dependent on HXK1, while its post-transcriptional glucose regulation is not.

Since sugars are extensively metabolized, downstream metabolites are also potential signaling molecules. Glucose induction of two pathogenesis-related genes (*PR1* and *PR2*) is higher in plants overexpressing *HXK1* but absent in *35S-antiHXK1* plants, consistent with the requirement of HXK for signaling. However, this induction is enhanced both in plants overexpressing *HXK1* and plants over-expressing yeast Sc*HXK2*, suggesting that the induction of *PR1* and *PR2* is dependent on HXK catalytic activity rather than on its regulatory function. This indicates that the regulatory signal is possibly generated by downstream metabolic intermediates. Genome-wide expression analyses with wild-type and *gin2* mutant plants will reveal more genes regulated by HXK1-dependent and HXK1-independent sugar signaling under various growth conditions.

Disaccharide Sensing

Sucrose is the main transported sugar in plants. Although sucrose can be substituted by the hexose products in many cases of sugar effects, a number of studies have provided compelling evidence for sucrose-specific regulation in gene expression and growth. Sucrose-induced expression has been reported for genes encoding UDP-glucose pyrophosphorylase, anthocyanin biosynthesis enzymes, and a putative chloroplast glucose-6-phosphate/phosphate translocator. Conversely, sucrose-specific repression has been shown for the transcription of a proton-sucrose symporter gene in sugar beet and for carrot somatic embryo radicle elongation. Also, the marked effects on leaf shape, caused by the overexpression of the homeodomain leucine zipper transcription factor gene *ATHB13*, are specifically induced by sucrose.

Specific disaccharide signaling, independent of metabolism, is also demonstrated using non-metabolizable sucrose analogs, such as palatinose and turanose. These analogs affect the expression of α-amylase in barley embryos and AGP in potato, and induce starch synthesis and sucrose degradation in potato tubers. However, they may activate novel pathways different from sucrose-signaling as they could be perceived as a more general stress-related signal rather than a specific non-metabolizable sucrose substitute.

Another line of fascinating research suggests that sucrose regulates the translation of the ATB2 protein. The *ATB2* gene encodes a group S basic region leucine zipper transcription factor (bZIP11), characterized by an unusually long 52 untranslated region (UTR). Point mutations and deletions in this 52 UTR have shown that a highly conserved small upstream open reading frame (uORF) encoding 42 amino acids is responsible for sucrose induced repression of translation. This uORF is also conserved in the 52 UTRs of other S-class bZIP transcription factors of Arabidopsis and many other plant species.

Expression of a 52 UTR-GUS (β-glucuronidase) fusion construct throughout the whole plant, indicates that this sucrose regulation mechanism is functional in all plant tissues. To demonstrate the specificity of sucrose in its repression mechanism, several other sugars, such as glucose, fructose, galactose, maltose and raffinose, were tested, but none induces repression of translation to the same extend as sucrose. Intriguingly, *ATB2* transcription is induced both by sucrose and glucose, providing a mechanism of rigorous fine-tuning of ATB2 protein levels in response to sucrose (Fig. 12).

Sucrose may also act as a signal, controlling specific developmental processes not affected by hexose sugars. In *Vicia faba* embryos, high hexose levels are associated with active division during early embryo development, while sucrose accumulation is correlated with the switch from active cell division to cell expansion, differentiation and reserve (starch) accumulation (Weber et al., 1996). It will be important to uncover how the sucrose signal is perceived and transduced to regulate the diverse responses.

Cell Surface Receptors

In yeast and mammals, extracellular sucrose and glucose can be perceived by G-protein-coupled-receptors (GPCRs) linked to heterotrimeric guanine nucleotide-binding proteins (G-proteins). In such a system, the extracellular signal molecule (ligand) specifically binds to the GPCR, resulting in a conformational change that stimulates the exchange of GTP for GDP at the guanine-nucleotide binding site on the interacting G-protein. This initiates intracellular signaling by dissociation of the α- and the β/γ-subunits and their interaction with a variety of downstream effectors. G-proteins act as molecular switches and intrinsic GTPase activity catalyses the return to the inactive heterotrimer. In many cases, GTPase activity and desensitization is accelerated by a new class of regulators of G-protein signaling (RGS) proteins.

One Arabidopsis gene encodes an unusual hybrid seven-transmembrane domain protein with a C-terminal RGS domain, RGS1. The rgs1 mutants are insensitive to 6% glucose, while overexpressors are hypersensitive. Growth assays with different sugars and sugar analogs suggested that RGS1 may function as a glucose sensor, independently of HXK (Fig. 12). Glucose appears to alter the interaction of RGS1 with GPA1 (the only Ga subunit in Arabidopsis) in vivo, leading to an increased GTP hydrolysis activity of GPA1 and sugar-specific signaling events.

In addition, Arabidopsis gpa1 mutants are hypersensitive to inhibition of germination by sugar and ABA, confirming their potential role in sugar signaling. Interestingly, THF1 (THYLAKOID FORMATION1), a protein found in the outer membrane and stroma of plastids, interacts with GPA1 in root hair cells when plasma membrane and plastid membrane are in close proximity. The thf1 mutant seedlings show altered sensitivity to glucose and THF1 protein levels are rapidly degraded by glucose, suggesting a role for this protein in the sugar response(Fig. 12).

Besides GPA1, three Arabidopsis genes encode one Gb (AGB1) and two Gg subunits (AGG1 and AGG2). A loss of function mutation in the Gb subunit (agb1) affects multiple processes, including hypersensitivity to glucose. Although the gpa1 and agb1 mutants often have opposite phenotypes, they seem to share the same function in glucose signaling. Interestingly, a mutation in a Golgi-localized hexose transporter (sgb1, suppressor of G beta1) can reverse glucose hypersensitivity in the agb1 mutants. In addition, agg1 and agg2 mutants are also hypersensitive to glucose. As agg1 mutants are sensitive to mannitol and more susceptible to infection, these proteins are likely involved in different cellular processes.

In budding yeast, novel hexose transporter-like genes, SNF3 and RGT2, encode extracellular glucose sensors that are responsible for the regulation of functional hexose transporter gene expression. These catalytically inactive transporters possess a long cytoplasmic domain for interaction with downstream signaling components. The plant disaccharide transporter SUT2, containing a similar extended cytoplasmic N-terminal and central loop domain, has been proposed to be a putative sucrose sensor. In addition, some monosaccharide transporters also have extended cytoplasmic loops. Further research is required to demonstrate that these putative receptors are actively involved in sugar sensing.

SUGAR SIGNALING MECHANISMS

After sugars have been detected by sensors at the plasma membrane or a specific intracellular location, the information is passed on through signal transduction and amplifying cascades, ultimately resulting in the appropriate responses. These responses vary from changes in gene expression to altered enzyme activities through the activity of protein kinases, protein phosphatases and other signal transduction mediators such as Ca^{2+} and calmodulin. The characterization of different sugar-signaling mutants isolated in diverse screens has also revealed extensive interactions with hormone biosynthesis and signaling pathways.

SUGAR-SIGNALING MUTANTS

Two major strategies have been used to screen for sugar signaling mutants in Arabidopsis. A simple approach is based on the effect of high levels of external sugars on seedling growth and development. For example, gin (glucose insensitive) mutants develop normally on MS medium supplemented with 6% (approximately 333 mM) glucose, while sis (sugar insensitive) and sig (sucrose insensitive growth) mutants are selected on media containing 300 mM and 350 mM sucrose, respectively (Fig. 13). Conversely, glo (glucose oversensitive), gss (glucose-supersensitive), and sss (sucrose-supersensitive) mutants show developmental arrest in the presence of otherwise non-inhibiting sugar concentrations. Remarkably, many mutants isolated in the different screens are allelic to mutants in hormone biosynthesis and signaling. Since these screens are done with seedlings, the information on sugar-hormone interactions in the signaling networks is necessarily limited to the early developmental stages.

To ensure uniform sugar responses, the initial mutant isolation screens were mostly performed on media containing high glucose or sucrose concentrations, raising concerns about the physiological relevance of sugar regulation. Moreover, the phenotypes could also be influenced by osmotic stress and other nutrients in the media. The analyses of wild-type and gin2 seedlings show that the presence of high nitrate concentrations (40 mM nitrate) in full MS medium can antagonize the glucose effects. When the nitrate effect can be minimized, the sugar effect can be observed at more physiological concentrations. Indeed, 2% (110 mM) glucose without MS medium is sufficient to cause glucose repression of RBCS, CAB and CAA (carbonic anhydrase) gene expression and chlorophyll accumulation in whole seedlings.

For HXK1-mediated growth promotion in seedlings, 0.2% (11 mM) glucose in a 0.1× MS medium is sufficient to provide the glucose signal requirement in vivo. In isolated maize and Arabidopsis mesophyll protoplasts, concentrations between 1 to 20 mM are sufficient to cause specific glucose but not sucrose repression in various assays. Recent microarray studies show that 15 mM sucrose can modulate the expression of a large number of Arabidopsis genes in carbon-deprived seedlings grown in liquid culture. Interestingly, changes in light intensity are correlated with endogenous glucose levels and plant growth phenotypes. The use of mutants (e.g. pgm, phosphoglucomutase) with different endogenous sugar levels has further substantiated the physiological relevance of the sugar sensing and signaling network in Arabidopsis. It would be interesting to examine the effects of pgm at various endogenous sugar levels without the presence of high nitrate.

Other more elaborate mutant screens are based on altered responses to sugar repression or activation of gene expression in transgenic reporter lines. For example, the sun (sucrose uncoupled) mutants show high activity of a sucrose-repressed plastocyanin (PC) promoter fused to a luciferase reporter in the presence of 3% sucrose. The isi (impaired sucrose induction) mutants loose their ability to induce ApL3 gene expression on sucrose-containing medium. The regulatory sequence of the ApL3 gene was fused to a negative selection marker. Hence, seedlings surviving this screen are impaired in sucrose induction of the starch biosynthesis gene promoter. Both sugar insensitive and hypersensitive mutants are also obtained based on the sugar-inducible expression of a b-amylase gene. Other screens include the use of the sugar inducible patatin promoter in front of a GUS reporter gene (rsr, reduced sucrose response mutants) or the ApL3 promoter in front of a luciferase reporter gene (hsr, high sugar response mutants).

Complex sugar signaling mutants are also identified in unrelated genetic screens. For example, the hypersenescence1 (hys1) mutant was isolated based on an early senescence screen and displays a sugar hypersensitive phenotype. Analyses of the double mutant hys1 gin2 reveal HXK1-dependent action of HYS1 in glucose-mediated developmental arrest but HXK1-independent function of HYS1 in senescence. The hys1 mutant is allelic to cpr5 (constitutive expression of pr genes5) with elevated salicylic acid (SA) levels. HXK1 and HYS1 act synergistically in the negative regulation of SA accumulation. Another unique sugar mutant, prl1 (pleiotropic regulatory locus1), displays complex hormone and sugar response phenotypes and is recently found to exhibit severe defects in plant defense.

PRL1 interacts with the C-terminal region of the Arabidopsis SnRK1 protein kinases KIN10 and KIN11, acting as a negative regulator. This C-terminal part of KIN10 and KIN11 also interacts with both the SCF ubiquitin E3 ligase subunit SKP1/ASK1 and the SKP1/ASK1-binding 26S proteasome. In addition, PRL1 is associated with CUL4, a subunit of the E3 ligase, in vivo. KIN10 degradation is slower both in the prl1 and cul4cs mutants, suggesting a link between SnRK1 and the ubiquitin/26S proteasome complex. Intriguingly, the glucose hypersensitive phenotypes found in the dark-grown hsr8 (high sugar response 8) mutant are suppressed by prl1, indicating that PRL1 acts as a positive regulator for the specific glucose effects observed in etiolated hsr8 seedlings.

Under light, hsr8 does not show overt phenotypes whereas prl1 displays hypersensitive glucose repression in developmental arrest, suggesting complex effects of glucose regulation in different growth conditions. Map-based cloning reveals that HSR8 is allelic to MUR4, encoding an enzyme involved in arabinose synthesis. The studies indicate that failure of arabinose synthesis and consequent changes in cell wall composition could have a strong impact on sugar signaling.

It remains an important challenge to link complex sugar phenotypes to specific protein-protein interactions and signal transduction pathways.

Interplay Between Abscisic Acid and Sugar Signaling

A first indication for crosstalk between sugar and ABA came from the observation that several glucose insensitive mutants (gin1/isi4/sis4 and gin5/isi2/sis3) are allelic to ABA biosynthesis mutants (aba2 and aba3) with lower endogenous ABA levels and reduced seed dormancy.

The extensive crosstalk is further corroborated by the inability of glucose to repress the CAB1 and plastocyanin (PC) genes in the gin1/aba2 and gin5/aba3 backgrounds. Also, overexpression of HXK1 in the gin1 and gin5 mutants does not alter glucose-insensitivity, indicating that ABA acts downstream of HXK1. Interestingly, ABA accumulation and transcript levels of several ABA biosynthesis genes are significantly increased by glucose, providing a mechanistic link at the molecular level.

Except for ABA2, which is specifically upregulated by glucose but not by ABA, several ABA biosynthesis genes (ABA1, ABA3, NCED3 and AAO3) show a synergistic effect of glucose and ABA in their induction. While expression of the APL3 gene is not induced by ABA, its sucrose-induction is significantly enhanced by ABA, indicating that sugar responses are dependent on an ABA-controlled responsiveness of the tissue. It is suggested that, through induction of ABA synthesis, high concentrations of sugars would make the seedlings also more responsive to the sugar signal.

More studies are needed to reveal whether ABA biosynthesis is directly involved in the sugar signal transduction cascade or indirectly stimulated by sugars, and modulating sugar-responsiveness. Since many of these mutants were isolated in screens on high amounts of glucose, glucose-specific (likely ABA-mediated) osmotic effects should also be considered.

In addition to mutants in ABA biosynthesis, several sugar-response mutants are affected in ABA signaling. The gin6 mutant contains a T-DNA insertion in the promoter of the ABI4 (ABSCISIC ACID INSENSITIVE 4) gene. In screens for mutants insensitive to sucrose, sun6, sis5 and isi3 were found to be allelic to ABI4. ABI4 encodes a transcription factor of the APETALA 2 (AP2) domain family, involved in seed development and germination. ABI4 expression is induced by high glucose levels and this induction is abolished in the gin6 mutant.

The gin6 mutant also shows insensitivity to ABA during seed germination, suggesting that cis-acting elements are required for the ABA and high glucose regulation of the ABI4 gene expression. Interestingly, the glucose-insensitive abi4 phenotype cannot be rescued by the addition of ABA, suggesting that ABI4 acts downstream of the ABA-mediated glucose signaling cascade.

However, detailed analysis of glucose effects on seed germination shows that there is a differential sensitivity to different glucose levels at distinct developmental stages. The same low glucose levels are inhibitory for seed germination but stimulatory for seedling growth and development. In addition, ABA is not required for sugar-dependent induction of leaf senescence, although leaf senescence can be triggered independently by application of ABA and sugars. The complexity of these data suggests caution when analyzing tissue-and developmental stage-specific crosstalk between sugars and hormones.

In addition to abi4, mutation of ABI5, encoding a transcription factor belonging to a large basic leucine zipper (bZIP) gene family, similarly confers a glucose-insensitive phenotype, albeit more moderate. Conversely, overexpression of ABI5 and some of the closely related bZIP transcription factors (ABF2, ABF3 and ABF4) confers hypersensitivity to sugars. All of these transcription factors bind to the ABA responsive cis-element ABRE.

However, abf3 and abf4 loss-of-function mutants are not insensitive to glucose, but display defects in salt and dehydration responses, suggesting that they may only play a minor role in glucose-mediated signaling. Interestingly, the ABI5 protein is only expressed in vegetative tissue shortly after seed germination, consistent with a role in monitoring environmental conditions during early seedling development. Furthermore, ABI5 gene expression is activated by glucose in an ABA-dependent manner, and, unlike ABI4, highly inducible by ABA and mannitol. Still, both ABI4 and ABI5 respond to the glucose analog 2-deoxy-glucose, suggesting a common HXK-dependent expression regulation.

Two other ABA signaling components may also be involved in the sugar response. Overexpression of the transcription factor ABI3 leads to glucose hypersensitivity at the seedling stage. Expression of ABI3 is induced by 6% glucose in an ABA-dependent manner, albeit to lower levels than ABI4 and ABI5, and loss-of-function abi3 mutants are insensitive to the glucose-induced inhibitory effect on seed germination. Based on glucose-induced ABI3 expression, altered glucose-mediated gene expression in the abi3 mutants and glucose insensitive phenotypes of abi3 mutant alleles, it is concluded that ABI3 is an essential component in the regulation of glucose insensitivity. In addition, loss-of-function mutation of ABI8, encoding a small plant-specific protein with unknown function, confers highly stunted root growth, which can be partially rescued by glucose.

While several ABA-signaling transcription factors appear to be involved in sugar-mediated responses of early seedling development, two dominant ABA-insensitive mutants, abi1 and abi2, respond normally to high concentrations of external glucose. Considering the complex nature of ABA signaling and the ABA-glucose interactions, more detailed molecular analyses will be necessary to determine the genetic and spatio-temporal network linking ABA with the mechanisms conducting sugar signaling.

Interaction with Ethylene Signaling

Interactions between the ethylene and sugar-signaling pathways were first suggested by the finding that the ethylene precursor 1-aminocyclopropane-1-carboxylic acid (ACC) prevents inhibition of cotyledon greening and expansion at high concentrations of glucose in wild type seedlings. In addition, the ethylene insensitive mutants *etr1*, *ein2* and *ein3* display glucose hypersensitivity. Conversely, the ethylene overproduction mutant (*eto1*) and the constitutive ethylene triple response mutant (*ctr1/gin4/sis1*) are glucose insensitive. However, consistent with the idea that the crosstalk between sugar and hormone-signaling pathways is strongly dependent on glucose levels and developmental stage, the *ctr1* mutant is sensitive to low glucose-mediated seed germination.

To further elucidate the role of GIN1 (later identified as ABA2) in the ethylene signal transduction cascade, etr1 gin1 and ein2 gin1 double mutants were generated and analyzed. These double mutants display a glucose insensitive phenotype, indicating that ethylene affects glucose signaling partially through ABA to promote germination and seedling development.

Moreover, protein stability of the transcription factor EIN3 (ETHYLENE INSENSITIVE 3), a key regulator in ethylene signalling acting downstream of ETR1, EIN2 and MKK9, is differentially regulated by glucose and ethylene. In maize and Arabidopsis mesophyll protoplasts, glucose enhances the proteasome-dependent degradation of EIN3 proteins in the nucleus. This glucose-induced degradation is dependent on HXK1, as it is no longer detected in gin2 mutants. Transgenic plants overexpressing EIN3 are insensitive to glucose, whereas ein3 mutants are hypersensitive. The degradation of EIN3 proceeds much faster in seedlings treated with glucose and the presence of the ethylene precursor ACC delays EIN3 degradation in Arabidopsis plants. The glucose hypersensitivity observed in etr1, ein2, ein6 and mkk9 mutants and the allelism of gin4 with ctr1 can also be explained by these findings.

Other Interactions between Sugar and Hormone Signaling

The observation that gin2 mutants show growth defects under high light suggests a link between glucose signaling and hormones important in cell elongation, such as auxin, gibberellins and brassinosteroids. Tissue cultures assays show that gin2 mutants are indeed defective in auxin-induced cell proliferation and root formation. Since auxin concentrations are similar in wild type and gin2 hypocotyl, and exogenously applied IAA (indole-3-acetic acid) does not induce cell proliferation in gin2 hypocotyls, gin2 mutants are likely deficient in auxin signaling and/or uptake.

Furthermore, the auxin-resistant mutants, axr1, axr2 and tir1, are insensitive to growth inhibition on 6% glucose. Remarkably, the catalytically inactive HXK1 alleles restore auxin-mediated cell proliferation in gin2 tissue cultures, demonstrating that glucose sensing by HXK acts synergistically with auxin signaling. Another aspect of sugar and auxin interactions is revealed by the hookless1 (hls1) mutant that enhances sugar-induced gene expression, which can be partially eliminated by IAA.

The delayed leaf senescence found in gin2 plants links sugars with another plant hormone, cytokinin. In tissue culture, gin2 mutants are hypersensitive to cytokinins for shoot regeneration. The addition of cytokinins to seedlings grown on 6% glucose also eliminates the developmental arrest. Because cytokinins promote ethylene biosynthesis and ethylene counteracts the effects of glucose, the effect of cytokinins could be indirect.

As cytokinins still promote greening in the ethylene insensitive mutants etr1 and ein2, ethylene and cytokinin act independently in response to glucose. Furthermore, constitutive cytokinin signaling transgenic plants can overcome the glucose repression response, supporting the antagonistic relationship between cytokinin and sugar signaling. A recent analysis of the hys1 mutant reveals that its hypersensitivity to glucose is HXK1-dependent but cannot be prevented by adding ethylene or cytokinin, suggesting broad HXK1 effects beyond ethylene and cytokinin signaling.

Several studies support a close link between sugars and hormones in the control of cell division and growth. Cyclin D3 (CycD3), important in the control of G1 to S-phase transition during the cell cycle, is induced by cytokinins in the presence of sucrose. In the absence of sucrose, addition of auxin or cytokinin no longer induces its expression. In contrast, sucrose alone is able to induce expression of this cyclin gene, suggesting that sucrose acts upstream of hormones or synergistically with hormones in regulating CycD3 expression.

Another hormone that interacts closely with sugars is gibberellin (GA). Glucose can affect GA-mediated a-amylase expression in barley embryos (Perata et al., 1997). Interestingly, several studies show that glucose and GA responses share the same promoter elements in rice. Three MYB transcription factors from rice bind the glucose and GA response element TATCCA, present in all promoters of a-amylase genes isolated from cereals, have been identified. All three MYBs are involved in the glucose-regulation and cooperate with a GA-regulated transcription factor to regulate expression in the absence of GA.

This suggests that these three MYB transcription factors play roles in both glucose- and GA-mediated regulation of a-amylase gene expression. In addition, RGL2 (RGA-like) and SPY (SPINDLY), two negative regulators of GA signaling may be involved in glucose repression of seed germination as rgl2 and spy mutants germinate normally in the presence of inhibiting glucose concentrations. Recently, it is shown that GA represses the sucrose activation of anthocyanin biosynthesis genes, whereas ABA and jasmonate promote the effect of sucrose.

Finally, several recent studies also link SA and brassinosteroids (BR) with sugar signaling. The sugar hypersensitivity of the bls1 (brassinosteroid, light and sugar1) mutant can be repressed by exogenous BR application, suggesting crosstalk between BR and glucose signaling. The sucrose- and glucose-induced expression of the PR2 and PR5 genes are abolished in NahG mutants with reduced SA levels, suggesting a role for SA in sugar-induction of PR2 and PR5 expression.

INTERACTIONS BETWEEN SUGARS AND OTHER NUTRIENTS

Carbon and nitrogen metabolism are tightly linked and need to be coordinated in order to maintain proper plant growth and development. Carbohydrates provide energy and the carbon backbone for nitrate and ammonium assimilation in the biosynthesis of amino acids. Nitrogen is essential for incorporation in nucleic acids, amino acids and chlorophyll. The integration of sugar and nitrogen metabolism as well as sensing and signaling enables plants to respond to acute changes in internal and external nitrogen concentrations or carbon supplies by regulating the expression of genes involved in nitrogen transport, assimilation and metabolism.

It has been clearly illustrated that Arabidopsis seedling growth, lipid mobilization and photosynthetic gene expression is profoundly influenced by the availability of sugars and nitrogen. Recently, the osu1 (oversensitive to sugar1) mutant, deficient in a putative methyl-transferase, was isolated based on enhanced anthocyanin accumulation and root inhibition under imbalanced sugar and nitrogen conditions. Interestingly, genes encoding diverse transporters for nitrogen and other nutrient uptake are up-regulated by sugars. Several studies have demonstrated the genome-wide expression regulation by nitrogen (nitrate and ammonium), sugars and their various combinations.

Although nitrogen and sugars each specifically regulate the expression of a subset of genes, sucrose or glucose are more potent signals, regulating the majority of the genome responses, and nitrogen appears to play an important role in modifying sugar responses. A recent microarray study supports the idea that nitrogen and sugars are sensed through sugar- and nitrogen-responsive pathways in Arabidopsis roots, illustrating an interconnection at the level of signaling or metabolism.

An informative system to study the effects of nitrogen and sugars on plant growth and development is based on the formation of lateral roots since their number and position are strongly dependent on nutritional cues. High sugar concentrations repress lateral root formation, and the lateral root initiation1 (lin1) mutant shows lateral root initiation even in repressing environmental conditions (high external sucrose/low external nitrate). The lin1 mutant is deficient in the high affinity nitrate transporter, NRT2. Since expression of NRT2 is induced by low exogenous nitrogen and by sucrose under low nitrogen conditions, NRT2 is a negative regulator of lateral root development in the signal transduction pathway sensing environmental status.

Another nutrient closely interacting with sugar signaling is inorganic phosphate (Pi). Pi is an essential component in many plant metabolites and plays an important role in energy balance and carbon assimilation. Since Pi is often limiting in soils, plants have evolved specific mechanisms to sense and adapt to Pi starvation. Interestingly, changes in internal Pi concentrations are often linked with altered source-sink distribution and carbohydrate content. Characterization of the phosphorus-deficient pho3 mutant (mutated in the SUC2 gene, encoding a major sucrose-proton symporter) shows that reduced levels of Pi in rosette leaves is associated with increased levels of sucrose, anthocyanins and starch. In addition, knock-out of the MYB-related transcription factor PHR1, involved in the response to Pistarvation, results in altered Pi allocation between roots and shoots, and reduced levels of starch and sugars.

Microarray analyses of Pi and/or sucrose treated Arabidopsis leaf segments show that nearly 150 genes are synergistically or antagonistically regulated by both stimuli. Interestingly, expression of several known sucrose-regulated genes including CHS and b-AMY (b-amylase) can be modulated by Pi starvation and vice versa, suggesting a tight crosstalk between Pi and sucrose-regulated gene expression. Sugars are essential in regulating Pistarvation mediated gene expression and root architecture, and it has been suggested that this response is partially mediated by HXK1. However, the sugar regulation of CHS and b-AMY expression is HXK-independent. These studies demonstrate the complexity of crosstalk between sugar and phosphate signaling and more analyses will be necessary to reveal the molecular nature of their interactions. An overview of Arabidopsis genes mediating sugar sensitivity is summarized in Table III.

TRANSCRIPTION CASCADES

One way for plants to respond to diverse sugar signals is to rapidly modulate gene expression. Using a variety of methods and assays, selected genes have been found to be specifically regulated by sugars. Microarray technology now enables genome-wide identification of specific sets of sugar-regulated genes in Arabidopsis seedlings and adult leaves. Unraveling the physiological functions of these Arabidopsis genes will bring new insight into our understanding of the sugar regulation and signaling network.

During the diurnal cycle, changes in sugar levels are accompanied by alterations in transcript levels. These alterations are strongly amplified in the starchless pgm mutant, which contains higher sugar levels than wild-type plants during the day and very low sugar levels at the end of the night, suggesting that the transcriptional changes observed during the diurnal cycle are caused by sugars. Using the pgm mutant, it was shown that these changes in sugar levels can even override circadian behavior. Interestingly, the extensive gene expression reprogramming promoted especially by carbon depletion can be effectively blocked by the addition of sucrose or glucose.

Genes involved in diverse metabolic pathways are activated or repressed by sugars. Importantly, many genes encoding transcription factors and signaling components are also highly regulated by sugars or sugar starvation, providing a molecular link of sugar signaling to multiple regulatory pathways and hormone signaling. It is important to note that some of these microarray experiments were carried out in the presence of high nitrogen levels and may not reveal a comprehensive coverage of sugar-regulated gene expression. Future studies aim at understanding the functional relations and kinetics of sugar-regulated gene expression will faciliate the elucidation of the transcriptional cascades and signaling networks central to the control of plant growth and development.

PROTEIN KINASES AND PHOSPHATASES

In all eukaryotic cell types, both protein kinases (PK) and phosphatases (PP) have established roles as central regulators of signal transduction. The first indication for the involvement of phosphatases in plant sugar signaling came from the use of PP1 and PP2A phosphatase inhibitors, mimicking sugar-induced repression of photosynthetic gene expression in maize protoplasts and Chenopodium cell cultures and blocking sucrose-induced expression of genes encoding sporamin, b-amylase and the small subunit of AGP in sweet potato petioles. Together with glucose- and stress-related stimuli, these PP inhibitors also activate the expression of a stress-inducible invertase and PAL genes, while rapidly activating a putative mitogen-activated protein kinase in Chenopodium cell cultures.

The differential effects of the PK inhibitor staurosporine on stress-and glucose-related responses suggest the involvement of different PKs in these processes. Similar pharmacological approaches have also shown the involvement of distinct PPs in dark- and sugar starvation-induced DIN (DARK INDUCED) gene expression. Studies with Ca2+ channel blockers (LaCl3), EGTA and calmodulin inhibitors have also shown the involvement of Ca2+ signaling in sugar induction of sporamin and b-amylase genes in sweet potato (Ohto and Nakamura, 1995). It is possible that Ca2+ signaling mediates protein phosphorylation in some of the sugar signaling pathways. For example, sucrose can induce cytosolic calcium fluxes and increases expression of a plasma membrane-associated calcium-dependent protein kinase (CDPK) from tobacco. The future challenges are to connect various sugar sensors and signaling pathways to the functions of specific PKs and PPs and sugar-regulated target genes.

ENERGY AND STRESS SIGNALING

Studies of sugar-regulated genes have mainly focused on establishing connections to specific sugar sensing and signal transduction pathways. However, more detailed investigation of some sugar-repressed genes reveals the interesting finding that these genes are activated in darkness, starvation or/and under diverse stress conditions), and are likely essential for plant survival under dark-induced starvation and stress conditions. Interestingly, sucrose enhances anoxia tolerance in Arabidopsis and partially relieves transcriptional changes under these conditions, suggesting a close relationship between more general energy-depriving stresses and sugars. Importantly, the activation of these genes can be HXK1-independent and is repressed by multiple metabolizable sugars. As sessile and autotrophic organisms, plants are constantly challenged by a changing environment that compromises photosynthesis and respiration for energy supplies. Quick and adequate responses are necessary for continued metabolic activity, growth and development. The studies suggest that plant cells may sense energy depletion in a cell-autonomous manner and trigger convergent gene activation in response to diverse and seemingly unrelated stress signals independent of light.

New findings now show that the evolutionarily conserved Arabidopsis protein kinases KIN10/11, catalytic subunits of the plant SnRK1 (Snf1-related protein kinase) heterotrimeric kinase complex, are master regulators of transcription networks and activate catabolism and repress anabolism under energy deprivation conditions. Consistent with a role in energy depleting conditions, sugar phosphates (especially glucose-6-phosphate) have been shown to inhibit SnRK1 activity. Genome-wide microarray analyses of KIN10 overexpressing cells and the characterization of *kin10/11* double mutant plants reveal that the energy sensors KIN10/11 control transcriptional responses to darkness, sugar and stress conditions.

In addition, KIN10/11 are important transcriptional regulators upon small changes in the carbon status during the night, suggesting high sensitivity of these components to energy status. At least part of the transcriptional reprogramming under these conditions is mediated through GBF (G-box binding factors)/bZIP transcription factors. In rice, the SnRK1A and SnRK1B orthologs activate the sugar-repressed a-amylase gene in germinating embryos, possibly through phosphorylation of the MYBS1 transcription factor. Consistently, antisense repression of SnRK1 in cultured wheat embryos represses an a-amylase gene promoter.

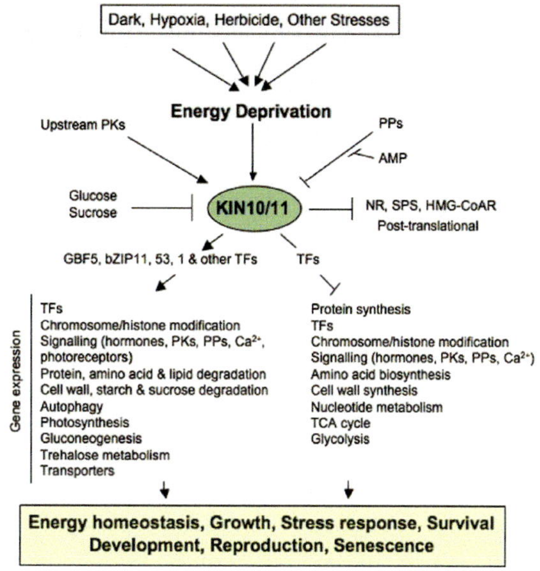

Fig. 14. Overview of the role of the Arabidopsis KIN10/11 protein kinases in the response to energy depletion.

Beside their role as global regulators of transcription networks, SnRK1 has been well known to modulate enzymes important for carbon and nitrogen metabolism through direct phosphorylation. For example, SnRK1 phosphorylates and inactivates the key enzymes 3-hydroxy-3-methylglutaryl CoA reductase, nitrate reductase and sucrose phosphate synthase in vitro, suggesting a pivotal role for SnRK1 in plant metabolism (Fig. 14). Analyses of an antisense transgenic potato line suggest that SnRK1 is important for sucrose and trehalose-6-phosphate activation of AGP and sucrose induction of SUS4 (encoding a sucrose synthase). However, unlike KIN10/11 and another potato gene (StubSNF1), the specific potato SnRK1 gene (PKIN1) used in these transgenic line does not complement the yeast snf1 mutant, and may have novel regulation and functions distinct from those of the more conserved yeast Snf1, mammalian AMPK and Arabidopsis KIN10/11 kinases.

To further study the physiological functions of KIN10/11 in plant growth and development, transgenic and mutant plants have been characterized. KIN10 overexpression plants display delayed senescence and flowering and an altered flower architecture under long-day conditions. Silencing of both KIN10 and KIN11 genes confers dramatic growth defects, with the most silenced plants senescing before the onset of flowering. In the moss Physcomitrella patens, the double knockout of both SnRK1 genes results in premature senescence and altered sensitivity to auxin and cytokinin. Interestingly, a Snf1-like kinase gene is expressed asymmetrically in tomato apical meristems, with higher expression in the part destined to form leaf primordia, suggesting a role for SnRK1 in linking metabolism and energy supplies with meristematic activity. Future studies will likely reveal more roles of KIN10/11 in plant growth and development and in antiviral defense.

In order to respond adequately to energy-mediated changes in metabolism, modulation of SnRK1 kinase activity could be important. Phosphorylation of the conserved threonine in the T-loop of the SnRK1 catalytic subunit has recently been demonstrated for Arabidopsis SnAK1 and SnAK2 (SnRK1-Activating Kinase). These two protein kinases can complement a yeast mutant deficient in Snf1 upstream kinase activity. Interestingly, using a similar approach, the same upstream kinases, dubbed Geminivirus Rep-interacting kinase 1 and 2 (GRIK1/2), were isolated and shown to be induced in infected leaves, suggesting a role for SnRK1 in the biotic stress response.

In addition to its conserved b, g and the plant-specific chimeric bg regulatory subunits, several more proteins were found to interact with SnRK1, including a putative Arabidopsis dual specificity phosphatase (PTPKIS1) (Fordham-Skelton et al., 2002), novel SnRK1-interacting proteins (SnIP1 and SnIP2), the heat shock protein BHSP17 from barley endosperm. In a screen for heterologous multi-copy suppressors of the yeast snf4 (Snf1 regulatory protein) deficiency, a plant casein kinase I orthologue, a novel hexose transporter (Mss1), and two Msn2/4-type zinc finger factors, AZF2 and ZAT10, involved in stress responses, have been isolated. These results suggest that a multitude of protein interactions are likely involved in the complex regulation of SnRK1 activity in plants.

TREHALOSE METABOLISM AND SIGNALING

Recently, trehalose and trehalose-6-phosphate (T6P) have emerged as novel regulators of carbon metabolism and development in plants. Synthesized from UDP-glucose and glucose-6-phosphate through the intermediate T6P, trehalose metabolism resides at the crossroads of major carbon fluxes in plants. Feeding sucrose to sucrose-starved seedlings results in a significant but transient increase in T6P levels and concomitant activation of AGP activity. Moreover, direct feeding of T6P to isolated chloroplasts was reported to increase AGP activity. Disruption of the outer membrane abolishes the response, suggesting perception of T6P signals at the chloroplasts.

Transgenic Arabidopsis plants overexpressing the E. coli trehalose-6-phosphate synthase (TPS; OtsA), the first enzyme of trehalose metabolism, exhibit better growth than wild-type seedlings on media supplemented with sugars. Conversely, overexpression of the E. coli trehalose-6-phosphate phosphatase (TPP; OtsB), the second enzyme in trehalose metabolism, causes a total arrest of seedling growth on sugar media, suggesting an important role for T6P in sugar utilization.

Furthermore, seedling growth is totally arrested on media with low amounts of exogenous trehalose that strongly induces APL3 gene expression. Further analyses show that expression of the starch breakdown genes, SEX1 and b-amylase BAM3, is significantly reduced under these conditions, leading to high amounts of starch at the end of the night. The effects on gene expression by exogenous trehalose are mediated by the ABI4 transcription factor). Interestingly, overexpression of Arabidopsis TPS1 confers glucose and ABA insensitivity, suggesting crosstalk between T6P and sugar signaling.

In the Arabidopsis genome, 21 genes with potential roles in trehalose and T6P metabolism are found. They are classified into three distinct subfamilies based on the sequence similarity with yeast TPS1 (class I, TPS1-4), TPS2 (class III, TPPA-J), or both (class II, TPS5-11). From the class I genes, only TPS1 has been studied in detail and it encodes an active TPS protein. Loss-of-function studies show that the embryo lethal tps1 mutant arrests growth at the torpedo stage of embryo development, at the transition from cell division to cell elongation, differentiation and carbohydrate storage.

Conditional TPS1 expression allows embryo maturation and further characterization of TPS1 functions in transgenic plants. Interestingly, TPS1 knock-out plants show shorter root length with a reduction in root apical meristem size, slow growth and strongly reduced flowering, suggesting an important role for TPS1 throughout the plant's life cycle.

Enzymes of the third class of trehalose metabolism proteins are active TPPs (Vogel et al., 1998), suggesting an important role for rapid T6P breakdown under sugar-depleting conditions. Importantly, RAMOSA3, a gene controlling inflorescence architecture in maize, encodes such a functional TPP enzyme. It will be important to determine the functions of related TPP genes in Arabidopsis and other plants.

The class II TPS genes are more enigmatic. Although extensive expression regulation under wide-ranging conditions suggests important regulatory roles, enzymatic activity has not been demonstrated for any of these proteins. Several of these proteins are phosphorylated by KIN10/11 to promote 14-3-3 protein binding. Interestingly, the tps5 mutant is thermosensitive and the tps6 mutant exhibits altered plant architecture, epidermal pavement cell shape and trichome branching. These data suggest potentially important roles for trehalose metabolism in coordinating metabolic status with plant growth and development.

Conclusions

Numerous investigations have provided data on the regulatory role of melatonin in multiple metabolic pathways in plants. In primary metabolism, its critical action on enzyme transcripts and regulatory factors in different organelles (chloroplasts, mitochondria, endoplasmic reticulum) and subcellular sites (cytosol, cell wall) stands out. Metabolic processes such as photosynthesis, the pentose phosphate shunt, gluconeogenesis, glycolysis as well as the Krebs cycle and the biosynthesis of amino acids and fatty acids are clearly under the influence of melatonin at several key steps.

Carbohydrate metabolism is one of the most studied, although much remains to be known. From the regulation of Rubisco to the processes of glycolysis and fermentation, melatonin appears to play a decisive role in the fate of carbohydrates synthesized in the chloroplast and cytosol. Thus, melatonin regulates the production of triose phosphate in the Calvin cycle, its transformation into hexoses and also the pool of starch in the chloroplastic stroma and that of sucrose in the cytosol and cell walls.

In general, melatonin activates the primary metabolism, both of carbohydrates and of other primary components such as lipids and amino acids. The result is an activation of the metabolic turnover such that it is adequate and conditioned to the physiological situation of the moment. In summary, melatonin has multiple regulation actions; for example, it influences photosynthesis, improving the efficiency of Rubisco and other Calvin cycle-related enzymes, Photosystem I and II, chlorophyll and carotenoid content and stomatal complex, with the result of a higher net photosynthesis, and, in specific carbohydrate metabolism, melatonin mobilizes some key pathways such as starch and sucrose biosynthesis, through SPS, SuS, and invertases upregulation.

In parallel, melatonin regulates many factors of the metabolism of plant hormones that, together with the modulation of the redox network, make melatonin an essential biostimulator or plant growth regulator, leading the plant through its functions to an adaptation to environmental situations against adverse effects and increasing tolerance to stressors.

Obviously, there are many aspects to be investigated relative to the influence of melatonin on carbohydrate metabolism, such as: the regulatory action of genes in the nucleus, chloroplasts and mitochondria; its interactions with other plant hormones; its functions in different organs (leaf, stem, root, flowers, fruits); its action on the accumulation and degradation of starch in amyloplasts; its ability to influence the metabolism of sucrose in source and sink tissues, thereby regulating phloem loading and unloading; and its action in the regulation of the biosynthesis of polyalcohols and proline, which is key to understanding the osmoregulatory response to stress. Finally, a complete understanding of its role in carbohydrates/fatty acid/amino acid balance has yet to be realized.

Every Suc molecule must be cleaved by SuSy or INV before it can be metabolized. The reason for the need of two Suc-catabolizing enzymes (SuSy and INV) in plants remains unclear. One possibility is suggested by their differential subcellular locations. The SuSy in plasma membranes and cell walls and their production of UDP-G may be important for directing carbon toward cellulose or callose synthesis; whereas INV may direct carbon to other metabolic pathways. However, SuSy and INV also co-localize to the cytosol. SuSy activity is feedback-inhibited by its product, Fru, and its activity is also reversible.

These features of SuSy may help to control the amount of Suc consumed in different organs, for example, in stems and petioles. Only controlled amounts of the transported Suc must be cleaved and metabolized to support the maintenance and development of vascular and other supporting tissues. It is likely that feedback inhibition of SuSy activity and substrate inhibition of fructokinase by Fru work together to impose a double break that controls the amount of Suc cleavage. That is, in case of excess cleavage of Suc by SuSy, the increased fructose Fru inhibits fructokinase activity so that fructose accumulates and that accumulated Fru inhibits further cleavage of Suc by SuSy.

Although plant SuSy proteins have been the subject of intensive study, we are still faced with major gaps in our understanding of the functions of these enzymes. The main issues that need to be further explored are:

- the regulation of SUS expression and intracellular localization of specific SuSy isozymes,
- the co-evolution of SuSy and INV and the division of Suc cleavage between these two group of enzymes,
- the differences between the SUS I, SUS II, and SUS III clades, and
- The roles of SuSy in meristem and leaf development.

Sugar sensing and signaling is mediated by a complex network comprising a multitude of interactions with metabolic and hormonal signals. Several key players involved in these processes have been identified and the complex genetic interactions are gradually being dissected and interpreted. Since many of the mutant screens are based on altered sugar sensitivity in germination and seedling growth assays, information about the spatial-temporal aspects of these interactions is still missing.

Defining sugar sensing and signaling cascades for specific sugar signals and concentrations at specific time points and developmental stages will facilitate the establishment of an integrated view, linking these mechanisms with normal plant growth and development.

Furthermore, the molecular and functional link between the different putative sugar sensing mechanisms is still elusive, and will require identification of the sensors, their exact location, expression and regulation, as well as an understanding of the dynamic carbon fluxes and associated signal generation during normal physiological development and in response to changing environmental stimuli. Further analyses will reveal specific sets of target genes and elucidate important protein-protein interactions and putative post-translational protein modifications in response to sugars.

Plant sugar sensing and signaling is a complex field of study, but within this complexity lay great potential and opportunities for scientific exploration and discovery. Novel approaches will reveal previously unidentified components and precisely define existing mechanisms, bringing together a dynamic and comprehensive sugar-signaling network central to the modulation of plant life and processes.

Abbreviations

- At Arabidopsis thaliana
- BD Brachipodium distachion
- Be Bamboo emeiensis
- Cit Citrus sinensis
- Ga Gossypium arboreum
- Gh Gossypium hirsutum
- Gm Glycine max
- Hb Hevea brasiliensis
- HV Hordeum vulgare
- MA Musa acuminata
- Md Malus domestica
- Os Oryza sativa
- PAB Picea abies
- PME Pseudotsuga menziesii
- Pp Prunus persica
- PPI Pinus pinaster
- PSI Pinus sitchensis
- PSY Pinus sylvestris
- PTA Pinus taeda
- Ptr Populus trichocarpa
- SB Sorghum bicolor

- SI Setaria italica
- Sl Solanum lycopersicum
- Vv Vitis vinifera
- Zm Zea mays.

CHAPTER

DIVERSITY OF ISOPRENOIDS HAS MULTIPLE FUNCTIONS IN PLANT METABOLISM

14

The isoprenoids play important ecological and physiological roles in plants. They also have a tremendous impact on human lives as food additives, medicines, and industrial raw materials, among others. Though some isoprenoids are highly abundant in nature, plants produce many at extremely low levels. Glandular trichomes (GT), which cover the aerial parts of more than 25% of vascular plants, have been considered as natural biofactories for the mass production of rare industrially important isoprenoids. In several plant genera (e.g., Lavandula and Mentha), GTs produce and store large quantities of the low molecular weight isoprenoids, in particular mono- and sesquiterpenes, as essential oil constituents. Within each trichome, a group of secretory cells is specialized to strongly and specifically express isoprenoid biosynthetic genes, and to synthesize and deposit copious amounts of terpenoids into the trichome's storage reservoir. Despite the abundance of certain metabolites in essential oils and defensive resins, plants, particularly those lacking glandular trichomes, accumulate small quantities of many of the biologically active and industrially important isoprenoids.

Therefore, there is a pressing need for technologies to enable the mass production of such metabolites, and to help meet the ever-increasing demand for plant-based bioproducts, including medicines and renewable materials. Considerable contemporary research has focused on engineering isoprenoid metabolism in GTs, with the goal of utilizing them as natural biofactories for the production of valuable phytochemicals. In this review, we summarize recent advances related to the engineering of isoprenoid biosynthetic pathways in glandular trichomes.

Isoprenoids are present in all living organisms, but with a remarkable diversity in plants. More than 40,000 different plant isoprenoids are known and new compounds are being constantly identified. These isoprenoids have many different functions. In primary metabolism, they func-tion as membrane constituents, photosynthetic pigments, electron trans-port carriers, growth substances, and plant hormones. They act as glucosyl carriers in glucosylation reactions and are involved in the regulation of cell growth. Plant isoprenoids (also known as terpenoids) are important commer-cially, for example as aroma substances for food, beverages, and cosmet-ics, as vitamins (A, D, and E), natural insecticides (e.g., pyrethrin), solvents (e.g., turpentine), and as rubber and gutta-percha. The plant isoprenoids also comprise important natural compounds, which are utilized as pharma-ceuticals or their precursors. Investigations are in progress to increase the ability of plants to synthesize isoprenoids by genetic engineering.

Plant ethereal oils have long been of interest to chemists. A number of mainly cyclic compounds containing 10, 15, 20, or correspondingly more C atoms have been isolated from turpentine oil. Such substances have been found in many plants and were given the collective name terpenes. Figure 1 shows some examples of terpenes. Limonene, an aromatic sub-stance from lemon oil, is a terpene with 10°C atoms and is called a monoterpene. Carotene, with 40°C atoms, is accordingly a tetraterpene. Rubber is a polyterpene with about 1500°C atoms. It is obtained from the latex of the rubber tree Hevea brasiliensis.

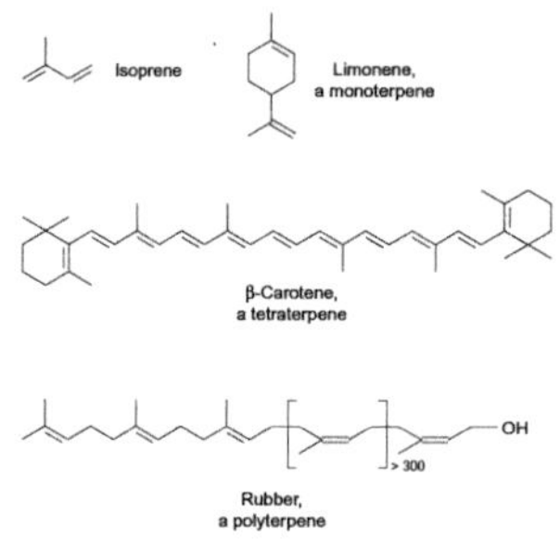

Fig. 1. Various Isoprenoids

Otto Wallach (Bonn, Göttingen), who in 1910 was awarded the Nobel Prize in Chemistry for his basic studies on terpenes, recognized that iso-prene is the basic constituent of terpenes (Fig. 1). Continuing these stud-ies, Leopold Ruzicka (Zürich) found that isoprene is the universal basic element for the synthesis of many natural compounds, including steroids, and for this he was awarded the Nobel Prize in Chemistry in 1939. He pos-tulated the biogenic isoprene rule, according to which all terpenoids (deriv-atives of terpenes) are synthesized via a hypothetical precursor, which he named active isoprene. This speculation was verified by Feodor Lynen in Munich (1964 Nobel Prize in Medicine), when he identified isopentenyl pyrophosphate to be the "active isoprene."

ISOPRENOID DIVERSITY AND BIOSYNTHESIS

The isoprenoids or terpenoids, make up the largest class of plant secondary, or specialized metabolites. They play crucial ecological roles as pollinator attractants and defensive agents, and have important physiological functions as plant hormones and photosynthetic pigments, among others. Isoprenoids impact human lives through imparting scent, flavor, and health-promoting properties to fruits, vegetables, and medicinal plants. Some terpenoids have potent biological activities, and have applications as prescription drugs and over-the-counter medicines. A few have also been used as sustainable replacements for petroleum-derived chemicals.

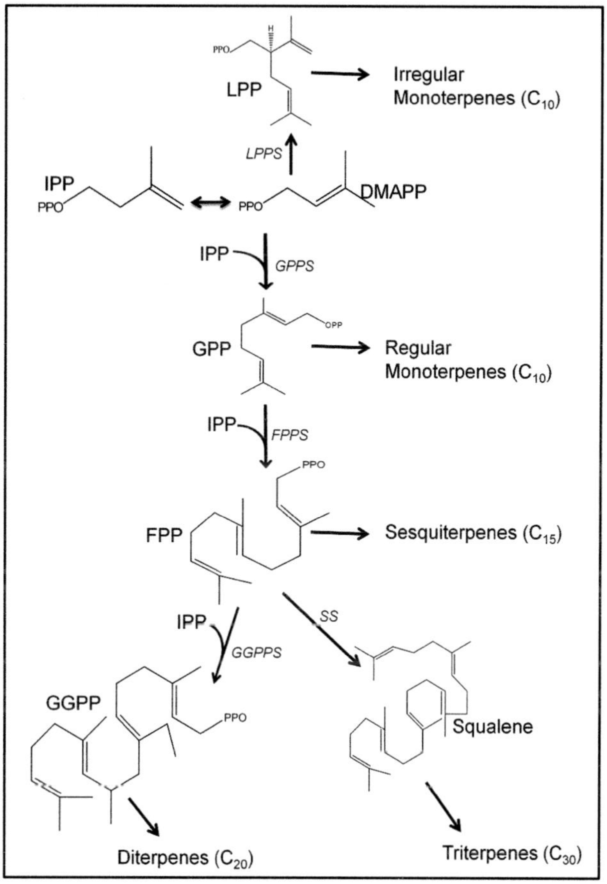

Fig. 2. Overview of the biosynthetic pathways for major plant isoprenoids.

Isoprenoids are synthesized through the condensation of isopentenyl diphosphate (IPP; C5) and its isomer dimethylallyl diphosphate (DMAPP; C5), and are classified by the number of five-carbon units present in the core structure. Major isoprenoid classes include monoterpenes (C10), sesquiterpenes (C15), diterpenes (C20), triterpenes (C30), and tetraterpenes (C40), although lower and higher-order isoprenoids (e.g., isoprene and natural rubber, respectively) also exist (Figure 1). In general, the biosynthesis of isoprenoids can be classified into the following four states: generation of general precursors IPP and DMAPP, production of specific isoprenyl diphosphates for various isoprenoid classes, the transformation of isoprenyl diphosphates to individual isoprenoids by terpene synthase (TPS) enzymes, and structural modifications catalyzed by other catalysts (Fig. 2).

In plants, IPP and DMAPP are generated through two distinct compartmentally separated pathways. In plastids, the 1-DXP pathway, aka MEP pathway, produces IPP/DMAPP for monoterpene, diterpene, and tetraterpene biosynthesis (Lois et al., 2000; Turner et al., 2000). In the cytosol, the classical mevalonic acid (MVA) pathway supplies IPP, which is isomerized to DMAPP by an isomerase. The cytosolic IPP/DMAPP pool is primarily used for the production of sesqui- and triterpenes. All biochemical steps of both pathways have been characterized, and the relevant genes cloned.

The head-to-tail condensation of IPP and DMAPP yields various linear isoprenoid precursors, including geranyl diphosphate (GPP; C10), neryl diphosphate (NPP; C10), farnesyl diphosphate (FPP; C15), and geranylgeranyl diphosphate (GGPP; C20). The head-to-middle condensation of two DMAPP units also occurs in some plants, and gives rise to lavandulyl diphosphate (LPP; C10) as a linear product. The condensation reactions are catalyzed by a group of enzymes known as isoprenyl diphosphate synthases (IDSs). IDSs are classificed into "cis" or "trans" based on their primary structures, and the stereochemistry of the products they generate.

The trans-IDSs catalyze the synthesis of the common terpene precursors, and are distinguished by two conserved aspartate-rich motifs, DDX2–4D and (N/D) DX2D, which serve as substrate and divalent metal ion cofactor binding sites. On the other hand, cis-IDSs lack these conserved motifs, and instead share five conserved regions designated as Regions I–V, including catalytically essential aspartate residue in Region IV and the glutamate residue in Region V. For example, GPPS, a typical trans-IDS, catalyze the head-to-tail condensation of DMAPP with IPP to produce GPP, while LPPS, a cis-IDS, generates LPP through the head-to-middle condensation of two DMAPP units in lavenders.

Other trans-IDSs include FPP, which gives rise to sesquiterpenes and triterpenes, and GGPP, which serves as a precursor for the synthesis of diterpenes and tetraterpenes. The prenyl diphosphates are transformed into various terpenoids by specific TPSs, of which hundreds have been cloned from a wide range of plants.

HIGHER PLANTS HAVE TWO DIFFERENT SYNTHESIS PATHWAYS FOR ISOPRENOIDS

Precursor for the synthesis of isoprenoids is isopentenyl pyrophosphate. Its synthesis proceeds in higher plants and some groups of algae in two different ways, one located in the cytosol and the other in the plastids.

Acetyl CoA is a precursor for the synthesis of isoprenoids in the cytosol

The basis for the elucidation of this isoprenoid biosynthesis pathway was the discovery by Konrad Bloch that acetyl CoA is a precursor for the biosyn-thesis of steroids. Figure 3 shows the synthesis of the intermediary product isopentenyl pyrophosphate: two molecules of acetyl CoA react to produce acetoacetyl CoA and then with another acetyl CoA yielding β-hydroxy-β-methylglutaryl CoA (HMG CoA). In yeast and animals, these reactions are catalyzed by two different enzymes, whereas in plants a sin-gle enzyme, HMG CoA synthase, catalyzes both reactions. The esterified carboxyl group of HMG CoA is reduced by two molecules of NADPH to a hydroxyl group, accompanied by hydrolysis of the energy-rich thioester bond.

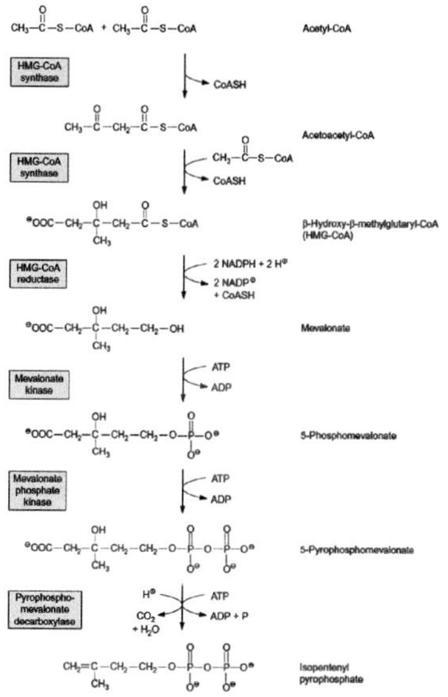

Fig. 3. Isopentenyl pyrophosphate synthesis in the cytosol proceeds via the acetatemevalonate pathway

Thus mevalonate is formed in an irreversible reaction. The forma-tion of mevalonate from HMG CoA is an important regulatory step of iso-prenoid synthesis in animals. It has not yet been resolved whether this also applies to plants. A pyrophosphate ester is formed in two successive phos-phorylation steps, catalyzed by two different kinases. With consumption of a third molecule of ATP, involving the transitory formation of a phosphate ester, a carbon-carbon double bond is generated and the remaining car-boxyl group is removed. Isopentenyl pyrophosphate thus formed is the basic element for the formation of an isoprenoid chain. This synthesis of isopentenyl pyrophosphate, termed the acetate-mevalonate pathway, is located in the cytosol. It is responsible for the synthesis of sterols, certain sesquiterpenes, and the side chain of ubiquinone.

Pyruvate and D-glyceraldehyde-3-phosphate are the precursors for the synthesis of isopentyl pyrophosphate in plastids

The acetate-mevalonate pathway can be blocked by mevilonin, a very specific inhibitor of HMG CoA reductase. Experiments with plants showed that mevilonin inhibits the isoprenoid synthesis in the cytosol, but not in the plastids. These findings led to the discovery that the synthesis of isopentenyl pyrophosphate in the plastids follows a different pathway from that in the cytosol. For the plastidal synthesis pathway, the precursors are pyruvate and D-glyceraldehyde-3-phosphate. As in the pyruvate dehydrogenase reaction, pyruvate is decarboxylated via thiamine pyrophosphate (TPP), and then, as in the transketolase reaction, is transferred to D-glyceraldehyde-3-phosphate to yield 1-deoxy-D-xylulose-5-phosphate (DOXP).

After isomerization and reduction by NADPH, 2-C-methyl-D-erythritol-4-phosphate (MEP) is synthesized. MEP is then activated by reacting with CTP to yield CDP methyl erythriol. Two further reduction steps, followed by dehydration and phosphorylation, finally yield isopentenyl pyro-phosphate. The MEP-synthase pathway for isoprenoids is present in bacte-ria, algae, and plants, but not in animals. A large part of plant isoprenoids, including the hemiterpene isoprene, monoterpenes like pinene and limonene, diterpenes (e.g., phytol chains, gibberellin, abietic acid as oleoresin constituent) as well as tetraterpenes (carotenoids) are synthesized via the MEP synthase pathway located in the plastids. Also, the side chains of chlorophyll and plastoquinone are synthesized by this pathway. Sesquiterpenes and triterpenes, on the other hand, according to our present knowledge are synthesized by the mevalonate pathway in the cytosol.

PRENYL TRANSFERASES CATALYZE THE ASSOCIATION OF ISOPRENE UNITS

Dimethylallyl pyrophosphate, which is formed by isomerization of isopen-tenyl pyrophosphate, is the acceptor for successive transfers of isopente-nyl moieties (Fig. 4). With the liberation of the pyrophosphate residue, dimethylallyl-PP condenses with isopentenyl-PP to produce geranyl-PP. In an analogous way, chain elongation is attained by further head-to-tail con-densations with isopentenyl-PP, and so farnesyl-PP and geranylgeranyl-PP are formed one after the other.

The transfer of the isopentenyl moieties is catalyzed by prenyl transferases. Prenyl residues are a collective term for isoprene or polyisoprene residues. A special prenyl transferase is required for the production of each of the pre-nyl pyrophosphates mentioned. For example, the prenyl transferase termed geranyl-PP synthase catalyzes only the synthesis of geranyl-PP. However, farnesyl-PP synthase synthesizes farnesyl-PP in two discrete steps: from dimethylallyl-PP and isopentenyl-PP, first geranyl-PP is formed, but this inter-mediate remains bound to the enzyme and reacts further with another isopen-tenyl-PP to produce farnesyl-PP. Analogously, geranylgeranyl-

PP synthase catalyzes all three steps of the formation of geranylgeranyl-PP. As these prenyl pyrophos-phates are synthesized by different enzymes, the synthesis of a certain prenyl pyrophosphate can be regulated by induction or repression of the corre-sponding enzyme. It appears that there is a synthesis pathway from isopente-nyl pyrophosphate to geranylgeranyl pyrophosphate, not only in the cytosol but also in the plastids. The differences between these two pathways have not yet been resolved in detail.

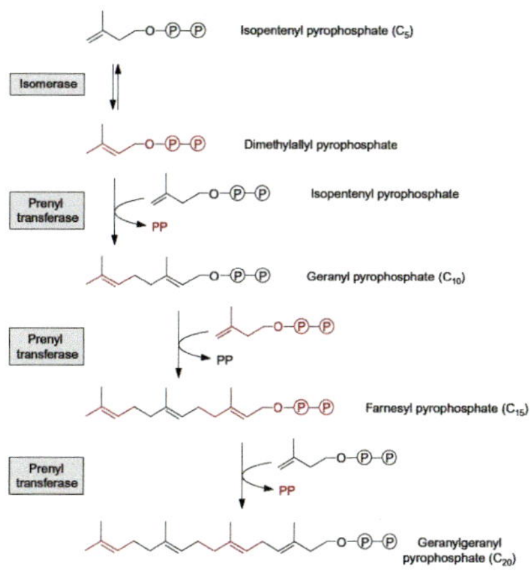

Fig. 4. Higher molecular prenyl phosphates are formed by head-to-tail addition of active isoprene units

The formation of a C-C linkage between two isoprenes proceeds by nucle-ophilic substitution: an Mg^{++} ion, bound to the prenyl trans-ferase, facilitates the release of the negatively charged pyrophosphate residue from the acceptor molecule, whereby a positive charge remains at the termi-nal C atom (C-1), which is stabilized by the neighboring double bond. The allyl cation thus formed reacts with the terminal C-C double bond of the donor molecule and a new C-C bond is formed with the release of a proton. According to the same reaction mechanism, not only isoprene chains, but also rings are formed, leading to the exceptional diversity of isoprenoids.

ENGINEERING ISOPRENOID BIOSYNTHESIS IN TRICHOMES

Several platforms, including bacteria (Farrokh et al., 2019), yeast (Rahmat and Kang, 2020), algae (Rico et al., 2017), and plants (Birchfield and McIntosh, 2020), have been successfully used for enhanced (or mass) production of specific isoprenoids of industrial value. Though traditionally microbial systems have been the platform of choice for the production of recombinant natural products, plants are emerging as important alternatives. This is in part due to the relative ease and cost-effectiveness of mass-producing plants in general, and in part due to the fact that genetic transformation of most plants is now routine. Plants bearing GTs are particularly suited for mass production of secondary metabolites, as GTs have specialized secretory cells, capable of production and secretion of large quantities of phytochemicals (that are often toxic to other cell types) into the GT's storage cavity. In this context, several GT-bearing plants have been investigated as potential biofactories for the mass production of specific metabolites, with most of the studies reported focussing on the plants discussed below.

Mints (Mentha)

Metabolic engineering to enhance isoprenoid (essential oil monoterpenes) metabolism has been reported in the two most commercially important mint species, Mentha x piperita (peppermint) and Mentha spicata (spearmint), which are widely grown for EO production. The EO in these plants is a complex mixture of mainly monoterpenes, many of which are highly abundant while others are present in trace quantities. In both species, GPP is initially converted to (–)-limonene through a reaction catalyzed by the enzyme (–)-limonene synthase. In peppermint, hydroxylation at the C-3 position of the limonene ring initiates a cascade of reactions that lead to the production of several monoterpenes, of which (–)-menthol is highly valued for its applications in food and alternative medicine. In spearmint, hydroxylation occurs on the C-6 position of limonene, leading to the production of (–)-trans-carveol, which is efficiently oxidized to (–)-carvone as the main EO constituent.

One of the earliest metabolic engineering efforts in mint was reported in 2001, in a study that attempted to increase flux toward monoterpene metabolism by overexpressing the coding sequence for the branch-point enzyme 4S-limonene synthase (4S-LimS) in transformed M. x piperita plants. The expression of 4S-LimS transgene was driven by the Cauliflower Mosaic Virus (CaMV) 35S promoter. Although monoterpene production was not altered in most plants, a few transformants accumulated higher levels of essential oil constituents, including 56% more pinene, 22–74% higher 1,8-cineole, and 18–40% more pulegone than control plants.

A few attempts followed this investigation in Professor Rodney Croteau's laboratory in the early 2000s. In one attempt, the DXP reductoisomerase (DXR) cDNA (Figure 1) was overexpressed in peppermint plants to enhance the output of the DXP pathway, and improve the production of precursor (IPP/DMAPP) for monoterpene biosynthesis. Several transgenic plants produced up to 50% more EO than the wild-type controls. In the same study, the menthofuran synthase gene, responsible for the production of the undesired oil monoterpene constituent (+)-Menthofuran, was overexpressed in separate peppermint plants in sense and antisense to evaluate the effects of these overexpressions on menthofuran production.

Most plants expressing the transgene in sense accumulated substantially more menthofuran than wild-type controls. Conversely, plants that overexpressed the gene in antisense orientation, accumulated significantly less menthofuran than control plants. Intriguingly, menthofuran production was almost entirely eliminated in plants in which the menthofuran synthase gene was co-suppressed. Surprisingly, growth and development in co-suppressed plants were compromised as some of the most affected plants appeared bleached, in particular, under stress conditions. Another study that aimed to enhance monoterpene production (oil yield) in peppermint, limonene synthase, and limonene-3-hydroxylase cDNAs were overexpressed in independently (separately) transformed peppermint plants.

Overexpression of either gene failed to improve oil yield significantly. However, limonene levels increased dramatically, surpassing 80% of total oil (compared to ~2% in wild-type plants), in the transformed peppermint lines, in which limonene-3-hydroxylase was co-suppressed. Taken together, studies in peppermint produced evidence that metabolic engineering may be employed to effectively engineer isoprenoid metabolism. However, not all metabolic steps respond to gene overexpression, while gene co-suppression can effectively eliminate unwanted metabolites (e.g., menthofuran) or enhance the accumulation of particular desired end products (e.g., limonene).

Modulating the expression of transcription factors that regulate other EO biosynthetic genes offers another avenue for enhancing isoprenoid production in GTs. A recent study investigated the effects of overexpressing and silencing the spearmint GT-specific transcription factor MsYABBY5 on the production of terpenes in spearmint. The TF was constitutively overexpressed in stably transformed plants, and silenced using RNA interference (RNAi). Silencing MsYABBY5 increased the levels of terpenes in spearmint from 20 to 77%, while its overexpression led to a 23–52% decrease in the overall levels of terpenes in spearmint plants. This study clearly demonstrated that the MsYABBY5 is a repressor of monoterpene metabolism in spearmint, and that manipulating the expression of transcription factors offers a viable method for enhancing terpene production in plant GTs. In a separate study, the expression of the spearmint GT-specific R2R3-MYB TF was suppressed by RNAi.

Likewise, suppression of MsR2R3-MYB expression led to a 2.3–4.5-fold increase in total monoterpene abundance in transgenic spearmint. It was further shown that MsR2R3-MYB suppresses the expression of the GPP synthase large subunit gene through binding of its cis-elements, and hence acts as a negative regulator of monoterpene metabolism in spearmint.

A recent study evaluated the potential of spearmint for the production of heterologous monoterpene through metabolic engineering. In this study, a transgenic spearmint line with reduced limonene and carvone was independently transformed with cDNAs of linalool synthase and myrcene synthase from Picea abies, and geraniol synthase from Cananga odorata, which were also controlled by 35S promoter. Silencing of the limonene synthase gene by RNAi led to a huge reduction in the production of limonene and carvone, and an increase in sesquiterpene, phytosterols, fatty acids, flavonoids, and phenolic metabolites. Surprisingly, overexpression of heterologous TPSs in these lines did not significantly increase the production of the related products, although small quantities of some heterologous terpenes were produced, which were oxygenated by the host plant.

Lavenders (*Lavandula*)

Lavenders (Lavandula) are perennial members of the mint family. To date, ca. 34 species have been described, three of which (L. angustifolia, L. latifolia, and their natural hybrid offspring L. x intermedia) are widely grown for EO production. Between 1,200 and 1,500 tons of lavender, EO is produced worldwide per annum for use in foods, cosmetics, and personal care products. As in other members of the Labiatae, lavender bears peltate GT that produce and store relatively large quantities of a monoterpene-rich essential oil.

The EO of a typical lavender species consists of over 50 monoterpenes, the most abundant of which include linalool, linalool acetate, borneol, camphor, and 1,8-cineole. Among these, camphor, linalool, and linalool acetate are key determinants of the quality (scent and bioactivity) of lavender EO. Lavender EOs with high linalool and linalool acetate, and low borneol, camphor, and 1, 8-cineole content are considered to be of "high quality," and are used in cosmetic products and aromatherapy. These oils are typically obtained from true lavender (L. angustifolia) species.

Oil marketed to the alternative medicine sector is typically obtained from L. latifolia plants, which accumulate high levels of linalool, borneol, camphor, and 1,8-cineole, but no linalool acetate. The EOs obtained from the hybrid L. x intermedia plants contain a mixture of monoterpenes present in both parents, and are mainly utilized in personal care and hygiene products, such as soaps, shampoos, mouthwashes, as well as industrial and household cleaners, among others. Lavandula x intermedia plants produce up to 10 fold more EO than either parent, and are hence widely grown for EO production.

Metabolic engineering to enhance monoterpene biosynthesis has been reported in two lavender species, L. latifolia (spike lavender) and L. x intermedia (lavandin). The earliest metabolic engineering study in a lavender species was reported in the early 2000s, when researchers overexpressed the Arabidopsis 1-DXP synthase (DXS) gene in spike lavender. Like the earlier results reported for the over-expression of DXR and DXS in peppermint, DXS overexpression increased EO (monoterpene) production compared to untransformed controls. In another study, researchers attempted to enhance EO quality (scent) in L. latifolia plants by increasing the biosynthesis of S-linalool, a sweet-scented monoterpene. In this study, the Clarkia breweri S-linalool synthase gene was transformed into L. latifolia plants via Agrobacterium-mediated transformation.

Transgenic lines overexpressing the transgene accumulated substantially (up to 1000 fold) more S-linalool compared to untransformed control plants. Interestingly double transgenic plants, which resulted from a cross between separate transgenic plants overexpressing either DXS or S-linalool synthase gene (LIS) did not yield the expected results. Both essential oil yield and linalool content in double DXS-LIS transgenic plants were lower than that of their parents. A separate study also aimed at enhancing the fragrance of L. x intermedia essential oil by reducing 1,8-Cineole biosynthesis by suppressing the expression of the 1,8-Cineole synthase gene using RNAi. The transcriptional expression of 1,8-Cineole synthase gene was suppressed in the transgenic plants compared to nontransgenic controls. RNAi was effective in reducing production of 1, 8-cineole significantly, and altering the overall scent of the EO of affected plants.

Finally, in a recent study, the L. x intermedia BPPS gene (LiBPPS) was placed under the control of the CaMV 35S promoter and stably expressed in transgenic L. latifolia plants via Agrobacterium-transformation in sense and antisense orientations.

CHAPTER

PRODUCTS OF NITRATE ASSIMILATION ARE DEPOSITED IN PLANTS AS STORAGE PROTEINS

15

INTRODUCTION

The macroelement Nitrogen (N) is an essential component in key molecules of living mater. However, dinitrogen, the most abundant form either in the atmosphere or dissolved in waters, can only be used by fixing bacteria but is inaccessible to eukaryotic algae in general. Nevertheless, an uncommon symbiotic process between a unicellular cyanobacteria and a single-celled eukaryotic alga was described. N can be incorporated in eukaryotic organisms from either organic or inorganic forms. The availability and concentrations of inorganic N sources, which in general are small, change depending on environments and usually limit growth and productivity. NH_4^+, NO_2^-, and NO_3^- the most frequent inorganic N sources assimilated by photosynthetic organisms show a different spatial distribution in oceans. In the euphotic zone of ocean waters (the area closer to the surface, which receives enough light for photosynthesis to be carried out), the estimated mean concentration of nitrate is about 7 μM being those of ammonium and nitrite much smaller, 0.3 and 0.1, respectively. Whereas in the aphotic zone (just below the euphotic zone and unable to support photosynthetic or autotrophic growth) these values are 31, 0.01, 0.006 for nitrate, ammonium, and nitrite, respectively (Gruber, 2008). However, in the soil nitrate concentrations can be very variable from 10 μM to 100 mM. In natural surface waters, nitrate concentration is usually < 1 μM, but it can increase several orders of magnitude specifically in underground waters mainly due to contamination from plant fertilizers or animal farms. Depending on colonized environments, microalgae own different adaptations for a proper N assimilation.

INTEREST OF USING CHLAMYDOMONAS IN NITRATE ASSIMILATION STUDIES

In this section, we will focus on nitrate assimilation in microalgae, which has generated basic knowledge and contributed significantly to the understanding of the pathway in crop plants. Algae are a group of polyphyletic organisms, which evolved along different endosymbiotic events, thus the several species developed different and distinctive cell structures, grouping organizations, and metabolisms. Here, we will only refer to algae whose ancestors derive from a primary endosymbiotic event. They correspond to individuals from the group of Glaucocystophytes, Rhodophytes, and Chlorophytes.

This latter group of chlorophytes is phylogenetically closely related to plants. Other algae groups add new layers of complexity, since they come from secondary and tertiary endosymbiotic events, with plastid loss, replacement, and gene transfers. Among those algae Chlamydomonas reinhardtii (referred herein as Chlamydomonas) has emerged as a very relevant experimental system, mostly due to the development of molecular and genetic tools useful in many different studies, such as the flagella structure and function, photosynthetic apparatus, chloroplast evolution, metabolism, or regulation under different stress conditions. Advances achieved in understanding of nitrate assimilation in this system were extrapolated to other algae and mostly to plants leading to remarkable progress.

Many of the methodological tools in Chlamydomonas rely on its efficient transformation system, development of vectors and the sequencing of its genome. In other algae, transformation methodology has been set up, which will allow efficient techniques to appear as in Chlamydomonas. We will refer here to 10 Chlorophytes (using Chlamydomonas as a reference), two Rhodophytes, and a Glaucophyte, for which their genomes are sequenced and available in the public databases.

NITRATE ASSIMILATION

Overview of Nitrate Assimilation

The nitrate assimilation pathway from nitrate to amino acid is relatively simple at structural level. By contrast, its regulation to ensure the efficient assimilation of nitrate coupled to that of other environmental factors, is complex.

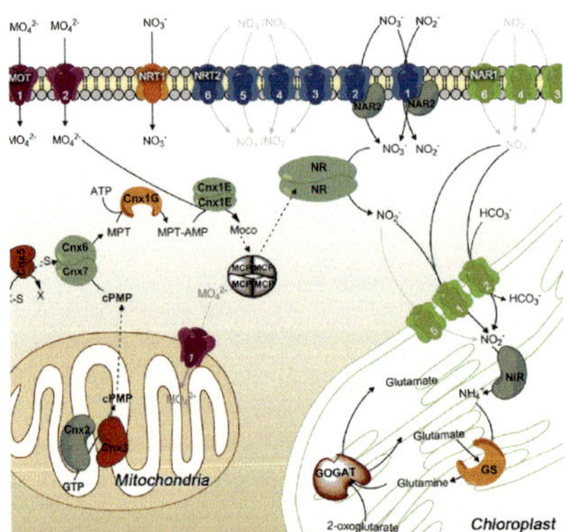

Fig. 1. general scheme of proteins involved in nitrate assimilation and molybdenum cofactor biosynthesis in *chlamydomonas*

Overview in Chlamydomonas/Algae

In photosynthetic eukaryotes, nitrate assimilation is performed by two transport and two reduction steps: First, nitrate is transported into the cell, then a cytosolic Nitrate Reductase (NR) catalyzes nitrate reduction to nitrite, which subsequently is transported into the chloroplast, where the enzyme Nitrite Reductase (NiR) catalyzes its reduction to ammonium. Finally, ammonium is incorporated to carbon skeletons by rendering glutamate, through the glutamine synthetase/glutamine oxoglutarate amino transferase or glutamate synthase (GS/GOGAT) cycle. First, ammonium is incorporated as the amide group of glutamine in a reaction involving glutamate and ATP (catalyzed by GS); then, the amide group is transferred reductively to α-oxoglutarate to form two molecules of glutamate.

Figure 1 summarizes the organization of enzymes/proteins for nitrate assimilation in Chlamydomonas. The assimilation of nitrate begins with its entry into the cell. The transport of nitrate and nitrite into the cell is a regulated and complex process in Chlamydomonas as suggested from the high number and types of proteins that participate. The transporters are classified, based on their substrate specificity, affinity and expression characteristics, in: specific or bispecific, high and low affinity, constitutive and inducible transporters. From the structural point of view three families of proteins are involved in nitrate and/or nitrite transport in Chlamydomonas. They are NRT1 (NPF), NRT2, and NAR1.

NRT1 Transporters Function

The NRT1 transporters belong to the PTR transporters (Peptide TRansporters), now renamed NPF for NRT1/PTR Family. There is a wide representation of these transporters in plant genomes, however its presence in algal genomes is in low number if any. Most of the plant NRT1 genes encode low-affinity nitrate transporters. The best-studied one is CHL1 from Arabidopsis (AtNRT1.1 or AtNPF2.6) and represents a paradigmatic protein. In fact, AtNRT1.1 is considered a transceptor with sensor and carrier functions. AtNRT1.1 is a dual affinity nitrate transporter that switches from high to low-affinity for nitrate, depending on the nitrate conditions and the phosphorylation/dephosphorylation of the threonine T101. At low nitrate conditions, AtNRT1.1 is phosphorylated by a CIPK23 protein kinase and functions as a high-affinity transporter. At high nitrate conditions AtNRT1.1 is dephosphorylated and is a low-affinity transporter. In addition AtNRT1.1 is involved in plant auxin transport and nitrate efflux.

NRT2 Transporters Function

The NRT2 proteins correspond to the NNP (Nitrate Nitrite Porter) family and belongs to the superfamily MFS (Major Facilitator Superfamily). NRT2 genes are present into genomes of organisms able to assimilate nitrate like plants, algae, fungi, yeast and bacteria. The NRT2 proteins can transport nitrate or nitrite and some of them are two-component systems, which require a second protein, NAR2, to be fully functional. NAR2 appears to be anchored to the membrane by a transmembrane domain. In Chlamydomonas there are six NRT2 members but NAR2 is only required for functionality of NRT2.1 and NRT2.2 but not for other NRT2 transporters. Thus, in Chlamydomonas NRT2.1/NAR2 constitutes a bispecific high-affinity nitrate/nitrite transporter and NRT2.2/NAR2 is a high-affinity nitrate transporter.

NAR1 Transporters Function

The family NAR1 (Nitrate Assimilation-Related component 1) is formed by transporters related to proteins called FNT. Chlamydomonas has six NAR1 genes (NAR1.1–6). NAR1.1 is a nitrite transporter to the chloroplast, and NAR1.2 (LCIA) shows NO2- and HCO3- transport activity when expressed in Xenopus oocytes. More recently, it has been demonstrated that NAR1.2 (LCIA) is a chloroplast envelope HCO3- transporter and part of a CO2-concentration mechanism (CCM) operating at very low CO2. The CCM is essential to accumulate CO2 close to RUBISCO and make efficient photosynthesis in aquatic microalgae.

Nitrate Reductase Function

The enzyme NR catalyzes the reduction of nitrate to nitrite with electrons from NAD(P)H. In eukaryotes, NR is a homodimeric protein with subunits of about 100–120 kDa containing each one three prosthetic groups: FAD, b_{557} heme and Molybdenum cofactor (Moco) (Zhou and Kleinhofs, 1996). Algae typically contain a single gene encoding NR (Fernαndez et al., 1989; Fernandez and Galvan, 2007) that, in addition to its physiological activity of nitrate reduction, has also two other activities. These partial activities can be assayed separately *in vitro*. The first one is the NAD(P)H-dehydrogenase or diaphorase, which catalyzes the transfer of electrons from pyridine nucleotides to artificial electron acceptors such as ferricyanide, dichlorophenol indophenol, or cytochrome c. This activity does not require the Moco domain. The second is the terminal activity, which catalyzes the transfer of electrons from artificial electron donors as flavins, bromophenol blue, or viologens (chemically reduced with dithionite) to nitrate (Solomonson et al., 1986; Caboche and Rouzé, 1990). This activity mostly requires the Moco domain.

Nitrite Reductase Function

Nitrite reduction to ammonium occurs in the stroma of chloroplasts and is catalyzed by the enzyme nitrite reductase (NiR). Reduced ferredoxin (Fdred) generated as a consequence of the photosynthetic electron transport is the electron donor for nitrite reduction. In the dark, nitrite reduction can also occur from NADPH, generated in the oxidative pentose phosphate cycle, by mediation of Fd-NADP+ oxidoreductase. Algal NiR, like in cyanobacteria and plants, is a monomer of approximately 63 kDa containing a [4Fe–4S] grouping and a siroheme as prosthetic groups. NiR is encoded by a single gene in Chlamydomonas, which is clustered with other genes essential for nitrate/nitrite assimilation.

INCORPORATION OF N INTO AMINO ACIDS

Inorganic N is incorporated finally in form of organic N as glutamate by the action of GS/GOGAT (see above). *Chlamydomonas* contains four GS (GLN) (two cytosolic and two plastid-targeted), and two plastidic GOGAT, one dependent on NADH and the other on Fd_{red}. GLN1 and GLN2 from *Chlamydomonas* have high sequence conservation and a similar octameric structure. Researchers found three forms of GS in *Chlamydomonas*, but the sequencing of its genome detected four.

However, function and localization of these two additional GS (GLN3 and GLN4) have not been studied in detail. In addition to GS/GOGAT, under certain stress conditions glutamate dehydrogenase (GDH) seems to be the predominant enzyme for ammonium incorporation in Chlamydomonas, catalyzing a reductive reaction of ammonium and a-oxoglutarate to yield glutamate. GDH is encoded by two genes (GDH1 and GDH2) and has three isoenzymes located in the mitochondria, resulting very probably from combination of the two different subunits.

Under stress conditions such as N-starvation or high ammonium, the aminating activity of GDH varies inversely to that of GS. The GDH function seems to be related to the maintenance on glutamate levels in the cell when GS/GOGAT is inefficient, appearing to be more connected to carbon than to nitrogen metabolism.

Since GDH is related to ammonium assimilation, which is not the topic of this review, it will not be mentioned any further.

PRODUCTS OF NITRATE ASSIMILATION ARE DEPOSITED IN PLANTS AS STORAGE PROTEINS

Whereas the products of CO_2 assimilation are deposited in plants in the form of oligo- and polysaccharides, the amino acids formed as products of nitrate assimilation are stored as proteins. These are mostly special storage proteins, which have no enzymatic activity and are often deposited in the cell within protein bodies. Protein bodies are enclosed by a single membrane that derived from the endomembrane system of the endoplasmic reticulum and the Golgi apparatus or the vacuoles. In potato tubers, storage proteins are also stored in the vacuole.

Storage proteins can be deposited in various plant organs, such as leaves, stems, and roots. They are stored in seeds and tubers and also in the cambium of tree trunks during winter to enable the rapid formation of leaves during seed germination and sprouting. Storage proteins are located in the endosperm in cereal seeds and in the cotyledons of most legume seeds.

Whereas in cereals the protein content amounts to 10% to 15% of the dry weight, in some legumes (e.g., soybean) it is as high as 40% to 50%. About 85% of these proteins are storage proteins. Globally, about 70% of the human demand for protein is met by the consumption of seeds, either directly or indirectly by feeding them to ani-mals for meat production. Therefore plant storage proteins are the impor-tant basis for human nutrition.

However, in many plant storage proteins the content of nutritionally essential amino acids is too low. In cereals, for example, the storage proteins are limited in threonine, tryptophan, and particularly in lysine, whereas in legumes there is a shortage of methionine. Since these amino acids cannot be synthesized by the human metabo-lism, humans have to take up essential amino acids through their food. In humans with an entirely vegetarian diet, the deficiency of essential amino acids can lead to irreparable physical and mental damage, especially in chil-dren. It can also be a serious problem in pig and poultry fodder.

A research goal in plant genetic engineering is to improve the amino acid composition of the storage proteins of harvest products. Scientists have long been interested in plant proteins. In 1745 the Italian Jacopo Beccari isolated proteins from wheat. In 1924, Thomas Osborne, at the Connecticut Agricultural Experimental Station, classified plant pro-teins according to their solubility properties. He fractionated plant proteins into albumins (soluble in pure water), globulins (soluble in diluted salt solutions), glutelins(soluble in diluted solutions of alkali and acids), and prolamins (soluble in aqueous ethanol).

Later, when the structures of theseproteins were determined, it turned out that glutelins and prolamins were structurally closely related. Therefore, in more recent literature, glutelins are regarded as members of the group of prolamins.

Globulins are the most abundant storage proteins

Storage globulins occur in varying amounts in practically all plants. The most important globulins are legumin and vicilin, both of which are encoded by a multigene family. These multigene families descend from a common ancestor. Legumin is the main storage protein of leguminous seeds. In broad bean, for instance, 75% of the total storage protein consists of legumin.

Legumin is a hexamer with a molecular mass of 300 to 400 kDa. The mono mers contain two different peptide chains (a, b), which are linked by a disulfide bridge. The large a-chain usually has a molecular mass of about 35 to 40 kDa, and the small b-chain has a molecular mass of about 20 kDa. Hexamers can be composed of different (a, b) monomers some of which contain methionine. In the hexamer, the protein molecules are arranged in a very regular package and can be deposited in this form in the protein bodies. Protein molecules, in which some of the protein chains are not properly folded, do not fit into this package and are degraded by peptidases.

Although it is easy nowadays to exchange amino acids in a protein by genetic engineering, it turned out to be difficult for storage proteins, most likely because the three-dimensional structure of the molecule was altered by such exchanges. Recent progress was made in obtaining protein crystals which enabled the analysis of the three-dimensional protein structure of the precursor trimers as well as of the mature storage proteins. These studies revealed that the stability of the storage proteins towards the proteases in the storage vacuoles is due to possible cleavage sites being hidden within the protein structure and in this way protected against proteolysis.

Vicilin shows similarities in its amino acid sequence to legumin, and primarily forms trimers, of which the monomers consist of only one pep-tide chain. Due to the lack of cysteine, the vicilin monomers are unable to form S-S bridges. In contrast to legumins, vicilins are often glycosylated; they contain carbohydrate residues, such as mannose, glucose, and N-acetylglucosamine.

Prolamins are formed as storage proteins in grasses

Prolamins are only present in grasses, such as cereals. They form polymor-phic mixtures of many different subunits of 30 to 90 kDa. Some of these subunits contain cysteine residues and are linked by S-S bridges. Also in glutenins which occur in the grains of wheat and rye, monomers are linked by S-S bridges. The glutenin molecules differ in size. Flour for baking bread depends on the content of high molecular glutenins. Flour from barley, oats or maize that lacks glutenin is not suitable for baking bread. Since the glutenin content is a critical factor in determining the quality of bread grain, investigations are in progress to improve the glutenin content of grains by genetic engineering.

2S-PROTEINS ARE PRESENT IN SEEDS OF DICOT PLANTS

2S-Proteins are also widely distributed storage proteins. They represent a heterogeneous group of proteins, of which the sole definition is their sedi-mentation coefficient of about 2 Svedberg (S). Structural investigations have revealed that most 2S-proteins have a related structure and therefore possibly derive, along with the prolamins, from a common ancestor pro-tein. Napin, the predominant storage protein in rape seed, is an example of a 2S-protein. This protein is of substantial economic importance since, after the oil has been extracted, the remainder of the rape seed is used as fodder. Napin and other related 2S-proteins consist of two relatively small polypeptide chains of 9 kDa and 12 kDa, which are linked by S-S bridges. So far, little is known about the packing of the prolamins and 2S-proteins in the protein bodies.

GENES FOR NITRATE ASSIMILATION IN ALGAE

The *NRT1 (NPF)* genes are represented in high number in plant genomes. In contrast, the algal genomes contain a very low number, two, one, or even none. For example, *Arabidopsis* has 53 NRT1 (NPF) but the number in plants varies between 51 (in *Capsella rubella*) and 139 (in *Malus domestica*). *Chlamydomonas* has a unique *NRT1*, like most of the algal genomes analyzed. *Coccomyxa subellipsoidea* and *Chlorella NC64A* contain two, but the *NRT1* gene was not found in the microalgae *Micromonas pusilla* or *Chlorella paradoxa*.

This is relevant since, provided that the analyzed genomes are complete, it suggests that along evolution some important function mediated by NRT1 has been maintained in the different algae species, and that in plants, the complexity of the new tissues that were appearing required new functions, which were assumed by this family of proteins.

The NRT2 proteins, responsible for high affinity nitrate/nitrite transport (HATS), are present in all algal species. However, NAR2, which with NRT2 forms the two-components HATS appears in Chlorophytes algae, like in plants, but not in Rhodophytes or Glaucophytes, suggesting that in these latter algae one-component nitrate HATS might be efficient enough to guarantee cell growth and performance. NAR2 is highly conserved between Chlamydomonas and Volvox but show a lower conservation with the other Chlorophytes.

So far, Chlamydomonas is the alga with the highest number of NRT2 genes, likewise for NAR1 genes (six of each). Like NAR2, NAR1 was not found in the genomes of Rhodophytes or Glaucophytes, and the number of NAR1 genes is just one in species of Micromonas, Ostreococcus, and Bathycoccus prasinos. This is interesting because in Chlamydomonas two members of this NAR1 family have specialized functions. NAR1.1 is a nitrite transporter to the chloroplast clustered with the nitrate assimilation genes. The gene encoding this transporter is expressed in the presence of nitrate under the control of NIT2, which is the major regulatory gene for nitrate assimilation. All this suggests a role of NAR1.1 in nitrate assimilation.

On the other hand, NAR1.2 (LCIA) is not nitrate-regulated nor NIT2-dependent. NAR1.2 is expressed under low CO2 conditions and is a HCO3- transporter to the chloroplast. The function of CrNAR1.3–6, as that of the single NAR1 in other microalgae species, has to be solved. The high number of these genes, NRT2 and NAR1, in Chlamydomonas might reflect an optimization in the utilization of those nutrients under many diverse environmental conditions.

Single genes encode NR and NiR in algae, excepting some particular cases as Coccomyxa subellipsoidea C-169, which has two NRs in its genome. It is interesting to point out that in Cyanidioschyzon merolae, which lacks NiR gene (NII1), one sulfite reductase gene CmSiRB of the two annotated in its genome, CmSiRA and CmSiRB, encodes really a nitrite reductase. Sulfite reductase and nitrite reductase have common structural and functional features. The nitrite reduction function of CmSiRB was deduced from its high NiR vs. its very low sulfite reductase activity, high expression in nitrate medium, and low in ammonium medium in contrast to CmSiRA, which is not affected by the presence of ammonium or nitrate. In addition, in the genome CmSiRB maps between a nitrate transporter and the NR genes.

Clustering of nitrate assimilation genes appears commonly in fungi, cyanobacteria and algae. In fungi the metabolic gene clusters are often associated with fungal virulence, and evolutionarily with pathways conferring some advantage over competitors.

However, the reasons for this clustering are not yet understood, but it is probable that this physical proximity of genes facilitates the coordination of their regulation and inheritance. Though the particular genes clustered change in its role, order or orientation from species to species, it is interesting to point out that most of Chlorophytes have a complete set of genes for nitrate assimilation (NRT2/NAR2/NIA1/NII1/NAR1) with the exception of Chlorella sp. NC64A, which has three NAR1 but none within the nitrate-cluster and Cocomyxa subellipsoidea in which genes are scattered in the genome.

A curiosity is that Chlorella sp. NC64A, even having a complete set of all the genes needed for nitrate assimilation (transporters, reductases and Moco genes), has developed an alternative adaptation strategy to acquire N.

This Chlorella strain is endosymbiotic with Paramecium bursaria and can use ammonium or amino acids but not nitrate or nitrite. In this alga NiR and not NR activity is considerably stimulated by nitrate but the mechanisms responsible for this silencing of nitrate utilization are unknown.

Another curiosity is that Ostreococcus species contain in addition to genes for nitrate/nitrite transporters and reductases others for molybdate transport (MOT1) and metabolism (CNX2 and CNX5). Thus, the complete cluster in Ostreococcus tauri is: NAR1-NIA1-NII1-Cytb5R-NAR2-NRT2-MOT1-CNX2-CNX5.

In Chlamydomonas two gene clusters are present: one consists on NII1-NRT2.2-NRT2.1-NAR2-NIA1-NAR1 and in the other NRT2.3-AOX1 (AOX1 encodes a mitochondrial alternative oxidase). In Micromonas species two different clusters were described: CNX2-NAR1-NII1-NAR2-NRT2-NIA1 (in M. pusilla CCMP1545), and NRT2.1-NAR2.1-NII1-NAR1-NIA1-NAR2.2-NRT2.2 (in RCC299).

A comparison between the different gene clusters for nitrate assimilation in microalgae is shown in Fig. 2. It is remarkable how essential genes for nitrate assimilation have been maintained clustered during the different genetic events needed to configure the present organization.

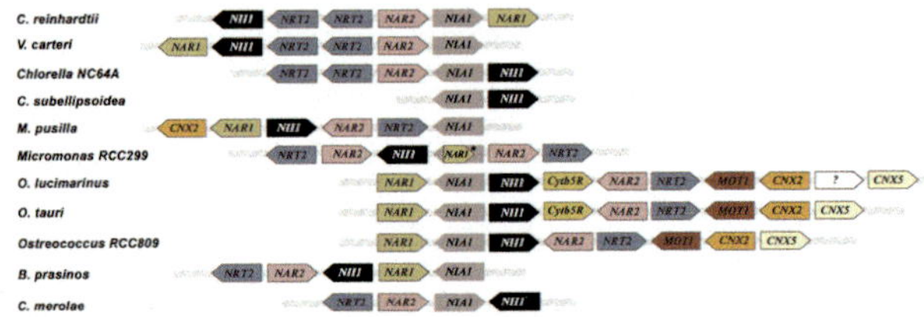

Fig 2. Clusters of nitrate assimilation genes in microalgae. NIA1 and NII1 (genes encoding NR and NiR)

Molybdate transporter genes (MOT), are present in the algal genomes highlighting the importance of this anion for a proper assimilation of nitrate. Accordingly, the complete set of genes for Moco biosynthesis is represented in all the species analyzed. Interestingly, the CNX1 gene, encoding the two domains (E and G) protein in plants, is split into two genes each one encoding a domain, CNX1G and CNX1E. However, some exceptions are found in Ostreococcus tauri, Ostreococcus RCC809, and Bathycoccus prasinos where these domains are merged in a single gene.

By fusing those two genes from Chlamydomonas reinhardtii in chimeric constructions CNX1E-G or CNX1G-E, it is shown that the orientation of the domains CNX1E and CNX1G does not affect the functionality of the chimeric CNX1 proteins (Llamas et al., 2007). This finding suggests that the fact of having two genes CNX1E and CNX1G, or just one CNX1 with the orientation E-G as in plants, or G-E as in mammals is irrelevant. In contrast to other Moco genes, ABA3 (Abscisic acid deficient 3) is not generally present in the analyzed algae (Table 1). ABA3 is required for functionality of the Mo-hydroxylases, Xanthine Dehydrogenase, and Aldehyde Oxidase, by the transfer of a sulfur atom to the Mo-center (sulfuration step). Thus, it will be interesting to understand what is the effect of this lack in the metabolism of those algae.

Similarly, MCP encoding a Moco Carrier Protein seems to be involved in an efficient mechanism of Moco storage and protection. MCP has a significant conservation with MBP (Moco Binding Protein) from plants, but it is only present in Volvocales and not in the other algal genomes analyzed.

Additionally, it is interesting to point out that the CNX5 gene for Moco biosynthesis was differentially conserved in the studied algae. CNX5 from Chlamydomonas and Volvox are closely related but markedly different from CNX5 genes from the other algae, which show a significant conservation among them. The reasons for this diversification are not clear.

Concerning the ammonium incorporation to amino acids, the corresponding genes were analyzed in algae. GS (GLN) genes are well-represented in Volvocales in a number of 3 or 4. However, their particular role and localization has only been studied in Chlamydomonas. The four GS genes encode cytosolic glutamine synthetases of type 1 (GLN1, GLN4) and plastidic GS of type 2 (GLN2, GLN3). The two isoforms of GS2 are highly conserved, and their genes located head to head in the genome suggesting a recent event of gene duplication in both Chlamydomonas and Volvox.

In the species of Ostreococcus and C. merolae, the single GLN gene is more related to bacterial GLN than to plant-type GLNs present in other algae. Genes for the plastidic NADH-GOGAT and Ferredoxin-GOGAT appear in Chlorophytes, however in Mamiellales, Rhodophytes, and Glaucophytes there appears only one (Ferredoxin-GOGAT).

REGULATION OF NITRATE ASSIMILATION

The regulation of nitrate acquisition is a complex process involving a network of proteins, many of them scarcely known. Majority of data available has been attained in Chlamydomonas. Below, we review some relevant regulatory aspects reported in microalgae.

In general, nitrate and ammonium have opposite effects over nitrate assimilation genes, which are referred to as positive (nitrate) and negative (ammonium) signals. Ammonium is the preferred N source due to its reduced state and energetically favorable assimilation. It is well-established that ammonium has a negative effect on nitrate assimilation at both transcriptional and posttranscriptional levels in Chlamydomonas, especially in high CO2 conditions.

NR gene (NIA1) expression, widely studied in many algal species, is induced in nitrate medium and strongly repressed in ammonium medium. Also nitrate transporters are expressed after ammonium depletion in Micromonas. In Chlamydomonas, when both positive and negative signals are present together, repression appears as a quantitative process that is not strictly sensitive to the ammonium concentration but to the nitrate/ammonium (N/A) balance. A determined ammonium concentration is less repressive in conditions of high nitrate (mM) than in low nitrate (μM).

Nevertheless, in medium without ammonium, NIA1 is similarly induced by low and high nitrate concentrations. This pattern points out to a finely tuned regulation mediated by the N/A balance. How this balance regulates NIA1 is scarcely known and a complex network of proteins and signal molecules are starting to be unraveled in Chlamydomonas. To understand these positive and negative signaling pathways, an insertional mutant collection of about 22,000 mutants was obtained and screened for aberrant sensing of nitrate or ammonium. The selection of these phenotypes was possible because the mutagenized strain bears a reporter gene that responds positively to nitrate and negatively to ammonium (pNIA1ARS, a chimeric gene: promoter of NIA1 fused to the arylsulfatase gene).

For the ammonium insensitive phenotype (AI), a broad range of insensitiveness is shown depending on the mutant but none of them is completely insensitive to ammonium. The numerous mutants isolated together with the absence of a mutant totally insensitive to ammonium highlight a regulation mechanism that probably involves several independent pathways. In contrast, the nitrate insensitive mutants (NI) reveal a master gene, NIT2, needed for nitrate assimilation genes expression. In addition, a number of NI mutants with partial phenotypes are identified that also illustrates the complexity of nitrate regulation.

POSITIVE REGULATION OF THE NITRATE ASSIMILATION PATHWAY

In photosynthetic organisms nitrate is not only an N source but also a signaling molecule. Nitrate induces expression of genes responsible for its assimilation. Proteins involved on nitrate sensing are scarcely known in microalgae. Several major transcription factors were identified in plants, fungi, yeast, and in some green and red algae. In *Chlamydomonas* the transcription factor NIT2 is essential for a proper upregulation of the main genes for nitrate assimilation (*NII1, NRT2.1, NRT2.2, NRT2.3, NAR2, NIA1, NAR1.1*, and *NAR1.6*).

Indeed, nit2 mutants are unable to grow in nitrate medium. The NIT2 gene is also repressed by ammonium and induced in nitrate and N-free medium, highlighting that other yet unidentified transcription factors work upstream to NIT2. Structural characteristics of NIT2 show an RWP-RK domain (so called by the consensus sequence) within a leucine zipper domain; this motif may serve in dimerization and DNA binding.

This domain is also found in algal minus dominance (MID) proteins and transcription factors from plants as the NIN-like proteins (NLP), some of them having a key role in regulating responses to nitrate availability. Actually, NLP7 seems to be a major player in the primary response to nitrate.

Transcription factors containing RWP-RK domains are numerous in green algae but many of them have an unknown function. Proteins with significant homology to NIT2 have not been identified. However, it is noteworthy an RWP-RK transcription factor along the cluster of N metabolism related genes in Micromonas pusilla. Though clustering is not a strong evidence for its function, this gene is a good candidate to study in future experiments in this alga. Furthermore, other RWP-RK transcription factors that are upregulated in N starvation in Chlamydomonas (RWP1 and RWP10) have an expression that correlates with those of LAO1 (encoding L-Amine Oxidase) and CNX1E, respectively.

In addition to the RWP-RK domain, NIT2 shows a GAF domain at the N-terminal region of the protein. This motif could bind small molecules such as cGMP, NO or a-oxoglutarate. However, the function of this domain remains poorly studied, although it is known that it does not bind nitrate and is not involved in DNA binding. Additionally, NIT2 possesses a glutamine-rich domain involved in protein-protein interactions, essential for NIT2 functionality, and a nuclear export sequence (NES). How NIT2 senses the nitrate signal for binding to different gene promoters is unknown. Nevertheless, a NES domain appears in the transcription factor NIRA involved in nitrate induction in Aspergillus and in NLP7 in Arabidopsis. This NES sequence is involved in NIRA and NLP7 exportation to the cytosol in the absence of nitrate. In Chlamydomonas the NES domain of NIT2 is close to the GAF motif, thus binding of small molecules to GAF domain might expose or hide the NES domain facilitating its nuclear exportation. However, in Chlamydomonas NIT2 is important to activate nitrate-inducible genes and to upregulate genes in nitrate plus ammonium medium, a condition under which the nitrate assimilation genes are repressed, as the truncated hemoglobin THB1.

Therefore, a nuclear export would not be favorable in those conditions. Furthermore, NIT2 is also involved in the repression of genes downregulated by nitrate like the ammonium transporter AMT1.1. It would be expected to find new proteins interacting with NIT2 to integrate the nitrate/ammonium balance and proportionate the specificity for a proper gene expression/regulation on each N condition.

In other green microalgae apart from Chlamydomonas, transcription factors involved in upregulating the nitrate assimilation genes are unknown. Nevertheless, in Chlorella vulgaris a unique GATA site is present in the promoter region of the NR gene. This sequence works as a nitrate response element (NRE) and can bind specifically a Neurospora crassa NIT2 zinc-finger domain/glutathione S-transferase fusion protein. Thus, similar GATA-binding transcription factors could be responsible for the transcriptional regulation in this alga. In the red alga Cyanidioschyzon merolae an R2R3-type MYB transcription factor, MYB1, has a key role in the nitrate upregulation of the main N assimilation genes (NIA1, NII1, NRT2, AMT, GLN). MYB1, as CrNIT2, is upregulated in nitrate and N-free medium and the transcription factor encoded binds directly to the promoter of all of these genes activating their expression.

This class of transcription factors is numerous in green algae but it has not been related to N metabolism. Nonetheless, two R2R3-type MYB type transcription factors directly bind to the promoter region of the glutamine synthetase gene in Pinus sylvestris and some MYB proteins are differentially regulated by the N source in Arabidopsis (Scheible, 2004). Thus, we could expect that MYB proteins would be also involved in N regulation in some algae.

In addition to NIT2, other regulatory protein NZF1 (nitrate zinc finger 1) involved in nitrate positive signaling was recently identified in Chlamydomonas. This is a tandem zinc finger protein CCCH-type that regulates NIT2 expression and, indirectly, the nitrate assimilation genes. NZF1 is proposed to have a role in NIT2 polyadenylation. The nzf1 knockout mutant shows aberrant forms of polyadenylated NIT2 mRNAs, which are not present in the wild type strain in nitrate medium. Thus, this nzf1 mutant has decreased levels of wild type NIT2 transcript. Analogous transcription factors were not found in other algae, but a posttranscriptional regulation related to the mRNA stability is reported for the NR gene in Chlorella vulgaris. In this case, two different mRNAs are synthesized in response to short and long times of induction in nitrate. The long mRNA is less stable than the short version, which is induced in minutes when the repressor (ammonium or derivatives) is removed or when the inductor (nitrate) appears. Differences between these mRNAs are located in the 52 UTR. This supposes a strategy to perform a rapid adaptation to the changing environmental conditions.

Transcriptomics analysis showed other transcription factors that are upregulated in N starvation. One of them is NRR1, which encodes a SQUAMOSA promoter-binding protein, SBP, responds early to N starvation, and has a relevant role in TAG accumulation. NRR1 is hypothesized as a "master" transcriptional regulator for reprogramming gene expression of N metabolism during N deprivation. However, more transcriptomic and proteomic studies in nitrate medium would be required to shed light about new regulatory factors.

ROLE OF INORGANIC NITROGEN UPTAKE AND PARTITIONING IN NUE

Root uptake of inorganic N forms often dominates and requires the function of nitrate and ammonium transporters, respectively. Nitrate uptake is achieved by transporters of the NRT1 and NRT2 family, respectively. Studies on natural variation and genetic manipulation of NRT1 or NRT2 expression have resolved that NRT function in roots influences N acquisition, plant growth, and seed development. To date, successes in improving NUE under both high and low N have mainly been achieved when NRT transporters were constitutively overexpressed or expressed in both root and leaf tissues.

While it remains to be determined if root-specific overexpression of NRTs may be sufficient for enhancing N uptake, research suggests that downstream nitrate transport processes may additionally or dominantly contribute to the observed increases in NUE. Indeed, recent work in Arabidopsis (Arabidopsis thaliana), tobacco (Nicotiana tabacum), and rice (Oryza sativa) showed that overexpression of the hyperactive chimeric nitrate transporter AtNC4N in the phloem of old leaves results in improved NUE under high and low N status.

This study also supports the fact that during leaf senescence, besides amino acids and peptides, nitrate recycling from the vacuole and subsequent source-to-sink partitioning contribute to sink N supply. In line with this, increased expression of the vacuolar OsNRT1.1A/OsNPF6.3 nitrate transporter either throughout rice plants, or in the epidermis and vascular tissues of roots and parenchyma cells of culms and leaf sheaths resulted in increased grain yield and NUE under high and low N supply, due to improved N utilization for grain development.

As nitrate transport, related N assimilation, and amino acid partitioning processes are highly intertwined, it would be interesting to examine if in AtNC4N and OsNRT1.1A overexpressors co-regulated by changes in amino acid pools and transport also add to the observed increases in seed yield, seed N content, and NUE.

NITROGEN ASSIMILATION AND NUE

Following uptake, nitrate is reduced in the cytosol to nitrite which is transported into the plastid for further reduction to ammonium. The two-step reduction is achieved by nitrate reductase (NR) and nitrite reductase (NiR), respectively. NR catalyzes the main regulatory step in the N assimilation process and its activity is highly regulated by factors such as nitrate, light, or water availability. In addition, amino acids (e.g., glutamine) present signals regulating uptake and assimilation by affecting both the activity of nitrate transporters and NR. Overexpression of NR or NiR has often resulted in increased N uptake, however improvements in seed yield, seed protein level, or NUE have generally not been achieved, potentially due to negative feedback of N-containing metabolites and related transcriptional or post-transcriptional regulation. Nevertheless, recent research in japonica rice showed that constitutive expression of indica OsNR2, which encodes a NADH/NADPH-dependent NR, confers increased tiller number, grain yield, and NUE at high N supply. These effects were further enhanced when indica OsNR2 and OsNRT1.1B were concurrently expressed, indicating that OsNR2 positively regulates OsNRT1.1B and thereby regulates nitrate uptake.

Ammonium is assimilated through the coordinated activity of glutamine synthetase (GS) and glutamate synthase (GOGAT) into glutamine and glutamate, respectively, which serve as N donors for the biosynthesis of other amino acids and amides. Two isoforms exist for GS, the cytosolic GS1 and plastidic GS2, which have distinct functions in N assimilation and are encoded by multigenic or single gene families. GS2 is involved in primary N assimilation (i.e., de novo amino acid synthesis) as well as re-assimilation of ammonium produced during photorespiration. GS1 isoforms are mainly involved in N re-assimilation and remobilization, with some exceptions such as AtGLN1;2 in Arabidopsis, which is important for N assimilation under high nitrate supply, and OsGS1;1 involved in ammonium assimilation in rice shoots and roots. Due to their fundamental role in N metabolism and to improve plant growth and seed yield, GS1 and GS2 genes have been overexpressed in numerous species, including in tobacco, Arabidopsis, Sorghum bicolor, Zea mays (maize), Triticum aestivum (wheat), Hordeum vulgare (barley), and rice. While in some cases no or negative effects on plant performance have been reported, other studies have shown improvements in seed yield, NUE, NUpE, or NUtE for both GS1 and GS2, and repeatedly at high and low N supply. For example, increased expression of HvGS1-1 under the control of its own promoter in barley led to improved grain yields and NUE under high and low N conditions. Researchers constitutively expressed bacterial GS I genes from Klebsiella sp. D1-5 and Lactococcus sp. Zjy3, respectively, in Arabidopsis and showed, for both genes, improved plant growth and NUE at low and sufficient N. Similarly, overexpression of TaGS2-2Ab in wheat using its endogenous promoter resulted in higher seed yield and NUE when N supply was high and low, due to increased N uptake and remobilization.

Compared to GS, relatively few studies have looked at altering the expression of genes encoding for plastid-localized ferredoxin-dependent (Fd-GOGAT) or NADH-dependent (NADH-GOGAT) GOGAT. While Fd-GOGAT is primarily involved in the assimilation of ammonium that originates either from nitrate reduction or leaf photorespiration, NADH-GOGAT is highly expressed in non-photosynthesizing cells and has a distinct function in non-photorespiratory ammonium reduction. Using Arabidopsis, barley, rice, or Medicago sativa (alfalfa) mutants, essential roles in plant growth and seed development have been demonstrated for both Fd-GOGAT and NADH-GOGAT.

However, relatively few studies manipulated GOGAT expression to enhance seed yield and NUE, and those focused on NADH-GOGAT with rather moderate outcomes. For instance, OsNADH-GOGAT overexpression in rice using its endogenous promoter only resulted in increased grain weight at low N fertilization. Further, constitutive expression of MsNADH-GOGAT showed higher biomass production in tobacco when N was sufficient, whereas in maize, overexpression of ZmNADH-GOGAT resulted in a lower shoot biomass and no changes in kernel yield. Interestingly, and as pointed out above, combined overexpression of OsAMT1;2 and OsNADH-GOGAT1 in rice improved NUE at both sufficient and low N conditions. At sufficient N, the transgenic lines achieved higher seed protein levels without altering seed yield, while at low N both seed protein and yield were improved. The results suggest that by increasing both N uptake and N assimilation, negative intrinsic effects that might happen when only AMT (see above) or GOGAT is overexpressed (e.g., through unbalanced pools of N metabolites) may be avoided.

AMINO ACID SYNTHESIS AND NUE

Relatively few studies have described improvements in NUE by altering the expression of genes related to amino acid synthesis, with the exception of asparagine synthetase (ASN) and alanine aminotransferase (AlaAT). Both asparagine and alanine are major long-distance transport forms of N in certain plants. Asparagine is synthesized through ASN, which functions in both N assimilation and remobilization. Constitutive overexpression of OsASN1 was somewhat successful in rice and showed increased seed N content and NUE under limiting and sufficient N treatments, as well as additional improvements in shoot biomass and seed yield at low N.

On the other hand, expression of Escherichia coli EcasnA in Brassica napus (canola) seems to only be successful at high N status. AlaAT is involved in alanine synthesis and catabolism, and the constitutive or root-specific expression of barley HvAlaAT resulted in improved seed yields in wheat, or in canola and rice. In addition, NUE was increased in rice, canola, and Saccharum officinarum (sugarcane) when HvAlaAT expression was targeted to roots and engineered plants were exposed to high N and/or low N. How AlaAT function in roots affects seed yield and NUE is unclear, as it seems to not be vital for plant N metabolism and source-to-sink partitioning.

REMOBILIZATION OF SOURCE NITROGEN AND NUE

A relatively high portion of N that is delivered to seeds or other storage sinks may derive from the recycling or remobilization of N in source leaves, rather than directly from root uptake. This includes nitrates, amino acids, and ureides that are temporarily stored in vacuoles during the vegetative stage, as well as N from proteins and organelles, especially from chloroplast degradation. Overall, the degree of N remobilization in source tissues and its allocation to sinks is assumed to have a significant impact on grain filling, yield, and NUE, while decreasing the residual N in vegetative tissues at harvest. N remobilization and reassimilation occurs especially during leaf senescence and mainly involves the activity of the cytosolic GS1, which has been shown to affect NUE.

Glutamate dehydrogenase (GDH) may also have some role in N assimilation, but it seems to predominantly function in glutamate deamination and ammonium supply for GS1 during N remobilization or when C is limited. Efforts to enhance plant performance and NUE by overexpressing GDH have been inconclusive. Studies have been unsuccessful when using constitutive overexpression of tomato Slgdh-NAD;B1 in tobacco, Nicotiana plumbaginifolia GDHA and GDHB in N. tabacum, and GDH from fungi Sclerotinia sclerotiorum and Magnaporthe grisea in rice. On the other hand, there is good evidence that constitutive overexpression of GDH can also lead to increased growth, seed yield, and/or NUE at different N regimes.

Specifically, expression of EcgdhA from E. coli in tobacco and maize resulted in increased biomass under sufficient N conditions (Ameziane et al., 2000) and increased seed yield under high N supply, respectively. Further, expression of AngdhA from Aspergillus nidulans in potato (Solanum tuberosum) caused increased photosynthesis rates, tuber numbers and tuber dry weight, as well as improved NUE at moderate and low N levels. In rice, increases in NUE and/or grain yield were achieved at low N when fungal GDH from Cylindrocarpon ehrenbergii (CeGDH) or Eurotium cheralieri (EcGDH) were expressed. In addition, improvements in rice grain yield and weight, or in grain storage protein levels were obtained at low and high N as well as without N fertilization by using fungal GDHs from Trichurus sp. (TrGDH) and Pleurotus cystidiosus (PcGDH), respectively. Why constitutive overexpression of GDH was successful in some but not all studies is unclear, but it could potentially result from the genes and plant species that were analyzed. Also, a leaf-specific engineering approach may help to more reliably promote source N remobilization through GDH and enhance source-to-sink N allocation.

Autophagy, an essential mechanism for the degradation of proteins, organelles, and cytosolic macromolecules, is enhanced during leaf senescence and strongly contributes to N remobilization and source-to-sink transport of N. Altering expression of autophagy-related proteins (ATGs) has been a successful strategy to enhance N remobilization in Arabidopsis and rice. Depending on the study, the outcome was increased plant growth, seed yield, or seed N and protein levels. In Arabidopsis, constitutive overexpression of AtATG8 resulted in higher seed protein levels when transgenic plants were grown with moderate N supply (Chen et al., 2019), while OsATG8b expression led to increased biomass, seed yield, and NUE at moderate and low N. In rice, shoot biomass, seed yield, and protein levels, as well as NUE were improved when OsATG8a was overexpressed and plants were exposed to high N nutrition, while at moderate N, positive effects were only observed for biomass and seed protein. Similarly, constitutive expression of OsATG8b in rice resulted in more biomass as well as higher grain yield and protein under sufficient N, and OsATG8c overexpression triggered improvements in shoot biomass, grain yield, and NUE under moderate and low N fertilization. To date, studies on autophagy-related proteins have focused on Arabidopsis and rice, and work is now needed to address their importance for N remobilization and subsequently yield and NUE improvements in other crop species.

Future Perspectives

The nitrate assimilation pathway is a relative simple process conserved in the majority of microalgae. This conservation highlights the importance of this pathway in the central metabolism of these organisms. This pathway owns an uptake step to incorporate nitrate into the cell, a first NR-mediated reduction followed by nitrite incorporation to chloroplasts and a second reduction step mediated by NiR. Finally N is incorporated into carbon skeletons by GS/GOGAT cycle. Additionally, the key enzyme for nitrate assimilation (NR) harbors molybdenum cofactor, and subsequently the genes responsible for Moco biosynthesis are also highly conserved in most studied algae. In spite of the low structural complexity of this pathway its regulation is intricate. A complex protein network controlling nitrate assimilation is starting to be unraveled in Chlamydomonas.

However, although structural genes for nitrate assimilation are well-conserved, being clustered in most algae, those involved in regulation show a lower conservation degree. Even so, a central role of NO and cGMP occurs in Chlamydomonas, and so the implication of these signal molecules in the regulation of other algae must be addressed. The higher diversification in the regulatory processes would allow a specific adaptation of each alga to its particular environment. At structural level, main differences are found in the number of transporters. Chlamydomonas is an alga with a high number of them, but other, algae with small genome sizes have transporter families with only one member. Probably, this would be a good reason to expect also a more complex regulation in Chlamydomonas. These features among others contribute to the high adaptability observed for this organism and have converted Chlamydomonas in a good model alga for basic science and biotechnological use. Nevertheless, future studies should improve our knowledge about the whole system of nutrient acquisition in algae, addressing not only assimilation of different N forms and its regulation but also their relationships with carbon, sulfur and phosphorus metabolisms. This understanding will allow us to develop many applications with algae and discover the real potential of these organisms on applied sciences.

The movement of N from the soil to final sink cells involves numerous transport and metabolic processes in source and sink, including (i) N uptake and partitioning within cells and over short and long distances, and (ii) N (re)assimilation, amino acid synthesis and catabolism, and protein synthesis. Since any of these processes could potentially present a bottleneck for the efficient use of N for biomass and seed production, associated key genes have been targets for improving plant performance and NUE, but, as reviewed above, with varying success.

The inconsistent outcomes might, in part, be caused by unwanted pleiotropic effects due to the use of constitutive promoters, an imbalance in N metabolite pools triggering negative feedback inhibition of N uptake or metabolism, substrate limitations for amino acid synthesis, or post-transcriptional and post-translational regulation. To circumvent such detrimental effects and to improve NUE, future research may consider the selection or design of "super promoters" for genetic manipulation that allow tissue or cell-specific, developmentally regulated and enhanced gene expression. Such promoters could, for example, be used for the overexpression of amino acid transporters or synthesis genes to drive N uptake from the soil throughout the vegetative and reproductive phases, or to accelerate source N remobilization and leaf export during leaf senescence. Other research studies could consider a "gene pyramiding approach". While selective metabolism and transporter genes have been successful targets to improve NUE in some plant species, at least under high N nutrition, simultaneous manipulation of several genes may be required for a broader approach to improve NUE at high and low N status.

For example, improvements in N assimilation or amino acid synthesis may only be successful if metabolic and transport processes are coordinated through concurrent manipulation of two or more genes. Further, one gene could be simultaneously manipulated in root and shoot tissues, or source and sink. For instance, to improve N flux from source to sink and seed N supply, leaf N export and phloem loading, as well as seed transport processes including embryo loading, will need to happen at a high rate and throughout sink development. Approaches to improve crop productivity and NUE should also take into account regulatory aspects of N assimilation and transport, as well as the sensing and signaling mechanisms involved. Consideration should further be given to the tight interaction between carbon, N, and sulfur metabolism and their transport processes.

NUE is a complex polygenic trait making its genetic dissection and engineering challenging. In fact, despite the notable progress that has been made in understanding plant NUE, successful agronomic applications of engineered crop plants with altered N transport and metabolism or other physiological functions have not yet been achieved. To better grasp the genetic basis of NUE for genetic manipulation, future strategies might take greater advantage of the available natural variation for NUE between and within crop species.

For example, using genetic association studies or genomic selection, in conjunction with high-throughput phenotyping, large crop populations, or different cultivars could be screened to discover further genes associated with improved NUE. Alternatively, or in addition, comparative transcriptome and proteome studies will allow for identification of fundamental NUE genes. Importantly, to develop improved crop cultivars for a range of growing regions, such analyses will have to be done with plants exposed to a variety of locations, latitudes, soil conditions, and varying N regimes and agricultural systems. These should include systems that do not solely rely on mineral N fertilizer but also consider organic matter for plant N nutrition, like organic farming or conservation agriculture, and the need of amino acid transporters for N uptake.

Further, to improve NUE for a sustainable agriculture, future climate and environmental changes will need to be addressed. For example, exposure of crop plants to drought stress will not only require adjustments with respect to N uptake, partitioning, and metabolic processes, but also other physiological as well as morphological changes. As it would be extremely challenging to conduct field experiments for conventional breeding, marker-assisted modern breeding, genomic selection, and genetic engineering under numerous scenarios, the use of crop modeling that integrates genetics, environmental factors, and management practices will be a critical tool for making progress in NUE. Indeed, interdisciplinary strategies and multiscale collaborations are essential to achieve a more robust and holistic understanding of NUE that can lead to global successes for crop production and a more sustainable agroecological system.

CHAPTER

A PLANT CELL HAS THREE DIFFERENT GENOMES 16

The cell was first discovered in 1665 by an English scientist named Robert Hooke. While looking through a microscope, he observed tiny box-like objects in a slice of cork (bark from an oak tree) and named these boxes cells. Cells are the basic units of life, which make up all living things. This idea forms the basis of the Cell Theory.

CELL THEORY

The three main parts of the cell theory are:

- All living things are made of cells.

Fig. 1. Cells seen in a plant stem cross-section

- The cell is the basic unit of structure and function in all living things.
- Cells only come from other pre-existing cells by cell division.

While some organisms are single-celled, others are made up of many cells. These organisms are called multicellular (having many cells). Cells differ in their size and complexity.

Eukaryotes are organisms which are made up of large and complex cells, whereas prokaryotes are organisms which are made up of small and simple cells. Animals and plants are examples of eukaryotes (have eukaryotic cells) while bacteria are examples of prokaryotes (have prokaryotic cells).

PLANT CELL STRUCTURE AND FUNCTION

In spite of the differences in size and complexity, all cells are mostly composed of the same substances and they all carry out similar life functions. These include growth and metabolism and reproduction by cell division.

Cells are made up of subcellular structures that are responsible for different and specific functions. These structures are known as organelles. A number of these organelles are common to both animal and plant cells. This section will focus on those parts which plants have.

Cell Structures (Cell Organelles)

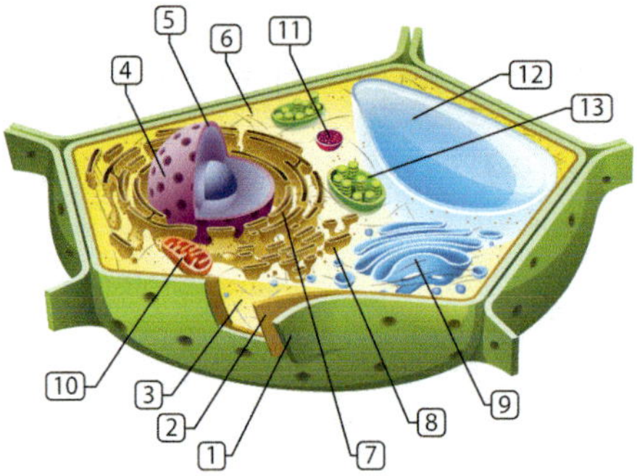

Fig. 2. Plant cell structures

See below for key to numbering:

- Cell Wall: This is the rigid outermost layer of a plant cell. It makes the cell stiff -providing the cell with mechanical support - and giving it protection. Animal cells do not have cell walls.
- Cell Membrane: This is a protective layer that surrounds every cell and separates it from its external environment. It is found just inside the cell wall and is made up of complex lipids (fats) and proteinsproteins .
- Cytoplasm: The cytoplasm is a thick, aqueous (water-based) solution in which the organelles are found. Substances such as salts, nutrients, minerals and enzymes (molecules involved in metabolism) are dissolved in the cytoplasm.
- NucleusNucleusThe nucleus is the 'control center' of the cell. It contains Deoxyribonucleic acid (DNA), the genetic material that directs all the activities of the cell. Only eukaryotic cells have nuclei (plural for nucleus), prokaryotic cells do not. The nucleus is separated from the cytoplasm by a specialized membrane called the
- Nuclear membrane.
- Ribosomes: These are little round structures that produce proteins. They are found in the cytoplasm or attached to the endoplasmic reticulum.

- Endoplasmic Reticulum (ER): The ER is a membrane system of folded sacs and tunnels. The ER helps move proteins within the cell as well as export them outside of the cell. There are two types of endoplasmic reticulum.
- Rough endoplasmic reticulum. The rough endoplasmic reticulum is covered with ribosomes. Smooth endoplasmic reticulum (no ribosomes)
- Golgi body: The Golgi body is a stack of membrane-covered sacs that prepares proteins for export from the cell.
- MitochondrionMitochondrion (plural mitochondria): This is the 'powerhouse' of the cell. It converts the energy stored in food (sugar and fat) into energy-rich molecules that the cell can use (Adenosine triphosphate – ATP for short).
- Lysosome: The lysosome is the digestive center of a cell that produces many different types of enzymes which are able to break down food particles and recycle worn out components of the cell.
- Vacuoles: These are large membrane-enclosed compartments that store toxic wastes as well as useful products such as water. These are mainly found in plants.
- ChloroplastChloroplast Chloroplasts contain a green pigment that traps sunlight and converts it into sugars by a process called photosynthesis. The sugars are a source of energy for the plants and the animals that eat them.

WHAT MAKES PLANT CELLS UNIQUE

Plant cells have a cell wall.

Plant cells are different from animal cells in a number of ways. Perhaps the most obvious difference is the presence of a cell wall. The cell wall provides strength and support to the plant, much like the exoskeletonexoskeleton of an insect or spider (our skeleton is on the inside of our body, rather than on the outside like insects or spiders).

The plant cell wall is mainly made up of the carbohydratescarbohydrates molecules cellulose and lignin. Cellulose is used extensively by humans for making paper. Cellulose can also be converted into cellulosic ethanol, a type of biofuel. Some animals, such as cows, sheep and goats, can digest cellulose with the help of bacteria in their stomachs. Humans cannot digest cellulose, which passes through our bodies and is better known as dietary fiber, something that we should eat to keep our waste moving as it should! Lignin fills in the spaces between cellulose and other molecules in the cell wall. Lignin also helps water molecules move from one side of the cell wall to the other – an important function in plants.

Plant cells contain vacuoles.

Most adult plant cells have one large vacuole that takes up more than 30% of the cell's volume. At certain times and conditions the vacuole takes up as much as 80% of the cell's volume! In addition to storing wastes and water, the vacuole also helps to support the cell because the liquid inside the vacuole exerts an outward pressurepressure on the cell, much like the water inside of a water balloon. This is called turgor pressure and keeps the cells from collapsing inward.

Plant cells contain chloroplasts.

Unlike animal cells, plant cells can harness the energy of the Sun, store it in the chemical bonds of sugar and later use this energy. The organelle which is responsible for this is the chloroplast. Chloroplasts contain chlorophyll, the green pigment that gives leaves their colour and absorbs light energy. Cyanobacteria, a type of prokaryote capable of photosynthesis, are considered to be the ancestors of chloroplasts!

PLANT CELL HAS THREE DIFFERENT GENOMES

A plant cell contains three genomes: in the nucleus, the mitochondria and the plastids. The size of the genomes is given in base pairs (bp). As discussed, the genetic information of the mitochondria and plastids is located on one or sometimes several circular double strand (ds) DNA, with many copies present in each organelle. During cell division and organelle reproduction, these copies are distributed ran domly between the daughter organelles. Each cell contains a large number of plastids and mitochondria, which are also randomly distributed during cell division between both daughter cells. The genetic information of mitochon dria and plastids is predominantly maternally inherited from generation to generation, and the large number of inherited gene copies protects against mutations.

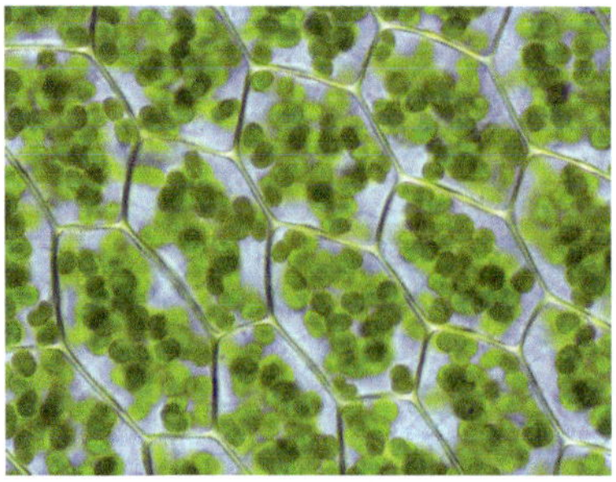

Fig. 3. Chloroplasts

IN THE NUCLEUS THE GENETIC INFORMATION IS DIVIDED AMONG SEVERAL CHROMOSOMES

For almost the whole of their developmental cycle, eukaryotic cells, as they are diploid, normally contain two chromosome sets, one set from the mother and the other set from the father. Only the generative cells (e.g., egg, pollen) are haploid, that is, they possess just one set of chromosomes. The DNA of the genome is replicated during the interphase of mitosis and gen erates in this way a quadruple chromosome set. During the anaphase, this set is distributed by the spindle apparatus to two opposite poles of the cell. Thus, after cell division, each daughter cell contains a double chromosome set and is diploid again. Colchicine, an alkaloid from the autumn crocus, inhibits the spindle apparatus and thus interrupts the distribution of the chromosomes during the anaphase. In such a case, all the chromosomes of the mother cell can end up in only one of the daughter cells, which then possesses four sets of chromosomes, to render it tetraploid. Infrequently this tetraploidy occurs spontaneously due to mitosis malfunction.

Tetraploidy is often stable and is then inherited by the following gene rations via somatic cells. Hexaploid plants can be generated by crossing tetraploid plants with diploid plants, although this is seldom successful. Tetraploid and hexaploid (polyploid) plants often show a higher growth rate, and this is why many polyploid crop plants are profitable in agriculture. Polyploid plants can also be generated by protoplast fusion, a method that can produce hybrids between two different breeding lines. When very differ ent species are crossed, the resulting diploid hybrids are often sterile due to incompatibilities in their chromosomes. In contrast, polyploid hybrids generally are fertile.

The often much higher number of chromosomes that can be found in other plants is in part the result of a gene combination. Hence, rape seed (Brassica napus) with 2×19 chromosomes is a crossing of Brassica rapa (2×10 chromosomes) and Brassica oleracea (2×9 chromosomes). In this case, a diploid genome is the result of crossing. In wheat, on the other hand, the successive crossing of three wild forms resulted in hexaploidy: Crossing of einkorn wheat and goat grass (2×7 chromosomes each) resulted in wild emmer wheat (4×7 chromosomes), and this crossed with another wild wheat resulted in the wheat (Triticum aestivum) with 6×7 chromosomes which is cultivated nowadays. Tobacco (Nicotiana tab-acum) is also a cross between two species (Nicotiana tomentosiformis and N. sylvestris), each with 2×12 chromosomes.

The nuclear genome of broad bean with 14.5×109 nucleotide pairs has a 200-fold higher DNA content than Arabidopsis. However, this does not mean that the number of protein-encoding genes (structural genes) in the broad bean genome is 200-fold higher than in Arabidopsis. Presumably the number of structural genes in both plants does not differ by more than a factor of two to three. The difference in the size of the genome is due to a different number of identical DNA sequences of various sizes arranged in sequence, termed repetitive DNA, of which a very large part may contain no encoding function at all. In broad bean, for example, 85% of the DNA represents repetitive sequences.

This includes the tandem repeats, a large number (sometimes thousands) of identical repeated DNA sequences (of a unit size of 170–180 bp, sometimes also 350 bp). The tandem repeats are spread over the entire chromosome, often arranged as blocks, especially at the beginning and the end of the chromosome, and sometimes also in the interior. This highly repetitive DNA is calledsatellite DNA. This also includes microsatel lite DNA.

Its sequence is genus or even species specific. In some plants, more than 15% of the total nuclear genome consists of satellite DNA. Perhaps it plays a role in the segregation of species. The sequence of the satellite DNA can be used as a species-specific marker in generating hybrids by protoplast fusion in order to check the outcome of the fusion in cell culture. The polymorphism of satellite DNA is also used for identifying human genomes in criminal cases.

The genes for ribosomal RNAs also occur as repetitive sequences and, together with the genes for some transfer RNAs, are present in the nuclear genome in several thousand copies.

In contrast, structural genes are present in only a few copies, sometimes just one (single-copy gene). Structural genes encoding for structurally and functionally related proteins with a high nucleotide identity often form a gene family. Such a gene family, for instance, is formed by the genes for the small subunit of the ribulose bisphosphate carboxylase, which exist several times in a slightly modified form in the nuclear genome (e.g., five times in tomato). So far 14 members of the light harvesting complex (LHC) gene family have been identified in tomato. Zein, which is present in maize kernels as a storage protein , is encoded by a gene family of about 100 genes. In Arabidopsis almost 40% of the proteins pre dicted from the genome sequence belong to gene families with more than five members, and 300 genes have been identified that encode P450 proteins.

THE DNA SEQUENCES OF PLANT NUCLEAR GENOMES HAVE BEEN ANALYZED

It was a breakthrough in the year 2000 when the entire nuclear genome of Arabidopsis thaliana was completely sequenced (157 Mb). The sequence data revealed that the nuclear genome contains about 26,800 structural genes, twice as many as in the insect Drosophila. A comparison with known DNA sequences from animals showed that about one-third of the Arabidopsis genes are plant specific. By now the rice, poplar and moss genome have been sequenced (rice: 490 Mb, ca. 32,000 genes, poplar: 485 Mb ca. 45,600 genes,Physcomitrella patens: 511 Mb, ca. 28,000 genes).

The function of many of these genes is not yet known. The elucidation of these functions is a great challenge. One approach to solve this is a comparison of sequence data with identified sequences from microorganisms, animals, and plants, available in data banks, by means of bioinformatics. Another way is to eliminate the function of a certain gene by mutation and then investigate its effect on metabolism. A gene function can be eliminated by random mutations (e.g., by Ti-plasmids) (T-DNA insertion mutant). In this case, the mutated gene has to be identified. Alternatively, defined genes can be mutated by the RNAi technique. All these investigations require an automated evaluation with a very high technical expenditure. The project Arabidopsis 2010 (USA) and the German partner program AFGN (Arabidopsis functional genomics network) aim to fully elucidate the function of the Arabidopsis genome by the year 2010.

NEXT-GENERATION SEQUENCING TECHNOLOGIES APPLIED TO DE NOVO ASSEMBLY OF PLANT GENOMES

Second-generation sequencing technologies

As indicated above, successful whole-genome sequencing projects have been achieved with the use of Sanger technology. However, such projects require dealing with several complicating factors, including high costs and relatively long turnaround times to completion. The emergence of next-generation sequencing (NGS) technologies has changed this paradigm, both by reducing costs and increasing sequencing throughputs, while at the same time introducing complexity related to the relative short reads of NGS reads. Several NGS technologies have emerged in the past 7 to 8 years. All follow a relatively uniform approach to library construction and sequencing. To complete sequencing:

- universal adapters are ligated at the end of single DNA molecule templates;
- adapter-ligated DNA templates are amplified via PCR to create a cluster of identical isoforms and
- clusters are loaded on sequencers and nucleotide incorporations occur in parallel on millions of clusters.

These generate an amplified signal that is recognized by the platform and translated into a base call.

The most widely used NGS technology nowadays is the one commercialized by Illumina, whose high-throughput instrument, the HiSeq4000, can produce up to 1.5 Tb of sequencing data in approximately 3.5 days.. In the Illumina sequencing platform, sequencing templates generated during library construction are immobilized on a solid surface, and a "bridge PCR" approach allows for the localized amplification of millions of single DNA molecules, thus generating millions of clusters, each containing thousands of copies of the original DNA molecules.

Sequencing then is performed using a sequencing-by-synthesis approach where single-base extension allows the incorporation of a fluorescently labelled nucleotide (a blocking chemical moiety at the 3' hydroxyl end allows the incorporation of one base only). Once incorporated, the label is detected and the resulting signal subsequently translated into a base call. Finally, the fluorescent dye and the blocking 3' agent are cleaved, allowing the next single base incorporation event to occur. Through the use of alternating cycles of base incorporation, image capture and dye cleavage, the Illumina sequencing technology can produce reads that are up to 300 bp in length.

The relatively high error rate (~0.1% or 10 times higher than Sanger sequencing) can be compensated by very high sequencing coverage, thus allowing random errors at any given base position to be ignored below a certain frequency threshold. The relative short read of Illumina sequencing reads can be explained by several noise factors accumulating after each cycle, including phasing, where imperfect single-base incorporation and imperfect cleavage of the dye and 3' hydroxyl blocking moiety lead to the accumulation of copies of various lengths within a cluster, and the subsequent increase of signal-to-noise ratio after each cycle.

Third-generation sequencing technologies

De novo assemblies of plant genomes have been performed with NGS reads only, either with reads generated on the Illumina platform alone or with reads generated with the Illumina platform combined with reads generated on the Roche 454 second-generation sequencing platform. However, those assemblies generally are fragmented, resulting in low N50 values and a high number of contigs, mostly because of the overall short read length, the complexity of the genome and the presence of conserved regions whose length exceeds the length of NGS reads and thus cannot be extended during the *de novo* assembly process. The emergence of third-generation sequencing technologies has started to address some of the inherent limitations of sequencing and assembling large and complex plant genomes.

Those technologies are characterized by the parallel sequencing of single molecules of DNA (rather than "clusters"), thus avoiding phasing issues, and the resulting sequences tend to be in the kb range, offering the opportunity to assemble genomes and generating longer contigs by encompassing complex and conserved genomic regions and allowing relatively high-confidence assemblies of overlapping reads. However, single sequencing reads tend to exhibit relatively high error rates (~15%–25% on average).

Deep sequencing coverage or repeated sequencing of the same DNA fragments therefore are required to offset the presence of a high number of sequencing errors [48, 49]. As of today, two companies have developed and commercialized third-generation sequencing technologies, namely, Pacific Biosciences [e.g., 50] (Menlo Park, CA) and Oxford Nanopore Technologies (Oxford, UK). Each company uses vastly different approaches to sequencing. The Pacific Biosciences (PacBio) RS II system uses a sequencing-by-synthesis approach to offer up to ~40-kb reads, where base incorporation is monitored in a real-time fashion. Nanoscale holes, described as Zero Mode Waveguides ("ZMW") are located on a chip, where individual polymerases are covalently attached to the surface of each ZMW.

Individual nucleotides with a fluorescent label attached to the phosphate chain are incorporated to the elongating strand and the excited dye emits a signal that is captured before diffusion of the released pyrophosphate, and translated into a specific base call. DNA fragments used as template are ligated to "bell-shaped" adapters at both ends, thus facilitating the sequencing of DNA fragments through multiple passes and the creation of a more accurate consensus sequence. The overall stability and activity of the polymerase remain limited by photo damage and the progressive dissociation of the polymerase/template complex from the surface of the ZMW. It is therefore expected that reads generated from smaller DNA fragments will exhibit higher consensus accuracy than reads from larger DNA fragments.

Oxford Nanopore Technologies released the MinION sequencing device in early access mode in 2014. Like the PacBio RS II system, the MinION delivers long reads in a real-time fashion, from single molecules of DNA. In that particular case, however, sequencing is performed by measuring the change in ionic current when a single DNA strand translocates through a protein nanopore located in an insulated membrane. The resulting signal is measured and translated into a base call. Because no enzyme is involved in the DNA sequencing process, it is expected that read length will be driven mostly by the physical length of the DNA strand being sequenced. Library construction involves the ligation of two types of adapters to DNA fragment, one "Y-shaped" adapter with a bound protein that unwinds the double-stranded DNA and facilitates the translocation of a single strand through the pore, and one "bell-shaped" adapter at the other end that allows the translocation, and sequencing, of both the sense and antisense strands. Sequencing reads then are generated by aligning base calls from the two strands and producing a higher quality consensus sequence.

Challenges in assembling plant genomes

De novo assembly of genomes has closely mimicked the trends and improvements in sequencing technologies and accompanying sequencing assembly software over the years. The emergence of next-generation sequencing technologies has allowed a much larger number of plant genomes to be sequenced and assembled than what would have been deemed possible with Sanger sequencing alone, mostly because of the costs and labor involved in such projects. However, the complexity of the majority of those genomes still makes it a challenge to resolve them with short reads alone.

As a result, most plant genome assemblies are highly fragmented, with large number of contigs and conserved regions of the genome in an unfinished state. The presence of highly conserved repeats often exceeding 10 kb in length represents a major challenge in assembling plant genomes. The most common types of repeats in plants are type II long-terminal repeat (LTR) retrotransposons and their proliferation within a genome often explains most of the structural variations between strains. Their movement also results in genome expansion, where repeats represent, in some instances, more than 80–90% of the structural content of a particular genome.

Repeat expansion also can lead to very large genome sizes. While NGS technologies can generate enough raw data to cover an entire genome in a relatively cost-effective manner, assembling such a large amount of data often represents a major computational challenge. For example, the assembly of the loblolly pine genome (~22 Gb), which represents the largest genome assembled to date, could be solved only using condensed sets and read pooling prior to assembly. Assembling large and repeat-rich genomes can also be facilitated by using supplemental layers of information, such as the physical distance between "paired" reads (end-sequences generated at both ends of a particular DNA fragment) in mate-pair libraries.

Another challenge for de novo assembly of plant genome is the issue of polyploidy. Polyploidy is an important force in plant genome evolution and it is estimated that ~80% of all living plants are polyploids, while close to 100% of all plant lineages have a paleo-polyploidy event in their history. As a consequence, some plants species, including economically important crop species like soybean, have entire gene families consisting of highly similar paralogs. Those gene families are the direct result of paleo-polyploidization events where the merger of genomes has been followed by extensive structural rearrangements, including gene loss, and the modification of gene expression for paralogs within a particular gene family. The diploid genomes of progenitor species can be used to determine the origin and structure of contigs when assembling large polyploid genomes.

Finally, heterozygosity may represent another important challenge when assembling plant genomes. Outcrossing species like grape, for instance, exhibit up to 13% sequence divergence between alleles, and the existence of such variation will impact contig assembly when both alleles are sequenced in a whole-genome assembly project.

Examples of plant genome assemblies

According to researchers, over 100 plants genomes have been sequenced since 2000, out of which 63% are genomes from various crop species. As indicated above, different Sanger sequencing strategies have been applied with varying degrees of success on several plant genomes. However, the most successful Sanger-based genome assemblies have been obtained from relatively small genomes (Arabidopsis, rice), while de novo assemblies for larger and complex genomes, such as maize, remains partial and unfinished (manual improvements of the maize genome were limited to nonrepetitive regions only).

In addition, due to the high costs and labor associated with such approaches, and the need for (in most cases) an international consortium to complete such projects, a vast majority of the most recent genomes have been sequenced using either a hybrid approach, complementing Sanger sequencing with NGS data, or using NGS data alone, from various NGS platforms. Such platforms include Illumina, 454/Roche, and more recently, Pacific Biosciences.

The domesticated tomato genome represents an example of Sanger/NGS hybrid genome assembly. A total of 30,800 BAC clones from three different BAC libraries were shotgun-sequenced and end-sequenced, generating a total of 3.3 Gb of Sanger reads. In addition, 454/Roche shotgun and mate-pair sequencing was performed, both on BAC pools and whole-genome DNA preparation, using different insert sizes and generating a total of 21 Gb of NGS data. The de novo assembly of Sanger and 454 data was performed using the Newbler assembly software and other sequence assembly and alignment tools.

Further scaffolding and polishing of the assembly were performed when integrating BAC end-sequence data and additional high-coverage Illumina and ABI/SOLiD data. Taken together, the de novo assembly resulted in 3761 scaffolds totalling to 782 Mb, with 95% of the assembled scaffold sequences present in 225 scaffolds. The predicted tomato genome size is approximately 900 Mb. The correctness and integrity of the assembly were validated through different means including the alignment of clone end-sequences, publicly available tomato EST sequences, and alignment of BAC contigs from a sequence-based physical BAC map. Interestingly comparison of the tomato, potato and grape genomes supported the existence of two successive whole-genome triplication events in common ancestors that added new gene family members that mediate important fruit functions, such as enzymes involved in ethylene biosynthesis (examples of whole genome duplication or triplication events abound among plant genomes that have been sequenced to date).

Because of the relatively cheap costs involved, a large number of plant genomes have been sequenced and assembled using NGS technologies alone. This includes the assembly of the complex tetraploid genome of cultivated cotton (Gossypium arboreum). The tetraploid cultivated cotton genome has a genome size of approximately 1.7 Gb and is thought to have appeared 1–2 million years ago through interspecific hybridization between diploid A (Gossypium arboretum) and D (Gossypium raimondii) subgenome progenitors. A total of 371.5 Gb of shotgun Illumina data was generated with various insert sizes ranging from 180 bp to 40 kb and complemented with 33,454 BAC end sequences. The assembly was performed with SOAPdenovo, which resulted in 40,381 contigs, anchored and oriented in 7,914 scaffolds, ranging in length from 140 kb to 5.9 Mb with 90% of the contigs included in 3740 scaffolds.

An example of a smaller, relatively less complex genome assembly is that of the crop species Brassica rapa. An estimated 72× sequencing coverage of the genome was generated, corresponding to Illumina shotgun paired-end data from NGS libraries with insert sizes ranging from 200 bp to 10 kb, and assembled using SOAPdenovo. The resulting assembly was made of 14,207 contigs larger than 2 kb, further assembled into 794 scaffolds, totalling approximately 283.8 Mb and estimated to cover more than 98% of the gene space, based on alignments of 214,425 B. rapa public EST sequences and 52,712 unigenes from the BrGP database. Further assessment of the integrity of the assembly was performed by aligning BAC clone Sanger sequences reported in previous studies.

While a large number of genomes have been sequenced with NGS technologies alone, the relatively short reads of the major NGS platforms that have been used in those assembly projects, combined with the general complexity of most of those genomes, generally require the use of alternative methods to facilitate the assembly or confirm its integrity. These methods rely on the use of various types of NGS libraries, such as mate-pair large inserts, or the use of Sanger-derived sequencing data such as EST or BAC-based shotgun reads.

However, scaffolding of NGS contigs, based on using pairing information between NGS reads originating from the same DNA fragment, generally leads to unresolved gaps between contigs, often due to the presence of large repeat regions whose size exceed the length and resolution of short NGS reads. As a result, significant portions of any given scaffold contain large batches of unknown sequences, and of unknown length. To address these issues and improve plant genome assemblies, researchers have developed a series of multifaceted solutions, combining alignment to known public data, such as ESTs or BAC ends, or, when available, reference genomes from related species, integration of physical and genetic map data, or new technologies.

COMPLEMENTARY APPROACHES TO DE NOVO ASSEMBLY OF PLANT GENOMES

Long-read assembly

NGS assembly strategies based on the use of short reads cannot solve long and identical transposable elements abundantly present in most plant genomes. The use of long reads is expected to address some of those shortcomings and improve the overall quality of de novo assembly by ordering contigs, closing gaps, and improving scaffolding. As a consequence, researchers have started to adopt the single-molecule long-read sequencing technology from Pacific Biosciences in plant genome assembling projects.

Spinach is an example of such genome assembly efforts. Spinach is a diploid species with a genome size estimated at 989 Mb. Researchers sequenced and assembled the Spinach genome using large fragment libraries of Pacific Biosciences sequence reads. They generated a 60× coverage of the genome, with 20% of the reads larger than 20 kb. Data were assembled using PacBio's Hierarchical Genome Assembly Process (HGAP), which showed that long-read assemblies exhibited a 63-fold improvement in contig size over an Illumina-only assembly, derived from multiple Illumina libraries.

A distinct strategy to long-read assembly, namely, the Illumina TruSeq Synthetic Long-Read (SLR) strategy, is also expected to improve the quality of assemblies generated with short reads only. In SLR libraries, genomic DNA is fragmented to ~10 kb and individual indexed Illumina libraries are generated in parallel from highly diluted pools of sheared DNA fragments. After Illumina sequencing and data deconvolution, the original ~10 kb fragments can be reassembled, effectively reducing the complexity level of the assembly and generating very-high quality synthetic long reads that can subsequently be assembled together or used for haplotype resolution.

The use of long reads in de novo assembly is bound to become more prevalent in the near future, reducing the number of scaffolds while at the same time increasing their average length. The use of PacBio in smaller genomes, such as microbial genomes, has already demonstrated that the assemblies often result in contigs corresponding in most cases to individual chromosomes or plasmids present in the microbial cells.

Likewise, it is likely that future plant studies will include such long reads, either alone or in combination with short-read NGS data to improve assembly and coverage in questionable regions, and to confirm the integrity of the assembly in a manner similar to Sanger data with current NGS assemblies.

Genetic anchoring

The emergence of NGS technologies has rapidly led researchers to develop methods and assays for variant discovery in various plant genomes. Some studies have shown that Single nucleotide polymorphisms (SNPs) can be discovered in parental inbred lines using next-generation sequencing. Entire mapping populations also have been simultaneously sequenced and genotyped, in a process known as "genotyping-by-sequencing" (GBS), discovering in the process extensive lists of segregating markers within the mapped population, that can be completed by using known reference maps or sequences to impute missing marker data from individual haplotypes. Various reduced-representation methods have been employed for NGS-derived SNP discovery in plant species where whole-genome shotgun sequencing still remains too expensive for sequencing more than a few individuals. These methods include the use of restriction enzyme digestion–based assays with methyl-sensitive restriction endonucleases, or methods based on sequence capture approaches, to sequence and map gene-rich portions of a genome, and allowing the anchoring of SNPs in a relatively unambiguous manner.

More recently, ultradense linkage maps have been created from genotyping by whole genome sequencing of a genetic mapping population. It has been used to place whole-genome sequencing contigs into a map, thus anchoring, and ordering, sequencing of contigs. Such an approach requires using a genetic linkage map as a framework, into which SNPs derived from the whole genome sequencing assembly can be integrated into a genetic framework derived from low coverage whole-genome sequencing data from a segregating population. The genetic position of the sequence-derived SNPs can then be used to assign chromosomal locations to the contigs harboring them. Such an approach has been used in the context of a whole-genome assembly project in barley where genetic anchoring was applied to a whole-genome assembly.

SNPs discovered by sequencing individuals from two mapping populations at low coverage (~1×) were placed into genetic maps that had been previously constructed through different means, including SNP array data and GBS, or made from the whole-genome shotgun sequencing data of the population. Their genetic positions then were used to assign chromosomal locations, and integrate into the combined physical and genetic genome framework, approximately two-thirds of all whole-genome shotgun sequencing contigs. While highly effective in plants, where mapping populations are often readily available, it must be noted that such an approach is limited by the overall recombination landscape, and the subsequent relationship between physical and genetic distance within a particular region of the genome.

Recombination events in plants often occur in distal regions of the chromosomes, and peri-centromeric regions may require very large mapping populations to improve their resolution. In addition, recent studies have suggested that specific features of the genome, such as chromosomal inversion, translocation and duplication varying between the two parents used to generate the mapping population, may lead to errors and potentially confound genome assemblies.

BAC pool sequencing in gene-rich regions

A large number of genome assemblies have been generated with the help of physical maps and the use of a BAC-by-BAC sequencing approach. While laborious and costly, this approach still remains relevant as it offers multiple advantages over a whole-genome sequencing approach, especially in terms of assembling sequencing reads conserved in the context of a whole-genome assembly but mapping exclusively to a defined portion of a genome in the context of an individual clone assembly.

Researchers proposed a modified version of clone sequencing to take advantage of the massive sequencing capacity offered by NGS platforms. In that study, subsets of overlapping genome-tiling BAC clones were selected and pooled according to a multidimensional grid design. Each pool then was sequenced on an Illumina HiSeq2000 instrument. The resulting paired-end reads were deconvoluted by determining, for each read the intersection between the pool it originates from and the individual BAC clone(s) within that same pool covering the portion of the genome the read corresponds to, based on physical map information. Once deconvolution is achieved, reads can be assembled using an NGS assembler (Velvet), to recreate the sequence of the original BAC clone.

Such an approach was successfully tested in barley BAC clones selected based on BAC-unigene associations described in that same study, thus suggesting that BAC pool sequencing can be used in correlation with existing physical maps to complement or correct whole-genome sequencing assemblies, offering in the process the likelihood of higher quality contig sequence assemblies in gene-rich regions of complex plant genomes.

Optical mapping

Optical mapping is a single-molecule approach that produces fingerprints using ordered restriction maps or specific nick sites. After enzymatic treatment and subsequent incorporation of fluorescent labels, the DNA molecules are stretched on a glass surface or in a nanochannel array and directly imaged to locate regions corresponding to the restriction sites or nick sites within the molecule. Distances between those sites are then inferred to produce an optical map of the DNA molecule. Two commercial platforms currently are available, namely, the Opgen Argus and the BioNano Genomics Irys systems. Using such techniques, very large DNA molecules, in the Mb range, can be interrogated for the presence and location of short recognition sites (whose sequence will vary with the enzyme being used to treat the DNA).

Consensus optical maps then can be created by determining the overlap, under highly redundant conditions, between optical maps of single DNA molecules. Such consensus maps have to take into account the possibility of errors inherent to this type of technology, including star activity and false enzyme cuts, or the possibility of chimeric maps when joining, for example, optically mapped molecules containing paralogous genomic regions.

Optical maps can be used for multiple applications, including comparative genomics and structural variation detection, as well as the development of optical map-guided genome assemblies, where the optical map is aligned and compared to in silico digested contig sequences. Optical map-guided genome assemblies can assist in building high-quality genome assemblies by providing evidence of the ordering of adjacent contigs and scaffolds, or by assessing the overall sequence accuracy of contigs and suggesting potential errors in an assembly, such as inversions, translocations or chimeric contig or scaffold sequences. The addition of optical maps to a genome assembly often results in a significant increase in the scaffold N50 value.

For example, researchers used the mapping of tiling BAC clones in a 2.1 Mb highly repetitive region of Aegilops tauschii (the D-genome donor of hexaploid wheat) to correct several misassemblies and improve the assembly from 75% to 95% complete. In another study, a high-resolution optical map, spanning 91% of the maize genome, was built, and used to characterize gaps within contigs, the maize genetic-physical (FPC) map and the reference pseudomolecules. Results also suggested that the placement of 12 FPC contigs on the maize genetic-physical map required re-evaluation.

Long-range Hi-C interactions

High-throughput Chromosome capture (Hi-C) is a method that uses cross-linking of DNA-binding protein to DNA followed by restriction digestion and self-ligation of protein-bound DNA fragments, to probe genome-wide three-dimensional chromatin interactions between chromosomal regions bound to the same proteins (such as enhancer and promoter regions). There is a statistically higher probability that those regions are located on the same chromosome rather than on different chromosomes, as expected within the context of chromosomes located in distinct three-dimensional spaces within the nucleus.

As a result, a vast majority of Hi-C read pairs (where each paired reads correspond to reads that may be millions of bases apart from each other on the same chromosome) can be used to determine what two contigs can be linked together on the same chromosome, based on the Hi-C paired reads they each contain. Researchers evaluated the use of Hi-C datasets for long-range scaffolding of *de novo* whole-genome assemblies. This approach works, first, by aligning Hi-C reads to *de novo* assembly contig sequences and indexing each contig to their respective chromosomes, ordering contigs within each respective chromosome group by using higher Hi-C interaction densities expected between closely located contigs, and orienting ordered contigs using the location and orientation of Hi-C reads within each contig. The approach tested on existing human and mouse contig datasets generated from next-generation shotgun and mate-pair sequencing reads showed that a vast majority of the contigs could be grouped (98.2% and 98% of all sequences, in human and mouse, respectively) and ordered (94.4% and 86.7% of all grouped sequences, in human and mouse, respectively) within individual chromosomes when combined with Hi-C sequencing reads.

Long-range scaffolding

Two companies, namely, 10X Genomics (Pleasanton, CA) and Dovetail Genomics (Santa Cruz, CA), recently presented new ways to assemble short reads delivered by the Illumina technology. The GemCode instrument from 10X Genomics is a microfluidic device used to partition very long DNA molecules (typically 50 kb or more) into oil-based droplets and to prepare Illumina-compatible libraries in combination with "gel beads", each containing a unique 14-bp indexing barcode. Once sequencing is performed, in-house software deconvolutes the barcodes and reconstructs the sequence of the original DNA subfragments as to where they originate from on the original long DNA molecule.

In contrast to 10X Genomics, Dovetail Genomics approach does not necessarily require an instrument but requires larger amount of starting material for preparing samples. Dovetail's approach works essentially by in vitro making a Hi-C library from chromatin-free purified DNA, thus recreating intramolecular interactions while reducing intermolecular ones. The resulting fragments can then be selected for mate-pair sets capturing long-range intramolecular interactions for genome scaffolding. While not yet applied on plant genome assemblies, it is presumed that the strategies and technologies highlighted above could potentially assist in grouping and ordering contigs and scaffolds from gene-rich regions of diploid plant genomes.

DNA MARKER TECHNOLOGIES IN PLANTS AND APPLICATIONS

In general, DNA markers are a fragment of DNA indicating (mutations/variations), which can be used to detect polymorphism between alleles of a gene for a particular sequence of DNA or different genotypes. Such frag- ments are linked with a definite location within the genome and may be detected by using certain molecular technology.

The marker systems that are now being progressively developed and also has shifted from the first and second generation marker systems including RFLPs, RAPDs, SSRs and AFLPs to the third and the fourth generation marker systems, which include SNPs, KASper, DArT assays, and Genotyping by Sequencing (GBS).

This review will address general principles and methodologies of different molecular markers in these categories with a major emphasis on emerging genotyping technologies in plants including SNPs and KASPer assays. Some issues related to applications of these methodologies in practical breeding will also be discussed.

Low-Throughput Marker Systems

Restriction Fragment Length Polymorphisms (RFLPs)

RFLP markers were mainly used in 1980s and 1990s in plant genetic studies, and are therefore, referred to as (first generation molecular markers). The polymorphisms detected by RFLPs are as a result of changes in nucleotide sequences in recognition sites of restriction enzymes, or due to mutation events (insertions ordeletions) of several nucleotides leading to obvious shift in fragment size. The main advantages of RFLP markers are co-dominance, high reproducibility, no need of prior sequence information, and high locus-specificity.

By using RFLP markers, genetic maps have been established in several crop species including rice maize, wheat. However, since the last decade fewer direct uses of RFLP markers in genetic research and plant breeding have been stated. Most plant breeders would think that RFLP is too time consuming procedure and it requires relatively large amounts of pure DNA, tedious experimental procedure. Additionally, each point muta- tion has to be analysed individually.

In the late 1980s, Colinearity across genomes was first reported between the three diploid genomes of hexaploid wheat and between potato and tomato. Soon after couple of years, cross-hybridization of RFLP markers which obtained from bread wheat with barley and rye revealed a few translocations of chromosome arms in the rye genome when compared to the wheat genomes, whereas most probes indicated that the order of the loci was preserved between those three species.

MEDIUM-THROUGHPUT MARKER SYSTEMS

Random Amplified Polymorphic DNA's (RAPDs)

RAPDs are based on the PCR amplification of random DNA segments with primers of random nucleotide sequences that were inexpensive and easy to use. The primers bind to complementary DNA sequences and where two primers bind to the DNA sample in close enough for successful PCR reaction. The amplified of DNA products can then be visualized by gel electrophoresis.

RAPD markers have been widely used in diverse plant species for assessment of genetic variation in populations and species, fingerprinting and study of phylogenetic relationships among species and subspecies. Nevertheless, disadvantages of RAPD markers are the fact that it predominantly provides dominant markers, and incapability to detect allelic differences in heterozygotes. Polymorphisms are detected only as the presence or absence of a band of a certain molecular weight, with no information on heterozygosity.

Additionally, because of their random nature of amplification and short primer length, they are not ideal for genome mapping. Moreover, these markers do not exhibit dependable amplification patterns and differ with the experimental conditions.

Simple Sequence Repeats (SSRs)

During 1990s, Simple sequence repeats (SSRs) which is also known as microsatellites were established and provided a choice for many genetic researches since they are amenable to low, medium and high-throughput approaches. They are randomly tandem repeats of short nucleotide motifs (2 - 6 bp). SSRs are frequently highly polymorphic sequences normally present in animal and plant species, and can be used to study the relationship between inherited traits within a species. Microsatellite markers are often derived from non- coding/anonymous genomic regions, such as bacterial artificial chromosomes (BACs) and genomic survey sequences (GSSs).

Therefore, development of SSR markers used to be expensive and laborious. This assay is easily detectable by gel electrophoresis for few to hundreds of samples, which could be inexpensive by researchers with limited resources. Polymorphism is based on the variation in the number of repeats in different genotypes. Since polymorphisms in longer penta-nucleotide and tetra repeats are easier to make a distinction in a variety of detection systems and longer repeats may be more robust.

In recent years, SSR markers can easily be developed in silico due to the availability of large-scale gene (expressed sequence tag) EST sequence information for many plant species. Since EST sequencing projects have provided sequence data that is available in online databases and can be scanned for identification of SSRs. The high degree of polymorphism as compared to RFLPs and RAPDs, their co-dominant nature and locus specific make them the markers of choice for a diversity of purposes including practical plant breeding. Therefore, (SSRs) have become a marker of choice for an array of applications in plants due to extensive genome coverage and hyper variable nature.

Amplified Fragment Length Polymorphism's (AFLPs)

AFLPs are PCR-based markers, simply RFLPs visualized by selective PCR amplification of DNA restriction fragments. Such a marker is a multi-locus marker technique that combines the techniques of selective PCR amplification of restriction and fragments restriction digestion and it is possible to be applied into DNA of any origin.

The technique involves three steps: Initially, oligonucleotide adapters are ligated to both ends of the resulting restriction fragments and genomic DNA is digested. Subsequently, the fragments are selectively amplified, using the adapter and restriction site sequences as primer binding sites for following PCR reactions. As the 3' ends of the primers extend into the restriction fragments by (1 to 4 bp), only those fragments are amplified, whose ends are absolutely complementary to the 3' ends of the selective primers. Therefore, only a certain amount of the restriction fragments is amplified. Finally, the amplified fragments are resolved by gel electrophoresis and visualized by either silver staining, autoradiography or fluorescence, resulting in a unique reproducible fingerprint for each individual.

The advantages of using this method are that it is cost-efficient, since a single assay allows detection of a large number of co-amplified restriction fragments and it requires moderate quantities of DNA. Additionally, higher levels of polymorphisms compared with RFLPs can be detected also AFLPs have much higher multiplex ratio (more markers per experiment) and better reproducibility than RAPDs. These features make this technology an attractive tool for saturating genomic region with low marker density and constricting genetic maps. However, AFLP assays have some limitations also. For instance, it requires the use of polyacrylamide gels for detection and it needs a greater technical skill. Moreover, most AFLP markers are dominant rather than co-dominant, because of the complex banding arrangements. In some cases the scoring of AFLP polymorphisms as co-dominant marker loci is possible, for the reason that diploid homozygous individuals cause more intense peak than heterozygous individuals.

By group of researchers, genetic relatedness and identity of the durum wheat Graziella Ra, four Italian commercial durum cultivars (Grazia, Cappelli, Flaminio and Svevo and Kamut) were evaluated using (AFLPs). Their results revealed that the percentage of polymorphic loci within accession ranged from 6.57% to 19.71% (mean, 12.77%) and molecular variance was partitioned into 80% (variance among accessions) and 20% (within accession).

HIGH-THROUGHPUT MARKER SYSTEMS

Single Nucleotide Polymorphisms (SNPs)

SNP is a single nucleotide base difference between two DNA sequences or individuals. SNPs are typically bi-allelic and arise either due to substitutions/point mutations (transversion and transition) or as a result of deletion/insertion of nucleotides and are detectable when similar genomic regions from different genotypes of different or same species are aligned.

SNPs provide the simplest and ultimate form of molecular markers as a single nucleotide base is the smallest unit of inheritance, and therefore they can provide a great marker density. SNPs happen frequently in animals and plants. The probability to find polymorphisms in a target gene are increases due to high density of SNP markers which provides a huge advantage over previous markers that are at best closely linked to a locus of interest and not within. In the case of linkage it can easily happen that a linkage is lost when a marker is applied to other populations with different recombination patterns. Typically, SNP frequencies are in a range of one SNP every (100-300) bp in plants. SNPs may present within coding sequences of genes, non-coding regions of genes or in the intergenic regions between genes at different frequencies in different chromosome regions.

Several methods are currently available for SNP discovery, either following the database approach, where SNPs are detected by following the experimental approach, or mining sequence databases, where genome regions of interest are screened for SNPs with one of various techniques established for the detection of SNPs. Moreover, it can be categorized into four reaction chemistries or principles: hybridization with allele-specific oligonucleotide probes, oligonucleotide ligation, enzymatic cleavage, and single nucleotide primer extension.

In principle, the SNP methods show differences between a probe of known sequence and a target DNA containing the SNP site. The target DNA sections are typically PCR products and mismatches with the probe reveal SNPs within the amplified target DNA segment. The mismatching DNA segments can be sequenced then as the most direct way to identify SNP polymorphisms.

SNP markers are likely to become the marker of choice for breeding in the near future, especially as the full sequences of more plant genomes will become available with the advantage of next generation sequencing tech- nologies (NGS).

More recently, it has become very-cost effective and easier to quickly identify a large number of SNPs in short time in any plant species. This was due to the emergence of the third generation sequencing. The advantage of this new sequence technology are expected to further reduce sequencing costs extremely to levels below $1 per mega base compared to $60, $2, and $1 expected costs for sequences generated by next generation sequencing.

SNPs in Wheat (Triticum aestivum)

The large size of wheat genome has led to various approaches to reduce the cost of data production. These include the targeted re-sequencing of captured exome fragments and the establishment of confederations to share the cost of genome sequence data generation. In a different study, by using illumina sequencing of cDNA libraries, 14,078 putative SNPs were recognized across representative samples of UK wheat germplasm, with a proportion of these SNPs validated using KASPar assays. In addition, in developed countries such as UK, France and Australia many attempts for large-scale SNP development in bread wheat were undertaken, leading to the development of millions of SNPs. These SNPs will be widely used for molecular breeding in wheat.

The KBioscience Competitive Allele?Specific PCR (KASPar)

KASPar genotyping may be of specific interest to researchers and breeders who are interested in analysing a small number of targeted SNPs in a large number of samples. This makes KASPar a cost-effective, simpleand flexible genotyping system; since the assays can be modified with a range of DNA samples and it does notrequire a hybridization step; as an alternative it includes real-time detection of the product.

The chemistry of KASPar assyas involves one common reverse primer and two competitive allele specific tailed forward primers. To determine the alleles at a specific locus, this assay system relies on the discrimination power of a novel pattern of competitive allele specific PCR. In such cases, this assay involves competitive allele-specific PCR for a given SNP, followed by SNP detection through Fluorescence Resonance Energy Transfer (FRET).

This assay for the target SNPs has been developed and used for genotyping commercially by Kbioscience UK (http://www.kbioscience.co.uk/). This company perfected this technique to improve the performance of the detection platform by incorporating a 5'-3' exonuclease cleaved Taq DNA polymerase and a homogeneous Fluorescence Resonance Energy Transfer (FRET) detection system. The two allele-specific primers of a SNP are designed so that they incorporate with a unique (18) bp tail to the respective allele specific products, which in later cycles allow incorporation of allele specific fluorescent labels to the PCR products (with the help of corresponding labelled primers).

In KASPar assays there is no need of sequencing to identify SNPs, instead SNP flanking sequences already known while developing different types of genotyping assays (for instance, illumina) can easily be used for primer design (one common and two allele-specific primers) for KASPar assays.

Despite the fact that KASPar genotyping assays have come to the market very recently, they have started to be used for a large number of commercial species. In maize, a set of 695 highly polymorphic gene-based SNPs from a total of 13,882 GG-validated SNPs were selected and converted into KASPar genotyping assay with a success rate of 98%. Additionally, in wheat, the technique has been used for constructing a linkage map containing several hundred SNPs.

Genotyping-by-Sequencing (GBS)

With the increased throughput of NGS platforms, re-sequencing for genome-wide surveys of genetic diversity became reasonable. However, this assay is bioinformatically challenging, impartial estimation of genetic diversity across the genome in both coding and non-coding regions could be determined. Additionally, it allows for the detection of various types of genetic variation; this detection capability contains not only SNPs and small indels, but also large mega-base scale indels.

This assay involves the use of restriction enzymes for reducing the complexity of genomes followed by targeted sequencing of reduced proportions, in that way each marker can be sequenced at high coverage across many individuals at low cost and high accuracy.

The main target for constructing GBS libraries was based on reducing genome complexity with restriction enzymes, which may reach important regions of the genome that are unreachable to sequence capture approaches. The procedure has been demonstrated with barley at the recombinant inbred lines populations and maize (IBM) where about (25,000 to 200,000) sequence tags were mapped, respectively. With this method, species that lack a complete genome sequence can have a reference map settled around the restriction sites, which can be done in the process of sample genotyping. This system has been adjusted for reducing missing data points and improved SNP calls.

APPLICATIONS AND STRATEGIES OF DNA MARKERS IN BREEDING PROGRAMS

Pyramiding Multiple Loci and Favourable Alleles

Gene pyramiding is defined as an assembly of multiple desirable genes which can be combined into a single genotype from multiple parents. This is referred as one of the major applications of marker assistant selection, as gene pyramiding via conventional plant breeding is difficult, if not impossible.

The methods for pyramiding favourable alleles can be used in the same way to accumulate QTL controlling different traits. A main difference in the model is that alleles at different trait loci to be accumulated may have different favourable directions, for instance negative alleles are preferable for some traits but positive alleles are favourable for others. As a result, to meet breeding objectives one may need to combine the positive QTL alleles of some traits with the negative alleles of others.

Selection for multiple traits may be completed in one cycle if the population size is large enough to allow desirable individuals to combine different traits. Nevertheless, the number of trait loci that can be manipulated in one cycle is restricted because the population size required covering the recombinants increases exponentially with the increase of the number of traits/loci.

To overcome this limitation, a strategy was proposed by Bonnett, Rebetzke. In this method, individuals has been selected by all target markers for both heterozygous and homozygous forms to gain a subset of population that contain higher frequencies of the target alleles so that to obtain the homozygotes at the target loci, a smaller population size is required in the following generation.

For Barley Yellow Mosaic Virus complex, a variety of markers have been developed for selection of the rym5 and rym4 resistance genes on chromosome. In another study, the regions of a typical Spanish barley line that have VRNH2 and VRNHl were introgressed into a winter variety.

A set of 12 lines introgressed with all four possible combinations of VRNHl and VRNH2 has been assessed for frost tolerance and vernalization requirement. In practical, what has to be taken into account when applying such strategies pyramiding has to be repeated after each crossing, since the pyramided resistance genes are segregating in the progeny.

Marker-Assisted Recurrent Selection (MARS)

In the 1990s, marker-assisted recurrent selection was proposed, which uses markers at each generation to target all traits of importance and for which genetic information can be achieved. Parents contribute different favourable alleles when the QTL mapping is conducted based on a bi-parental population. Therefore, the perfect genotype is a mosaic of chromosomal segments produced by recombination between the two parents.

MARS refers to the improvement of an (F2) population by one cycle of marker-assisted selection (for instance, based on data marker scores and phenotypic) followed usually by two or three cycles of marker-based selection (for instance, based on marker scores only). It is possible today to define an ideal genotype as a pattern of QTLs, all QTLs carrying favourable alleles from various parents. After several successive generations of crossings, it might be possible to get close to the ideal genotype. In addition, this system can start without any QTL information, and selection can be based on significant marker-trait association established during the MARS process.

Studies have revealed that, in accumulating favourable alleles, (MARS) was superior to phenotypic selection. Moreover, through MARS in maize, the rates of genetic gain accomplished for complex traits were about twice those of phenotypic selection in some reference populations. The usefulness of including prior knowledge of QTL under genetic models has been studied that included QTL number, gene effects, heritability, epistasis and linkage. It is concluded that with known QTL, MARS is most favourable for traits controlled by a large number of QTL.

Conclusions

As it is evident from the discussion above, that different levels of throughput are available. Therefore, an appropriate marker system can be selected based on the need. The earlier types of molecular markers include neutral markers. For instance, (RFLPs), and which were later followed by based on the (PCR) reaction, a faster and less expensive technology. A PCR-based DNA marker includes (RAPDs), (AFLPs).

RFLPs offer the best marker type for many purposes. The main disadvantages of using RFLPs are low through- put and high cost of genotyping. RAPDs and AFLPs have also been widely used in genetic diversity studies and gene mapping. Both technologies are particularly useful when there is a necessity to assay loci across the entire genome. However, their lack of reproducibility, dominant nature of RAPDs compared with AFLPs and the lack of specificity in both cases, are limiting factors for their application in precise MAS breeding approaches.

Recently, the availability of whole genome sequences of a few selected crops and the sequence information has also led to the development of a new generation of markers, such as KASpar, SNP assays. SNP markers which are transferable through different genotyping chemistries will offer as flexible selection tools for plant breeders in (MAS). In addition, the focus should be placed in the identification of SNPs in as many genes as possible and the parallel analysis of many different lines.

It is very likely that improvement of complex traits will depend on the ability to manipulate genes, which have minor effects, and show interaction with each other.

As shown by some simulation studies, genotype by sequencing seems to be the best approach for improvement of complex traits. However, the parameters and conditions included in simulations may not fully reflect the complex situations of diverse plant breeding programs, the genetic gain per unit time and cost that has been achieved in simulations needs to be supported by the long-term selection response by comparing with other breeding approaches. In addition, this approach can technically lead to the discovery of thousands of SNPs in one single experiment. Moreover; it can be used of those plants that do not have the reference genome available

GLOSSARY

Absorption spectrum: the specific pattern of absorption for a substance that absorbs electromagnetic radiation

Alginate: component of the cell walls of many rhodophytes and kelps. Alginates have an affinity for water, and so help to slow dessication when the algae are exposed to the air; they are commercially important in the production of paper, toothpaste, beer, and frozen foods.

Amino acid: unit molecule from which proteins are constructed by polymerization.

Anaphase: the stage of mitosis during which sister chromatids are separated from each other

Atp: "adenosine triphosphate". A relatively stable, high energy molecule used to fuel chemical reactions within cells.

Autotroph: an organism capable of producing its own food

Biochemistry: the study of those molecules used and manufactured by living things.

Bioluminescence: the production of light by a chemical reaction within an organism. The process occurs in many bacteria and protists, as well as certain animals and fungi.

Brevitoxin: neurotoxin produced by the dinoflagellate *ptychodiscus brevis.*

Calcite: a common crystalline form of natural calcium carbonate, $caco_3$, that is the basic constituent of limestone, marble, and chalk. Also called calcspar.

Calcium carbonate: a "salt" used by many marine invertebrates, such as corals and echinoderms, and by protists, such as coccolithophorids, to construct their exoskeletons.

Calvin cycle: the reactions of photosynthesis that use the energy stored by the light-dependent reactions to form glucose and other carbohydrate molecules

Carbohydrates: class of biochemical compounds which includes sugars, starch, chitin, and steroids.

Carbon fixation: the process of converting inorganic co_2 gas into organic compounds

Cell cycle checkpoints: mechanisms that monitor the preparedness of a eukaryotic cell to advance through the various cell cycle stages

Cell cycle: the ordered sequence of events that a cell passes through between one cell division and the next

Cell plate: a structure formed during plant-cell cytokinesis by golgi vesicles fusing at the metaphase plate; will ultimately lead to formation of a cell wall to separate the two daughter cells

Cellulose: carbohydrate polymer of the simple sugar glucose. It is found in the cell walls of plants and green algae, as well as dinoflagellates. Cellulose is the most abundant compound on earth that is manufactured by living things.

Centriole: a paired rod-like structure constructed of microtubules at the center of each animal cell centrosome

Chitin: a carbohydrate polymer found in the cell walls of fungi and in the exoskeletons of arthropods, which provides strength for support and protection.

Chlorophyll: the green pigment that captures the light energy that drives the reactions of photosynthesis

Chlorophyll a: the form of chlorophyll that absorbs violet-blue and red light

Chlorophyll b: the form of chlorophyll that absorbs blue and red-orange light

Chloroplast: the organelle where photosynthesis takes place

Cleavage furrow: a constriction formed by the actin ring during animal-cell cytokinesis that leads to cytoplasmic division

Collagen: long proteins whose structure is wound into a triple helix. The resulting fibers have a high tensile strength. Collagen is a primary component of mammalian hair.

Cytokinesis: the division of the cytoplasm following mitosis to form two daughter cells

Dinosteranes/dinosteroids: chemicals found in dinoflagellates, which have been useful in documenting their existence early in the fossil record.

Dna: "deoxyribonucleic acid". The nucleic acid which carries the genetic code of an organism. It is the primary component of chromosomes.

Electromagnetic spectrum: the range of all possible frequencies of radiation

Enzyme: complex protein which helps to speed biochemical reactions. Enzymes are important in the construction and degradation of other molecules.

Flagellin: protein which is the primary component of prokaryotic flagella.

Fucoxanthin: yellowish-brown pigment found in some members of the chromista, including kelps and diatoms.

G_0 phase: a cell-cycle phase distinct from the g_1 phase of interphase; a cell in g_0 is not preparing to divide

G_1 phase: (also, first gap) a cell-cycle phase; first phase of interphase centered on cell growth during mitosis

G_2 phase: (also, second gap) a cell-cycle phase; third phase of interphase where the cell undergoes the final preparations for mitosis

Glucose: simple sugar, and the primary product of photosynthesis. It is polymerized to make cellulose and chitin.

Glycoprotein: a membrane-bound protein which has attached branching carbohydrates. These may function in cell-cell recognition, such as in human blood groups and immune system response, as well as in resisting compression of cells.

Granum: a stack of thylakoids located inside a chloroplast

Hemoglobin: protein complex found in the blood of most chordates and the roots of certain legumes. It binds oxgen molecules, and in chordates serves as the means by which the oxygen is supplied to the cells of the body.

Heterotroph: an organism that consumes other organisms for food

Histones: proteins attached to the dna of eukaryotes which allows it to be packaged into chromosomes.

Hydrophilic: "water loving". Hydrophilic compounds dissolve easily in water, and are usually polar.

Hydrophobic: "water fearing". Hydrophobic compounds do not dissolve easily in water, and are usually non-polar. Oils and other long hydrocarbons are hydrophobic.

Integrin: adhesive protein of the extracellular matrix in animals.

Interphase: the period of the cell cycle leading up to mitosis; includes g_1, s, and g_2 phases; the interim between two consecutive cell divisions

Ion: an atom or small molecule which carries a positive or negative charge.

Kinetochore: a protein structure in the centromere of each sister chromatid that attracts and binds spindle microtubules during prometaphase

Laminarin: a beta-glucan polysaccharide produced by many chromists through photosynthesis.

Light-dependent reaction: the first stage of photosynthesis where visible light is absorbed to form two energy-carrying molecules (atp and nadph)

Lipids: a class of biochemical compounds which includes fats, oils, and waxes.

Luciferase: enzyme which activates luciferin to produce bioluminescence.

Luciferin: compound whose activated form emits light.

Mesophyll: the middle layer of cells in a leaf

Metaphase plate: the equatorial plane midway between two poles of a cell where the chromosomes align during metaphase

Metaphase: the stage of mitosis during which chromosomes are lined up at the metaphase plate

Mitosis: the period of the cell cycle at which the duplicated chromosomes are separated into identical nuclei; includes prophase, prometaphase, metaphase, anaphase, and telophase

Mitotic phase: the period of the cell cycle when duplicated chromosomes are distributed into two nuclei and the cytoplasmic contents are divided; includes mitosis and cytokinesis

Mitotic spindle: the microtubule apparatus that orchestrates the movement of chromosomes during mitosis

Neurotoxin: poison which interferes with nerve function, usually by affecting the flow of ions through the cell membrane.

Nucleic acid: class of biochemical compounds which includes dna and rna. They are among the largest molecules known

Nucleotide: unit from which nucleic acids are constructed by polymerization. It contains a sugar, a phosphate group, and an organic base. Atp is a nucleotide.

Peptidoglycan: carbohydrate polymer cross-linked by proteins. It is found in the cell wall of gram positive bacteria, where it stains with the dye crystal-violet.

Peridinin: carotenoid pigment found in dinoflagellates.

Phosphate: an ion consisting of a phosphorus atom and four oxygen atoms. Among other things, it is used in the constuction of nucleic acids.

Photoautotroph: an organism capable of synthesizing its own food molecules (storing energy), using the energy of light

Photon: a distinct quantity or "packet" of light energy

Photosynthesis: biochemical process in which light energy is absorbed by chlorophyll, and is used to fuel the building of sugar molecules.

Photosystem: a group of proteins, chlorophyll, and other pigments that are used in the light-dependent reactions of photosynthesis to absorb light energy and convert it into chemical energy

Phycocyanin: blue, water-soluble pigment found in the cyanobacteria and the red algae.

Phycoerythrin: red, water-soluble pigment found in the cyanobacteria and red algae.

Pigment: any colorful compound, used by living things to absorb or block sunlight, and in sexual displays.

Polymer: a large molecule constructed from many smaller identical units. These include proteins, nucleic acids, and starches.

Prometaphase: the stage of mitosis during which mitotic spindle fibers attach to kinetochores

Prophase: the stage of mitosis during which chromosomes condense and the mitotic spindle begins to form

Protein: class of biochemical compounds constructed from amino acids. Proteins may be structural, such as those that make up hair and cartilage, or they may be reactive, such as the enzymes.

Proteinaceous: describes any structure which is composed of protein.

Quiescent: describes a cell that is performing normal cell functions and has not initiated preparations for cell division

Rbcl: a gene which is located in the chloroplast of photosynthetic organisms. It codes for the large subunit of the protein rubisco, and its sequence has been useful in plant phylogenies.

Rna: "ribonucleic acid". The nucleic acid which carries the dna message into parts of the cell where it is interpreted and used. The 18s ribosomal rna sequence has been used in many groups of organisms to reconstruct phylogeny.

Rubisco: protein which fixes carbon in photosynthetic organisms. It binds molecules of carbon dioxide to a five-carbon molcule. Rubisco is the most common protein on earth.

S phase: the second, or synthesis phase, of interphase during which dna replication occurs

Saxitoxin: neurotoxin found in a variety of dinoflagellates. If ingested, it may cause respiratory failure and cardiac arrest.

Silica: amorphous silicon dioxide (glass). It is a structural component in many organisms, such as diatoms and horsetails.

Spongin: proteinacous compound of which the spicules in demospongiae are composed.

Starch: a complex polymer of glucose, used by plants and green algae to store surplus sugar for later use.

Stoma: the opening that regulates gas exchange and water regulation between leaves and the environment; plural: stomata

Sugar: any of several small carbohydrates, such as glucose, which are "sweet" to the taste.

Telophase: the stage of mitosis during which chromosomes arrive at opposite poles, decondense, and are surrounded by new nuclear envelopes

Thylakoid: a disc-shaped membranous structure inside a chloroplast where the light-dependent reactions of photosynthesis take place using chlorophyll embedded in the membranes

Wavelength: the distance between consecutive points of a wave

BIBLIOGRAPHY

Ainsworth EA, Long SP. What have we learned from 15 years of free-air CO_2 enrichment (FACE)? A meta-analytic review of the responses of photosynthesis, canopy properties and plant production to rising CO2, *New Phytologist*, 2005, vol. 165 (pg. 351-371)

Ainsworth EA, Rogers A, Blum H, Nosberger J, Long SP. Variation in acclimation of photosynthesis in Trifolium repens after eight years of exposure to Free Air CO2 Enrichment (FACE), *Journal of Experimental Botany*, 2003, vol. 54 (pg. 2769-2774)

Ainsworth EA, Rogers A, Vodkin LO, Walter A, Schurr U. The effects of elevated CO_2 concentration on soybean gene expression. An analysis of growing and mature leaves, *Plant Physiology*, 2006, vol. 142 (pg. 135-147)

Ainsworth EA, Rogers A. The response of photosynthesis and stomatal conductance to rising [CO2]: mechanisms and environmental interaction, *Plant, Cell and Environment*, 2007, vol. 30 (pg. 258-270)

Allen JF, Santabarbara S, Allen CA, Puthiyaveetil S. Discrete redox signaling pathways regulate photosynthetic light harvesting and chloroplast gene transcription, *PLoS One*, 2011, vol. 6 pg. e26372

Aro EM, Virgin I, Andersson B. Photoinhibition of PS II. Inactivation, protein damage and turnover, *Biochimica et Biophysica Acta*, 1993, vol. 1143 (pg. 113-134)

Asada K. The water–water cycle as alternative photon and electron sinks, Philosophical Transaction of the Royal Society B: *Biological Sciences*, 2000, vol. 355 (pg. 1419-1431)

Atienza SG, Faccioli P, Perrotta G, Dalfino G, Zschiesche W, Humbeck K, Stanca AM, Cattivelli L. Large scale analysis of transcripts abundance in barley subjected to several single and combined abiotic stress conditions, *Plant Science*, 2004, vol. 167 (pg. 1359-1365)

Avenson TJ, Cruz JA, Kanazawa A, Kramer DM. Regulating the proton budget of higher plant photosynthesis, *Proceedings of the National Academy of Sciences, USA*, 2005, vol. 102 (pg. 9709-9713)

Avenson TJ, Kanazawa A, Cruz JA, Takizawa K, Ettinger WE, Kramer DM. Integrating the proton circuit into photosynthesis: progress and challenges, *Plant, Cell and Environment*, 2005, vol. 28 (pg. 97-109)

Badger MR, von Caemmerer S, Ruuska S, Nakano H. Electron flow to oxygen in higher plants and algae: rates and control of direct photoreduction (Mehler reaction) and rubisco oxygenase, Philosophical Transaction of the Royal Society B: *Biological Sciences*, 2000, vol. 355 (pg. 1433-1446)

Bae MS, Cho EJ, Choi EY, Park OK. Analysis of the Arabidopsis nuclear proteome and its response to cold stress, *The Plant Journal*, 2003, vol. 36 (pg. 652-663)

Baena-Gonzalez E, Baginsky S, Mulo P, Summer H, Aro EM, Link G. Chloroplast transcription at different light intensities. Glutathione-mediated phosphorylation of the major RNA polymerase involved in redox-regulated organellar gene expression, *Plant Physiology*, 2001, vol. 127 (pg. 1044-1052)

Baginsky S, Kleffmann T, von Zychlinski A, Gruissem W. Analysis of shotgun proteomics and RNA profiling data from Arabidopsis thaliana chloroplasts, *Journal of Proteome Research*, 2005, vol. 4 (pg. 637-640)

Baginsky S, Tiller K, Pfannschmidt T, Link G. PTK, the chloroplast RNA polymerase-associated protein kinase from mustard (Sinapis alba), mediates redox control of plastid *in vitro* transcription, *Plant Molecular Biology*, 1999, vol. 39 (pg. 1013-1023)

Baier M, Dietz KJ. Chloroplasts as source and target of cellular redox regulation: a discussion on chloroplast redox signals in the context of *Plant Physiology*, *Journal of Experimental Botany*, 2005, vol. 56 (pg. 1449-1462)

Baker A, Graham IA, Holdsworth M, Smith SM, Theodoulou FL. Chewing the fat: â-oxidation in signalling and development, *Trends in Plant Science*, 2006, vol. 11 (pg. 124-132)

Baker NR, Harbinson J, Kramer DM. Determining the limitations and regulation of photosynthetic energy transduction in leaves, *Plant, Cell and Environment*, 2007, vol. 30 (pg. 1107-1125)

Baker NR, Ort DR. Baker NR, Thomas H. *Light and crop photosynthetic performance, Topics in photosynthesis. Crop photosynthesis: spatial and temporal determinants*, 1992, AmsterdamElsevier (pg. 289-312)

Baldet P, Alban C, Douce R. Biotin synthesis in higher plants: purification and characterization of bioB gene product equivalent from Arabidopsis thaliana overexpressed in Escherichia coli and its subcellular localization in pea leaf cells, FEBS Letters, 1997, vol. 419 (pg. 206-210)

Bannai H, Tamada Y, Maruyama O, Nakai K, Miyano S. Extensive feature detection of N-terminal protein sorting signals, *Bioinformatics*, 2002, vol. 18 (pg. 298-305)

Baroja-Fernandez E, Munoz FJ, Zandueta-Criado A, Moran-Zorzano MT, Viale AM, Alonso-Casajus N, Pozueta-Romero J. Most of ADP-glucose linked to starch biosynthesis occurs outside the chloroplast in source leaves, *Proceedings of the National Academy of Sciences, USA*, 2004, vol. 101 (pg. 13080-13085)

Beale SI. Chloroplast signaling: retrograde regulation revelations, *Current Biology*, 2011, vol. 21 (pg. R391-R393)

Beckles DM, Smith AM, ap Rees T. A cytosolic ADP-glucose pyrophosphorylase is a feature of graminaceous endosperms, but not of other starch-storing organs, *Plant Physiology*, 2001, vol. 125 (pg. 818-827)

Bedhomme M, Hoffmann M, McCarthy EA, Gambonnet B, Moran RG, Rébeillé F, Ravanel S. Folate metabolism in plants: an Arabidopsis homolog of the mammalian mitochondrial folate transporter mediates folate import into chloroplasts, *Journal of Biological Chemistry*, 2005, vol. 280 (pg. 34823-34831)

Beevers H. Stumpf PK. The role of the glyoxylate cycle, The biochemistry of plants, 1980, vol. Vol. IV New York, NYAcademic Press(pg. 117-130)

Bendall DS, Manasse RS. Cyclic photophosphorylation and electron transport, *Biochimica et Biophysica Acta*, 1995, vol. 1229 (pg. 23-38)

Bilgin DD, Zavala JA, Zhu J, Clough SJ, Ort DR, DeLucia EH. Biotic stress globally downregulates photosynthesis genes, *Plant, Cell and Environment*, 2010, vol. 33 (pg. 1597-15613)

Björkman O, Demmig B. Photon yield of O2 evolution and chlorophyll fluorescence characteristics at 77 K among vascular plants of diverse origins, *Planta*, 1987, vol. 170 (pg. 489-504)

Bläsing OE, Gibon Y, Günther M, Höhne M, Morcuende R, Osuna D, Thimm O, Usadel B, Scheible WR, Stitt M. Sugars and circadian regulation make major contributions to the global regulation of diurnal gene expression in Arabidopsis, *The Plant Cell*, 2005, vol. 17 (pg. 3257-3281)

Bloom AJ, Burger M, Asensio JSR, Cousins A. Carbon dioxide enrichment inhibits nitrate assimilation in wheat and Arabidopsis, Science, 2010, vol. 328 (pg. 899-902)

Borner GH, Sherrier DJ, Stevens TJ, Arkin IT, Dupree P. Prediction of glycosylphosphatidylinositol-anchored proteins in Arabidopsis: a genomic analysis, *Plant Physiology*, 2002, vol. 129 (pg. 486-499)

Bowes G. Growth at elevated CO2: photosynthesis responses mediated through Rubisco, *Plant, Cell and Environment*, 1991, vol. 14 (pg. 795-806)

Bowsher CG, Tobin AK. Compartmentation of metabolism within mitochondria and plastids, *Journal of Experimental Botany*, 2001, vol. 52 (pg. 513-527)

Bräutigam K, Dietzel L, Pfannschmidt T. Hypothesis: a binary redox control mode as universal regulator of photosynthetic light acclimation, *Plant Signaling and Behavior*, 2010, vol. 5 (pg. 81-85)

Breyton C, Nandha B, Johnson G, Joliot P, Finazzi G. Redox modulation of cyclic electron flow around PS I in C3 plants, *Biochemistry*, 2006, vol. 45 (pg. 13456-13475)

Brugière S, Kowalski S, Ferro M, et al. The hydrophobic proteome of mitochondrial membranes from Arabidopsis cell suspensions, *Phytochemistry*, 2004, vol. 65 (pg. 1693-1707)

Bukhov N, Carpentier R. Alternative PS I driven electron transport routes: mechanisms and functions, *Photosynthesis Research*, 2004, vol. 82 (pg. 17-33)

Calikowski TT, Meulia T, Meier I. A proteomic study of the arabidospis nuclear matrix, *Journal of Cell Biochemistry*, 2003, vol. 90 (pg. 361-378)

Cape JL, Bowman MK, Kramer DM. Understanding the cytochrome bc complexes by what they don't do. The Q-cycle at 30, *Trends in Plant Sciences*, 2006, vol. 11 (pg. 46-55)

Carqueijeiro I, Martins V, Noronha H, Gerós H, Sottomayor M (2016) Analytical and fluorimetric methods for the characterization of the transmembrane transport of specialized metabolites in plants. *Methods Mol Biol* 1405: 121-135.

Carter C, Pan S, Zouhar J, Avila EL, Girke T, Raikhel NV. The vegetative vacuole proteome of Arabidopsis thaliana reveals predicted and unexpected proteins, *The Plant Cell*, 2004, vol. 16 (pg. 3285-3303)

Casati P, Walbot V. Gene expression profiling in response to ultraviolet radiation in maize genotypes with varying flavonoid content, *Plant Physiology*, 2003, vol. 132 (pg. 1739-1754)

Chabregas SM, Luche DD, van Sluys M-A, Menck CFM, Silva-Filho MC. Differential usage of two in-frame translational start codons regulates subcellular localization of Arabidopsis thaliana THI1, *Journal of Cell Science,* 2003, vol. 116 (pg. 285-291)

Cheeseman JM, Herendeen LB, Cheeseman AT, Clough BF. Photosynthesis and photoprotection in mangroves under field conditions, *Plant, Cell and Environment*, 1997, vol. 20 (pg. 579-588)

Chen GY, Yong ZH, Liao Y, Zhang DY, Chen Y, Zhang HB, Chen J, Zhu JG, Xu DQ. Photosynthetic acclimation in rice leaves to free-air CO2 enrichment related to both ribulose-1,5-bisphosphate carboxylation limitation and ribulose-1,5-bisphosphate regeneration limitation, *Plant and Cell Physiology*, 2005, vol. 46 (pg. 1036-1045)

Coxon KM, Chakauya E, Ottenhof HH, Whitney HM, Blundell TL, Abell C, Smith AG. Pantothenate biosynthesis in higher plants, *Biochemical Society Transactions*, 2005, vol. 33 (pg. 743-746)

Croce R, Müller MG, Bassi R, Holzwarth AR. Carotenoid-to-chlorophyll energy transfer in recombinant major light-harvesting complex (LHCII) of higher plants. I. Femtosecond transient absorption measurements, *Biophysical Journal*, 2001, vol. 80 (pg. 901-915)

Crofts AR, Hong S, Ugulava N, Barquera B, Gennis R, Guergova-Kuras M, Berry EA. Pathways for proton release during ubihydroquinone oxidation by the bc1 complex, *Proceedings of the National Academy of Sciences, USA,* 1999, vol. 96 (pg. 10021-10026)

Cruz JA, Sacksteder CA, Kanazawa A, Kramer DM. Contribution of electric field (ÄØ) to steady-state transthylakoid proton motive force (pmf) in vitro and *in vivo. Biochemistry*, 2001, vol. 40 (pg. 1226-1237)

Cure JD, Acock B. Crop responses to carbon-dioxide doubling—a literature survey, *Agricultural and Forest Meteorology*, 1986, vol. 38 (pg. 127-145)

Cutler SR, Ehrhardt DW, Griffitts JS, Somerville CR. Random GFP::cDNA fusions enable visualization of subcellular structures in cells of Arabidopsis at a high frequency, *Proceedings of the National Academy of Sciences, USA,* 2000, vol. 97 (pg. 3718-3723)

DalCorso G, Pesaresi P, Masiero S, Aseeva E, Schunemann D, Finazzi G, Joliot P, Barbato R, Leister D. A complex containing PGRL1 and PGR5 is involved in the switch between linear and cyclic electron flow in Arabidopsis, *Cell*, 2008, vol. 132 (pg. 273-285)

Dashty M (2013) A quick look at biochemistry: Carbohydrate metabolism. *Clin Biochem*. 46: 1339-1352.

Davey PA, Parsons AJ, Atkinson L, Wadge K, Long SP. Does photosynthetic acclimation to elevated CO2 increase photosynthetic nitrogen-use efficiency? A study of three native UK grassland species in open-top chambers, *Functional Ecology*, 1999, vol. 13 (pg. 21-28)

de Groot CC, van den Boogaard R, Marcelis LF, Harbinson J, Lambers H. Contrasting effects of N and P deprivation on the regulation of photosynthesis in tomato plants in relation to feedback limitation, *Journal of Experimental Botany*, 2003, vol. 54 (pg. 1957-1967)

De la Torre A, Delgado B, Lara C. Nitrate-dependent O2 evolution in intact leaves, *Plant Physiology*, 1991, vol. 96 (pg. 898-901)

de Vries SC, Weijers D (2017) Plant embryogenesis. Curr Biol 27: R870-R873

de Weerd FL, Kennis JTM, Dekker JP, van Grondelle R. â-Carotene to chlorophyll singlet energy transfer in the PS I core of *Synechococcus elongatus* proceeds via the â-carotene S2 and S1 states, *Journal of Physical Chemistry B*, 2003, vol. 107 (pg. 5995-6002)

Delucia EH, Sasek TW, Strain BR. Photosynthetic inhibition after long-term exposure to elevated levels of atmospheric carbon-dioxide, *Photosynthesis Research*, 1985, vol. 7 (pg. 175-184)

Demmig-Adams B, Adams WW. Photoprotection in an ecological context: the remarkable complexity of thermal energy dissipation, *New Phytologist*, 2006, vol. 172 (pg. 11-21)

Denyer K, Dunlap F, Thorbjørnsen T, Keeling P, Smith AM. The major form of ADP-glucose pyrophosphorylase in maize endosperm is extra-plastidial, *Plant Physiology*, 1996, vol. 112 (pg. 779-785)

Desikan R, Mackerness SAH, Hancock JT, Neill SJ. Regulation of the Arabidopsis transcriptome by oxidative stress, *Plant Physiology*, 2001, vol. 127 (pg. 159-172)

Dietz K-J, Pfannschmidt T. Novel regulators in photosynthetic redox control of plant metabolism and gene expression, *Plant Physiology*, 2011, vol. 155 (pg. 1477-1485)

Dong HL, Deng Y, Mu JY, Lu QT, Wang YQ, Xu YY, Chu CC, Chong K, Lu CM, Zuo JR. The Arabidopsis Spontaneous Cell Death1 gene, encoding a zeta-carotene desaturase essential for carotenoid biosynthesis, is involved in chloroplast development, photoprotection and retrograde signalling, *Cell Research*, 2007, vol. 17 (pg. 458-470)

Dunkley TPJ, Watson R, Griffin JL, Dupree P, Lilley KS. Localization of organelle proteins by isotope tagging (LOPIT), *Molecular and Cellular Proteomics*, 2004, vol. 3 (pg. 1128-1134)

Dutilleul C, Driscoll S, Cornic G, De Paepe R, Foyer CH, Noctor G. Functional mitochondrial complex I is required by tobacco leaves for optimal photosynthetic performance in photorespiratory conditions and during transients, *Plant Physiology*, 2003, vol. 313 (pg. 264-275)

Edwards J, ap Rees T, Wilson PM, Morrell S. Measurement of the inorganic pyrophosphate in tissues of Pisum sativum L., *Planta*, 1984, vol. 162 (pg. 188-191)

Eisenreich W, Rohdich F, Bacher A. Deoxyxylulose phosphate pathway to terpenoids, *Trends in Plant Science*, 2001, vol. 6 (pg. 78-84)

Emanuelsson O, Nielsen H, Brunak S, von Heijne G. Predicting subcellular localization of proteins based on their N-terminal amino acid sequence, *Journal of Molecular Biology*, 2000, vol. 300 (pg. 1005-1016)

Escoubas JM, Lomas M, Laroche J, Falkowski PG. Light intensity regulation of cab gene transcription is signaled by the redox state of the plastoquinone pool, *Proceedings of the National Academy of Sciences, USA*, 1995, vol. 92 (pg. 10237-10241)

Evans JR, Terashima I. Photosynthetic characteristics of spinach leaves grown with different nitrogen treatments, *Plant and Cell Physiology*, 1988, vol. 29 (pg. 157-165)

Evans JR. Photosynthetic acclimation and nitrogen partitioning within a lucerne canopy. II. Stability through time and comparison with a theoretical optimum, Australian Journal of *Plant Physiology*, 1993, vol. 20 (pg. 69-82)

Evron Y, Avron M. Characterisation of an alkaline pH-dependent proton 'slip' in the ATP synthase of lettuce thylakoids, *Biochimica et Biophysica Acta*, 1990, vol. 1019 (pg. 115-120)

Farage PK, Blowers D, Long SP, Baker NR. Low growth temperatures modify the efficiency of light use by PS II for CO2 assimilation in leaves of two chilling-tolerant C4 species, *Cyperus longus* L. and *Miscanthus × giganteus*, *Plant, Cell and Environment*, 2006, vol. 29 (pg. 720-728)

Farquhar GD, Caemmerer S, Berry JA. A biochemical model of photosynthetic CO2 assimilation in leaves of C3 species, *Planta*, 1980, vol. 149 (pg. 78-90)

Farré EM, Tiessen A, Roessner U, Geigenberger P, Trethewey RN, Willmitzer L. Analysis of the compartmentation of glycolytic intermediates, nucleotides, sugars, organic acids, amino acids, and sugar alcohols in potato tubers using a nonaqueous fractionation method, *Plant Physiology*, 2001, vol. 127 (pg. 685-700)

Ferro M, Salvi D, Brugière S, Miras S, Kowalski S, Louwagie M, Garin J, Joyard J, Rolland N. Proteomics of the chloroplast envelope membranes from Arabidopsis thaliana., *Molecular and Cellular Proteomics*, 2003, vol. 2 (pg. 325-345)

Ferro M, Salvi D, Rivière-Rolland H, Vermat T, Seigneurin-Berny D, Grunwald D, Garin J, Joyard J, Rolland N. Integral membrane proteins of the chloroplast envelope: identification and subcellular localization of new transporters, *Proceedings of the National Academy of Sciences, USA*, 2002, vol. 99 (pg. 11487-11492)

Fey V, Wagner R, Brautigam K, Pfannschmidt T. Photosynthetic redox control of nuclear gene expression, *Journal of Experimental Botany*, 2005, vol. 56 (pg. 1491-1498)

Finazzi G. Redox-coupled proton pumping activity in cytochrome b6f, as evidenced by the pH dependence of electron transfer in whole cells of *Chlamydomonas reinhardtii, Biochemistry*, 2002, vol. 41 (pg. 7475-7482)

Fischer BB, Wiesendanger M, Eggen RIL. Growth condition-dependent sensitivity, photodamage and stress response of *Chlamydomonas reinhardtii* exposed to high light conditions, *Plant and Cell Physiology*, 2006, vol. 47 (pg. 1135-1145)

Foyer CH, Bloom A, Queval G, Noctor G. Photorespiratory metabolism: genes, mutants, energetics, and redox signaling, *Annual Review of Plant Biology*, 2009, vol. 60 (pg. 455-484)

Foyer CH, Furbank RT, Harbinson J, Horton P. The mechanisms contributing to photosynthetic control of electron transport by carbon assimilation in leaves, *Photosynthesis Research*, 1990, vol. 25 (pg. 83-100)

Foyer CH, Lelandais M, Harbinson J. Control of the quantum efficiencies of photosystems I and II, electron flow, and enzyme activation following dark-to-light transitions in pea leaves, *Plant Physiology*, 1992, vol. 99 (pg. 979-986)

Foyer CH, Noctor G, Hodges M. Respiration and nitrogen assimilation: targeting mitochondria-associated metabolism as a means to enhance nitrogen use efficiency, *Journal of Experimental Botany*, 2011, vol. 62 (pg. 1467-1482)

Foyer CH, Noctor G. Redox regulation in photosynthetic organisms: signaling, acclimation and practical implications, *Antioxidants and Redox Signaling*, 2009, vol. 11 (pg. 861-905)

Foyer CH, Shigeoka S. Understanding oxidative stress and antioxidant functions to enhance photosynthesis, *Plant Physiology*, 2011, vol. 155 (pg. 93-100)

Fridlyand L, Backhausen J, Scheibe R. Flux control of the malate valve in leaf cells, *Archives of Biochemistry and Biophysics*, 1998, vol. 349 (pg. 290-298)

Friso G, Giacomelli L, Ytterberg AJ, Peltier JB, Rudella A, Sun Q, van Wijk KJ. In-depth analysis of the thylakoid membrane proteome of Arabidopsis thaliana chloroplasts: new proteins, new functions, and a plastid proteome database, *The Plant Cell*, 2004, vol. 16 (pg. 478-499)

Froehlich JE, Wilkerson CG, Ray WK, McAndrew RS, Osteryoung KW, Gage DA, Phinney BS. Proteomic study of the Arabidopsis thaliana chloroplastic envelope membrane utilizing alternatives to traditional two-dimensional electrophoresis, *Journal of Protein Research*, 2003, vol. 2 (pg. 413-425)

Fryer MJ, Andrews JR, Oxborough K, Blowers DA, Baker NR. Relationship between CO2 assimilation, photosynthetic electron transport, and active O2 metabolism in leaves of maize in the field during periods of low temperature, *Plant Physiology*, 1998, vol. 116 (pg. 571-580)

Fryer MJ, Ball L, Oxborough K, Karpinski S, Mullineaux PM, Baker NR. Control of Ascorbate Peroxidase 2 expression by hydrogen peroxide and leaf water status during excess light stress reveals a functional organisation of Arabidopsis leaves, *The Plant Journal*, 2003, vol. 33 (pg. 691-705)

Fukao Y, Hayashi M, Hara-Nishimura I, Nishimura M. Novel glyoxysomal protein kinase, GPK1, identified by proteomic analysis of glyoxysomaes in etiolated cotyledons of Arabidopsis thaliana, *Plant Cell Physiology*, 2003, vol. 44 (pg. 1002-1012)

Fukao Y, Hayashi M, Nishimura M. Proteomic analysis of leaf peroxisomal proteins in greening cotyledons of Arabidopsis thaliana., *Plant Cell Physiology*, 2002, vol. 43 (pg. 689-696)

Fukayama H, Fukuda T, Masumoto C, Taniguchi Y, Sakai H, Cheng WG, Hasegawa T, Miyao M. Rice plant response to long term CO2 enrichment: gene expression profiling, *Plant Science*, 2009, vol. 177 (pg. 203-210)

Fukayama H, Sugino M, Fukuda T, et al. Gene expression profiling of rice grown in free air CO2 enrichment (FACE) and elevated soil temperature, *Field Crops Research*, 2011, vol. 121 (pg. 195-199)

Fuks B, Homblé F. Mechanism of proton permeation through chloroplast lipid membranes, *Plant Physiology*, 1996, vol. 112 (pg. 759-766)

Gadjev I, Vanderauwera S, Gechev TS, Laloi C, Minkov IN, Shulaev V, Apel K, Inze D, Mittler R, van Breusegem F. Transcriptomic footprints disclose specificity of reactive oxygen species signaling in Arabidopsis, *Plant Physiology*, 2006, vol. 141 (pg. 436-445)

Genty BG, Harbinson J. Baker NR. The regulation of light utilisation for photosynthetic electron transport, Environmental stress and photosynthesis, 1996, Dordrecht Kluwer (pg. 67-99)

Gerhardt R, Heldt HW. Measurement of subcellular metabolite levels in leaves by fractionation of freeze-stopped material in non-aqueous media, *Plant Physiology*, 1987, vol. 75 (pg. 542-547)

Gester S, Wuest F, Pawelke B, Bergmann R, Pietzsch J. Synthesis and biodistribution of an 18F-labelled resveratrol derivative for small animal positron emission tomography, *Amino Acids*, 2005, vol. 29 (pg. 415-428)

Ghassemian M, Lutes J, Tepperman JM, Chang HS, Zhu T, Wang X, Quail PH, Lange BM. Integrative analysis of transcript and metabolite profiling data sets to evaluate the regulation of biochemical pathways during photomorphogenesis, *Archives of Biochemistry and Biophysics*, 2006, vol. 448 (pg. 45-59)

Giegé P, Heazlewood JL, Roessner-Tunali U, Millar AH, Fernie AR, Leaver CJ, Sweetlove LJ. Enzymes of glycolysis are functionally associated with the mitochondrion in Arabidopsis cells, *The Plant Cell*, 2003, vol. 15 (pg. 2140-2151)

Gout E, Bligny R, Douce R. Regulation of intracellular pH values in higher plant cells, *Journal of Biological Chemistry*, 1992, vol. 267 (pg. 13903-13909)

Gueguen V, Macharel D, Jaquinod M, Douce R, Bourguignon J. Fatty acid and lipoic acid biosynthesis in higher plant mitochondria, *Journal of Biological Chemistry*, 2000, vol. 275 (pg. 5016-5025)

Gunderson CA, Norby RJ, Wullschleger SD. Foliar gas-exchange responses of 2 decidious hardwoods during 3 years of growth in elevated CO2—no loss of photosynthetic enhancement, *Plant, Cell and Environment*, 1993, vol. 16 (pg. 797-807)

Gupta P, Duplessis S, White H, Karnosky DF, Martin F, Podila GK. Gene expression patterns of trembling aspen trees following long-term exposure to interacting elevated CO_2 and tropospheric O_3, *New Phytologist*, 2005, vol. 167 (pg. 129-142)

Harbinson J, Foyer CH. Relationships between the efficiencies of PS I and II and stromal redox state in CO2-free air: evidence for cyclic electron flow in vivo, *Plant Physiology*, 1991, vol. 97 (pg. 41-49)

Harbinson J, Genty B, Foyer C. Relationship between photosynthetic electron transport and stromal enzyme activity in pea leaves: toward an understanding of the nature of photosynthetic control, *Plant Physiology*, 1990, vol. 94 (pg. 545-553)

Harbinson J, Hedley CL. The kinetics of P-700+ reduction in leaves—a novel in situ probe of thylakoid functioning, *Plant, Cell and Environment*, 1989, vol. 12 (pg. 357-369)

Hatch MD. C4 photosynthesis: a unique blend of modified biochemistry, anatomy and ultrastructure, *Biochimica et Biophysica Acta*, 1987, vol. 895 (pg. 81-106)

Havaux M, Niyogi K. The violaxanthin cycle protects plants from photooxidative damage by more than one mechanism, *Proceedings of the National Academy Sciences, USA*, 1999, vol. 96 (pg. 8762-8767)

Hays SG, Yan LLW, Silver PA, Ducat DC (2017) Synthetic photosynthetic consortia define interactions leading to robustness and photoproduction. J Biol Eng 11: 4.

Heazlewood JL, Tonti-Filippini JS, Gout AM, Day DA, Whelan J, Millar AH. Experimental analysis of the Arabidopsis mitochondrial proteome highlights signaling and regulatory components, provides assessment of targeting prediction programs, and indicates plant-specific mitochondrial proteins, *The Plant Cell*, 2004, vol. 16 (pg. 241-256)

Heazlewood JL, Tonti-Filippini JS, Verboom RE, Millar AH. Combining experimental and predicted datasets for determination of the subcellular location of proteins in Arabidopsis, *Plant Physiology*, 2005, vol. 139 (pg. 598-609)

Hendrickson L, Furbank RT, Chow WS. A simple alternative approach to assessing the fate of absorbed light energy using chlorophyll fluorescence, *Photosynthesis Research*, 2004, vol. 82 (pg. 73-81)

Hibberd J, Quick WP. Characteristics of C4 photosynthesis in stems and petioles of C3 flowering plants, *Nature*, 2002, vol. 415 (pg. 451-454)

Hill R, Rich P. A physical interpretation for the natural photosynthetic process, *Proceedings of the National Academy of Sciences, USA*, 1983, vol. 80 (pg. 978-982)

Hodges M, Dellero Y, Keech O, Betti M, Raghavendra AS, et al. (2016) Perspectives for a better understanding of the metabolic integration of photorespiration within a complex plant primary metabolism network. *J Exp Bot* 67: 3015-3026.

Hope AB, Valente P, Matthews DB. Effects of pH on the kinetics of redox reactions in and around the cytochrome bf complex in an isolated system, *Photosynthesis Research*, 1994, vol. 42 (pg. 111-120)

Horton P, Wentworth M, Ruban A. Control of the light harvesting function of chloroplast membranes: the LHCII-aggregation model for non-photochemical quenching, *FEBS Letters*, 2005, vol. 579 (pg. 4201-4206)

Hosler JP, Yocum CF. Regulation of cyclic photophosphorylation during ferredoxin-mediated electron transport: effect of DCMU and the NADPH/NADP ratio, *Plant Physiology*, 1987, vol. 83 (pg. 965-969)

Hu H, Boisson-Dernier A, Israelsson-Nordström M, Böhmer M, Xue S, Ries A, Godoski J, Kuhn JM, Schroeder JI. Carbonic anhydrases are upstream regulators in guard cells of CO_2-controlled stomatal movements, *Nature Cell Biology*, 2010, vol. 12 (pg. 87-18)

Hua S, Sun Z. Support vector machine approach for protein subcellular localization prediction, *Bioinformatics*, 2001, vol. 17 (pg. 721-728)

Idso SB, Kimball BA. Effects of atmospheric CO2 enrichment on photosynthesis, respiration, and growth of sour orange trees, *Plant Physiology*, 1992, vol. 99 (pg. 341-343)

Im CS, Eberhard S, Huang KY, Beck CF, Grossman AR. Phototropin involvement in the expression of genes encoding chlorophyll and carotenoid biosynthesis enzymes and LHC apoproteins in *Chlamydomonas reinhardtii*, *The Plant Journal*, 2006, vol. 48 (pg. 1-16)

Im CS, Zhang ZD, Shrager J, Chang CW, Grossman AR. Analysis of light and CO2 regulation in *Chlamydomonas reinhardtii* using genome-wide approaches, *Photosynthesis Research*, 2003, vol. 75 (pg. 111-125)

Inada K. Action spectra for photosynthesis in higher plants, *Plant and Cell Physiology*, 1976, vol. 17 (pg. 355-365)

Jabaut JM, Dudum R, Margulies SL, Mehta A, Han Z (2016) Teaching and learning of medical biochemistry according to clinical realities: A case study. *Biochem Mol Biol Educ* 44: 95-98.

Kim J-W, Dang CV. Multifaceted roles of glycolytic enzymes, *Trends in Biochemical Sciences*, 2005, vol. 30 (pg. 142-150)

Kleffmann T, Hirsch-Hoffmann M, Gruissem W, Baginsky S. plprot: a comprehensive proteome database for different plastid types, *Plant and Cell Physiology*, 2006, vol. 47 (pg. 432-436)

Kleffmann T, Russenberger D, von Zychlinski A, Christopher W, Sjolander K, Gruissem W, Baginsky S. The Arabidopsis thaliana chloroplast proteome reveals pathway abundance and novel protein functions, *Current Biology*, 2004, vol. 14 (pg. 354-362)

Köckenberger W. Nuclear magnetic resonance micro-imaging in the investigation of plant cell metabolism, *Journal of Experimental Botany*, 2001, vol. 52 (pg. 641-652)

Kolbe A, Tiessen A, Schluepmann H, Paul M, Ulrich S, Geigenberger P. Trehalose 6-phosphate regulates starch synthesis via post-translational redox activation of ADP-glucose pyrophosphorylase, *Proceedings of the National Academy of Sciences, USA*, 2005, vol. 102 (pg. 11118-11123)

Koroleva OA, Tomlinson ML, Leader D, Shaw P, Doonan JH. High-throughput protein localization in Arabidopsis using Agrobacterium-mediated transient expression of GFP-ORF fusions, *The Plant Journal*, 2005, vol. 41 (pg. 162-174)

Kruft V, Eubel H, Jansch L, Werhahn W, Braun HP. Proteomic approach to identify novel mitochondrial proteins in Arabidopsis, *Plant Physiology*, 2001, vol. 127 (pg. 1694-1710)

Lichtenthaler HK. The 1-deoxy-D-xylulose-5-phosphate pathway of isoprenoid biosynthesis in plants, *Annual Review of Plant Physiology and Plant Molecular Biology*, 1999, vol. 50 (pg. 47-65)

Ludwig A, Stolz J, Sauer N. Plant sucrose-H+ symporters mediate the transport of vitamin H, *The Plant Journal*, 2000, vol. 24 (pg. 503-509)

Lunn JE, Douce R. Transport of inorganic pyrophosphate across the spinach chloroplast envelope, *Biochemical Journal*, 1993, vol. 290 (pg. 375-379)

Lunn JE, Droux M, Martin J, Douce R. Localization of ATP sulfurylase and O-acetylserine(thiol)lyase in spinach leaves, *Plant Physiology*, 1990, vol. 94 (pg. 1345-1352)

Lunn JE, Feil R, Hendriks JHM, Gibon Y, Morcuende R, Osuna D, Scheible WR, Carillo P, Hajirezaei MR, Stitt M. Sugar-induced increases in trehalose 6-phosphate are correlated with redox activation of ADPglucose pyrophosphorylase and higher rates of starch synthesis in *Arabidopsis thaliana*, *Biochemical Journal*, 2006, vol. 397 (pg. 139-148)

Lunn JE, Furbank RT. Localisation of sucrose-phosphate synthase and starch in leaves of C4 plants, *Planta*, 1997, vol. 202 (pg. 106-111)

Lunn JE. Evolution of sucrose synthesis, *Plant Physiology*, 2002, vol. 128 (pg. 1490-1500)

MacRae EA, Lunn JE. Plaxton W, McManus MT. *Control of sucrose biosynthesis, Annual plant reviews - control of primary metabolism in plants*, 2006, Oxford Blackwell Publishing (pg. 234-257)

Majeran W, Cai Y, Sun Q, van Wijk KJ. Functional differentiation of bundle sheath and mesophyll maize chloroplasts determined by comparative proteomics, *The Plant Cell*, 2005, vol. 17 (pg. 3111-3140)

Marmagne A, Rouet M-A, Ferro M, Rolland N, Alcon C, Joyard J, Garin J, Barbier-Brygoo H, Ephritikhine G. Identification of new intrinsic proteins in Arabidopsis plasma membrane proteome, *Molecular and Cellular Proteomics*, 2004, vol. 3 (pg. 675-691)

Martin W, Rujan T, Richly E, Hansen A, Cornelsen S, Lins T, Leister D, Stoebe B, Hasegawa M, Penny D. Evolutionary analysis of Arabidopsis, cyanobacterial, and chloroplast genomes reveals plastid phylogeny and thousands of cyanobacterial genes in the nucleus, *Proceedings of the National Academy of Sciences, USA*, 2002, vol. 99 (pg. 11996-11997)

Maurer-Stroh S, Eisenhaber F. Refinement and prediction of protein prenylation motifs, *Genome Biology*, 2005, vol. 6 pg. R55 doi:10.1186/gb-2005-6-6-r55

Mendes J, Borges N, Santos A, Padrão P, Moreira P, et al. (2018) Nutritional status and gait speed in a nationwide population-based sample of older adults. *Sci Rep* 8: 4227.

Mettler IJ, Beevers H. Oxidation of NADH in glyoxysomes by a malate-aspartate shuttle, *Plant Physiology*, 1980, vol. 66 (pg. 555-560)

Millar AH, Heazlewood JL, Kristensen BK, Braun H-P, Møller IM. The plant mitochondrial proteome, *Trends in Plant Science*, 2005, vol. 10 (pg. 1360-1385)

Millar AH, Sweetlove LJ, Giege P, Leaver CJ. Analysis of the Arabidopsis mitochondrial proteome, *Plant Physiology*, 2001, vol. 127 (pg. 1711-1727)

Miras S, Salvi D, Ferro M, Grunwald D, Garin J, Joyard J, Rolland N. Non-canonical transit peptide for import into the chloroplast, *Journal of Biological Chemistry*, 2002, vol. 277 (pg. 47770-47778)

Moulin M, Smith AG. Regulation of tetrapyrrole biosynthesis in higher plants, *Biochemical Society Transactions*, 2005, vol. 33 (pg. 737-742)

Neuhaus HE, Häusler RE, Sonnewald U. No need to shift the paradigm on the metabolic pathway to transitory starch in leaves, *Trends in Plant Science*, 2005, vol. 10 (pg. 154-156)

Niittyla T, Messerli G, Trevisan M, Chen J, Smith AM, Zeeman SC. A previously unknown maltose transporter essential for starch degradation in leaves, *Science*, 2004, vol. 303 (pg. 87-89)

Patron NJ, Keeling PJ. Common evolutionary origin of starch biosynthetic enzymes in green and red algae, *Journal of Phycology*, 2005, vol. 41 (pg. 1131-1141)

Paunov M, Koleva L, Vassilev A, Vangronsveld J, Goltsev V (2018) Effects of different metals on photosynthesis: Cadmium and zinc affect chlorophyll fluorescence in durum wheat. *Int J Mol Sci* 19: pii E787.

Peltier JB, Emanuelsson O, Kalume DE, et al. Central functions of the lumenal and peripheral thylakoid proteome of Arabidopsis determined by experimentation and genome-wide prediction, *The Plant Cell*, 2002, vol. 14 (pg. 211-236)

Pendle AF, Clark GP, Boon R, Lewandowska D, Lam YW, Andersen L, Mann M, Lamond AI, Brown JW, Shaw PJ. Proteomic analysis of the Arabidopsis nucleolus suggests novel nucleolar functions, *Molecular Biology of the Cell*, 2005, vol. 16 (pg. 260-269)

Pinon V, Ravanel S, Douce R, Alban C. Biotin synthesis in plants: the first committed step of the pathway is catalyzed by a cytosolic 7-keto-8-aminopelargonic acid synthase, *Plant Physiology*, 2005, vol. 139 (pg. 1666-1676)

Pracharoenwattana I, Cornah JE, Smith SM. Arabidopsis peroxisomal citrate synthase is required for fatty acid respiration and seed germination, *The Plant Cell*, 2005, vol. 17 (pg. 2037-2048)

Ratcliffe RG, Shachar-Hill Y. Measuring multiple fluxes through plant metabolic networks, *The Plant Journal*, 2006, vol. 45 (pg. 490-511)

Rea PA, Poole RJ. Vacuolar H+-translocating pyrophosphatase, *Annual Review of Plant Physiology and Plant Molecular Biology*, 1993, vol. 44 (pg. 157-180)

Richly E, Leister D. An improved prediction of chloroplast proteins reveals diversities and commonalities in the chloroplast proteomes of Arabidopsis and rice, *Gene*, 2004, vol. 329 (pg. 11-16)

Rubio MÁ, Napolitano M, Ochoa de Alda JA, Santamaría-Gómez J, Patterson CJ, et al. (2015) Trans-oligomerization of duplicated aminoacyl-tRNA synthetases maintains genetic code fidelity under stress. *Nucleic Acids Res* 43: 9905-17.

Sahr T, Ravanel S, Rébeillé F. Tetrahydrofolate biosynthesis and distribution in higher plants, *Biochemical Society Transactions*, 2005, vol. 33 (pg. 758-762)

Sandoval FJ, Roje S. An FMN hydrolase is fused to a riboflavin kinase homolog in plants, *Journal of Biological Chemistry*, 2005, vol. 280 (pg. 38337-38345)

Schad M, Lipton MS, Giavalisco P, Smith RD, Kehr J. Evaluation of two-dimensional electrophoresis and liquid chromatography–tandem mass spectrometry for tissue-specific protein profiling of laser-microdissected plant samples, *Electrophoresis*, 2005, vol. 26 (pg. 2729-2738)

Schubert M, Petersson UA, Haas BJ, Funk C, Schroder WP, Kieselbach T. Proteome map of the chloroplast lumen of Arabidopsis thaliana, *Journal of Biological Chemistry*, 2002, vol. 277 (pg. 8354-8365)

Schwender J, Goffman F, Ohlrogge JB, Shachar-Hill Y. Rubisco without the Calvin cycle inproves the carbon efficiency of developing green seeds, *Nature*, 2004, vol. 432 (pg. 779-782)

Servillo L, Castaldo D, Giovane A, Casale R, D'Onofrio N, et al. (2018) Ophthalmic acid is a marker of oxidative stress in plants as in animals. *Biochim Biophys Acta* 1862: 991-998.

Shimaoka T, Ohnishi M, Sazuka T, Mitsuhashi N, Hara-Nishimura I, Shimazaki K, Maeshima M, Yokota A, Tomizawa K, Mimura T. Isolation of intact vacuoles and proteomic analysis of tonoplast from suspension-cultured cells of Arabidopsis thaliana., *Plant Cell Physiology*, 2004, vol. 45 (pg. 672-683)

Sikka VK, Choi S-B, Kavakli IH, Sakulsingharoj C, Gupta S, Ito H, Okita TW. Subcellular compartmentation and allosteric regulation of the rice endosperm ADPglucose pyrophosphorylase, *Plant Science*, 2001, vol. 161 (pg. 461-468)

Small I, Peeters N, Legeai F, Lurin C. Predotar: a tool for rapidly screening proteomes for N-terminal targeting sequences, *Proteomics*, 2004, vol. 4 (pg. 1581-1590)

Smirnoff N, Conklin PL, Loewus FA. Biosynthesis of ascorbic acid in plants, *Annual Review of Plant Physiology and Plant Molecular Biology*, 2001, vol. 52 (pg. 437-467)

Smith AL, Purcell RJ, Vaughan JM (2015) Guided inquiry activities for learning about the macro- and micro-nutrients in introductory nutrition courses. *Biochem Mol Biol Educ* 43: 449-459.

Stitt M, Lilley RM, Gerhardt R, Heldt HW. Metabolite levels in specific cells and subcellular compartments of plant leaves, *Methods in Enzymology*, 1989, vol. 174 (pg. 518-550)

Szponarski W, Sommerer N, Boyer JC, Rossignol M, Gibrat R. Large-scale characterization of integral proteins from Arabidopsis vacuolar membrane by two-dimensional liquid chromatography, *Proteomics*, 2004, vol. 4 (pg. 397-406)

Tambasco-Studart M, Titiz O, Raschle T, Forster G, Amrhein N, Fitzpatrick TB. Vitamin B6 biosynthesis in higher plants, *Proceedings of the National Academy of Sciences, USA*, 2005, vol. 102 (pg. 13687-13692)

Tetlow IJ, Davies EJ, Vardy KA, Bowsher CG, Burrell MM, Emes MJ. Subcellular localization of ADP glucose pyrophosphorylase in developing wheat endosperm and analysis of the properties of a plastidial isoform, *Journal of Experimental Botany*, 2003, vol. 54 (pg. 715-725)

Thorbjørnsen T, Villand P, Denyer K, Olsen O-A, Smith AM. Distinct isoforms of ADPglucose pyrophosphorylase occur inside and outside the amyloplasts in barley endosperm, *The Plant Journal*, 1996, vol. 10 (pg. 243-250)

Tian G-W, Mohanty A, Chary SN, et al. High-throughput fluorescent tagging of full-length Arabidopsis gene products in *Planta, Plant Physiology*, 2004, vol. 135 (pg. 25-38)

Tiessen A, Hendriks JH, Stitt M, Branscheid A, Gibon Y, Farre EM, Geigenberger P. Starch synthesis in potato tubers is regulated by post-translational redox modification of ADP-glucose pyrophosphorylase: a novel regulatory mechanism linking starch synthesis to the sucrose supply, *The Plant Cell*, 2002, vol. 14 (pg. 2191-2213)

Tobin AK, Bowsher CG. Nitrogen and carbon metabolism in plastids: evolution, integration and coordination with reactions in the cytosol, *Advances in Botanical Research*, 2005, vol. 42 (pg. 113-165)

Tobin AK, Thorpe JR, Hylton CM, Rawsthorne S. Spatial and temporal influences on the cell-specific distribution of glycine decarboxylase in leaves of wheat (Triticum aestivum L.) and pea (Pisum sativum L.), *Plant Physiology*, 1989, vol. 91 (pg. 1219-1225)

Tobin AK, Yamaya T. Cellular compartmentation of ammonium assimilation in rice and barley, *Journal of Experimental Botany*, 2001, vol. 52 (pg. 591-604)

van Wijk KJ. Plastid proteomics, *Plant Physiology and Biochemistry*, 2004, vol. 42 (pg. 963-977)

Vargas W, Cumino A, Salerno GL. Cyanobacterial alkaline/neutral invertases: origin of sucrose hydrolysis in the plant cytosol, *Planta*, 2003, vol. 216 (pg. 951-960)

Villarejo A, Burén S, Larsson S, et al. Evidence for a protein transported through the secretory pathway en route to the higher plant chloroplast, *Nature Cell Biology*, 2005, vol. 7 (pg. 1224-1232)

von Zychlinski A, Kleffmann T, Krishnamurthy N, Sjolander K, Baginsky S, Gruissem W. Proteome analysis of the rice etioplast: metabolic and regulatory networks and novel protein functions, *Molecular and Cellular Proteomics*, 2005, vol. 4 (pg. 1072-1084)

Voznesenskaya EV, Franceschi VR, Kiirats O, Artusheva EG, Freitag H, Edwards GE. Proof of C4 photosynthesis without Kranz anatomy in *Binertia cycloptera* (Chenopodiaceae), *The Plant Journal*, 2002, vol. 31 (pg. 649-662)

Voznesenskaya EV, Franceschi VR, Kiirats O, Freitag H, Edwards GE. Kranz anatomy is not essential for terrestrial C4 plant photosynthesis, *Nature*, 2001, vol. 414 (pg. 543-546)

Walker RP, Chen Z-H, Johnson KE, Famiani F, Tecsi L, Leegood RC. Using immunohistochemistry to study plant metabolism: the examples of its use in the localization of amino acids in plant tissues, and of phosphoenolpyruvate carboxykinase and its possible role in pH regulation, *Journal of Experimental Botany*, 2001, vol. 52 (pg. 565-576)

Weiner H, Stitt M, Heldt HW. Subcellular compartmentation of pyrophosphate and alkaline pyrophosphatase in leaves, *Biochimica et Biophysica Acta*, 1987, vol. 893 (pg. 13-21)

Werhahn W, Braun HP. Biochemical dissection of the mitochondrial proteome from Arabidopsis thaliana by three-dimensional gel electrophoresis, *Electrophoresis*, 2002, vol. 23 (pg. 640-646)

Winkel BSJ. Metabolic channeling in plants, Annual Review of Plant Biology, 2004, vol. 55 (pg. 85-107)

Winter H, Robinson DG, Heldt HW. Subcellular volumes and metabolite concentrations in spinach leaves, *Planta*, 1994, vol. 4 (pg. 530-535)

Yasuno R, Wada H. Biosynthesis of lipoic acid in Arabidopsis: cloning and characterization of the cDNA for lipoic acid synthase, *Plant Physiology*, 1998, vol. 118 (pg. 935-943).

INDEX